Lutz Breuer and Philipp Kraft (Eds.)

Hydro-Ecological Modeling

MDPI

This book is a reprint of the Special Issue that appeared in the online, open access journal, *Water* (ISSN 2073-4441) from 2015 (available at: http://www.mdpi.com/journal/water/special_issues/hydro-eco_model).

Guest Editors
Lutz Breuer and Philipp Kraft
Justus Liebig University Gießen
Germany

Editorial Office
MDPI AG
Klybeckstrasse 64
Basel, Switzerland

Publisher
Shu-Kun Lin

Managing Editor
Cherry Gong

1. Edition 2016

MDPI • Basel • Beijing • Wuhan • Barcelona

ISBN 978-3-03842-239-6 (Hbk)
ISBN 978-3-03842-212-9 (PDF)

Table of Contents

List of Contributors VII

About the Guest Editors.. XI

Preface to "Hydro-Ecological Modeling"..XIII

Chapter 1: Ecological Controls on Water Resources

Valentin Aich, Stefan Liersch, Tobias Vetter, Jafet C. M. Andersson,
Eva N. Müller and Fred F. Hattermann
Climate or Land Use?—Attribution of Changes in River Flooding in the
Sahel Zone
Reprinted from: *Water* **2015**, 7(6), 2796–2820
http://www.mdpi.com/2073-4441/7/6/2796 3

Anastassi Stefanova, Cornelia Hesse and Valentina Krysanova
Combined Impacts of Medium Term Socio-Economic Changes and Climate
Change on Water Resources in a Managed Mediterranean Catchment
Reprinted from: *Water* **2015**, 7(4), 1538–1567
http://www.mdpi.com/2073-4441/7/4/1538 32

Mana Gharun, Mohammad Azmi and Mark A. Adams
Short-Term Forecasting of Water Yield from Forested Catchments after Bushfire:
A Case Study from Southeast Australia
Reprinted from: *Water* **2015**, 7(2), 599–614
http://www.mdpi.com/2073-4441/7/2/599 67

Cornelia Hesse, Anastassi Stefanova and Valentina Krysanova
Comparison of Water Flows in Four European Lagoon Catchments under a Set of
Future Climate Scenarios
Reprinted from: *Water* **2015**, 7(2), 716–746
http://www.mdpi.com/2073-4441/7/2/716 85

Jieqiong Su, Xuan Wang, Shouyan Zhao, Bin Chen, Chunhui Li and Zhifeng Yang
A Structurally Simplified Hybrid Model of Genetic Algorithm and Support Vector Machine for Prediction of Chlorophyll *a* in Reservoirs
Reprinted from: *Water* **2015**, *7*(4), 1610–1627
http://www.mdpi.com/2073-4441/7/4/1610 119

Berit Arheimer, Johanna Nilsson and Göran Lindström
Experimenting with Coupled Hydro-Ecological Models to Explore Measure Plans and Water Quality Goals in a Semi-Enclosed Swedish Bay
Reprinted from: *Water* **2015**, *7*(7), 3906–3924
http://www.mdpi.com/2073-4441/7/7/3906 138

Chapter 2: Water Flux Impact on Ecosystem Functions

Lei Huang and Zhishan Zhang
The Stability of Revegetated Ecosystems in Sandy Areas: An Assessment and Prediction Index
Reprinted from: *Water* **2015**, *7*(5), 1969–1990
http://www.mdpi.com/2073-4441/7/5/1969 161

Kellie Vaché, Lutz Breuer, Julia Jones and Phil Sollins
Catchment-Scale Modeling of Nitrogen Dynamics in a Temperate Forested Watershed, Oregon. An Interdisciplinary Communication Strategy
Reprinted from: *Water* **2015**, *7*(10), 5345–5377
http://www.mdpi.com/2073-4441/7/10/5345 185

Mikołaj Piniewski, Paweł Marcinkowski, Ignacy Kardel, Marek Giełczewski, Katarzyna Izydorczyk and Wojciech Frątczak
Spatial Quantification of Non-Point Source Pollution in a Meso-Scale Catchment for an Assessment of Buffer Zones Efficiency
Reprinted from: *Water* **2015**, *7*(5), 1889–1920
http://www.mdpi.com/2073-4441/7/5/1889 220

Marc Vis, Rodney Knight, Sandra Pool, William Wolfe and Jan Seibert
Model Calibration Criteria for Estimating Ecological Flow Characteristics
Reprinted from: *Water* **2015**, 7(5), 2358–2381
http://www.mdpi.com/2073-4441/7/5/2358 256

Huijie Li, Jun Yi, Jianguo Zhang, Ying Zhao, Bingcheng Si, Robert Lee Hill, Lele Cui and Xiaoyu Liu
Modeling of Soil Water and Salt Dynamics and Its Effects on Root Water Uptake
in Heihe Arid Wetland, Gansu, China
Reprinted from: *Water* **2015**, 7(5), 2382–2401
http://www.mdpi.com/2073-4441/7/5/2382 282

Dejian Zhang, Xingwei Chen and Huaxia Yao
Development of a Prototype Web-Based Decision Support System for
Watershed Management
Reprinted from: *Water* **2015**, 7(2), 780–793
http://www.mdpi.com/2073-4441/7/2/780 305

List of Contributors

Mark A. Adams Faculty of Agriculture and Environment, University of Sydney, 1 Central Avenue, Eveleigh, NSW 2015, Australia.

Valentin Aich Potsdam Institute for Climate Impact Research (PIK), Telegraphenberg, 14473 Potsdam, Germany.

Jafet C. M. Andersson Swedish Meteorological and Hydrological Institute (SMHI), SE-601 76 Norrköping, Sweden.

Berit Arheimer Swedish Meteorological and Hydrological Institute (SMHI), Norrköping 60176, Sweden.

Mohammad Azmi Faculty of Engineering, Monash University, Clayton Campus, VIC 3800, Australia.

Lutz Breuer Institute for Landscape Ecology and Resource Management, University of Giessen, Giessen 35392, Germany.

Bin Chen State Key Laboratory of Water Environment Simulation, School of Environment, Beijing Normal University, Beijing 100875, China.

Xingwei Chen College of Geographic Sciences, Fujian Normal University, Fuzhou 350007, China; Cultivation Base of State Key Laboratory of Humid Subtropical Mountain Ecology, Fujian Normal University, Fuzhou 350007, China.

Lele Cui Key Laboratory of Plant Nutrition and the Agri-Environment in Northwest China, Ministry of Agriculture, Northwest Agriculture and Forestry University, Yangling 712100, China; Suide Test Station of Water and Soil Conservation, Yellow River Conservancy Committee of the Ministry of Water Resources, Yulin, Shanxi 719000, China.

Wojciech Frątczak European Regional Centre for Ecohydrology of the Polish Academy of Sciences, Tylna str. 3, Łódź 90-364, Poland; Regional Water Management Authority in Warsaw, 13B Zarzecze, Warszawa 03-194, Poland.

Mana Gharun Faculty of Agriculture and Environment, University of Sydney, 1 Central Avenue, Eveleigh, NSW 2015, Australia.

Marek Giełczewski Department of Hydraulic Engineering, Warsaw University of Life Sciences (WULS-SGGW), Nowoursynowska str. 159, Warszawa 02-776, Poland.

Fred F. Hattermann Potsdam Institute for Climate Impact Research (PIK), Telegraphenberg, 14473 Potsdam, Germany.

Cornelia Hesse Potsdam Institute for Climate Impact Research, Post Box 601203, Potsdam 14412, Germany.

Robert Lee Hill Department of Environmental Science and Technology, University of Maryland, College Park, MD 20742, USA.

Lei Huang Shapotou Desert Research and Experimental Station, Cold and Arid Regions Environmental and Engineering Research Institute, Chinese Academy of Sciences, Lanzhou 730000, China; Key Laboratory of Stress Physiology and Ecology in Cold and Arid Regions Gansu Province, Lanzhou 730000, China.

Katarzyna Izydorczyk European Regional Centre for Ecohydrology of the Polish Academy of Sciences, Tylna str. 3, Łódź 90-364, Poland.

Julia Jones Department of Geosciences, Oregon State University, Corvallis, OR 97331, USA.

Ignacy Kardel Department of Hydraulic Engineering, Warsaw University of Life Sciences (WULS-SGGW), Nowoursynowska str. 159, Warszawa 02-776, Poland.

Rodney Knight Geological Survey Lower Mississippi—Gulf Water Science Center, 640 Grassmere Park, Suite 100, Nashville, TN 37211, USA.

Valentina Krysanova Potsdam Institute for Climate Impact Research, Post Box 601203, Potsdam 14412, Germany.

Chunhui Li Key Laboratory for Water and Sediment Sciences of Ministry of Education, School of Environment, Beijing Normal University, Beijing 100875, China.

Huijie Li Key Laboratory of Plant Nutrition and the Agri-Environment in Northwest China, Ministry of Agriculture, Northwest Agriculture and Forestry University, Yangling 712100, China.

Stefan Liersch Potsdam Institute for Climate Impact Research (PIK), Telegraphenberg, 14473 Potsdam, Germany.

Göran Lindström Swedish Meteorological and Hydrological Institute (SMHI), Norrköping 60176, Sweden.

Xiaoyu Liu Key Laboratory of Plant Nutrition and the Agri-Environment in Northwest China, Ministry of Agriculture, Northwest Agriculture and Forestry University, Yangling 712100, China.

Paweł Marcinkowski Department of Hydraulic Engineering, Warsaw University of Life Sciences (WULS-SGGW), Nowoursynowska str. 159, Warszawa 02-776, Poland.

Eva N. Müller Institute of Earth and Environmental Science, University of Potsdam, Karl-Liebknecht-Staße 24-25, 14476 Potsdam-Golm, Germany.

Johanna Nilsson Swedish Meteorological and Hydrological Institute (SMHI), Norrköping 60176, Sweden.

Mikołaj Piniewski Department of Hydraulic Engineering, Warsaw University of Life Sciences (WULS-SGGW), Nowoursynowska str. 159, Warszawa 02-776, Poland; Current address: Potsdam Institute for Climate Impact Research (PIK), P.O. Box 60 12 03, Potsdam 14412, Germany.

Sandra Pool Department of Geography, University of Zurich, Winterthurerstrasse 190, 8057 Zurich, Switzerland.

Jan Seibert Department of Geography, University of Zurich, Winterthurerstrasse 190, 8057 Zurich, Switzerland; Department of Earth Sciences, Uppsala University, Villavägen 16, 752 36 Uppsala, Sweden.

Bingcheng Si Department of Soil Science, University of Saskatchewan, Saskatoon, SK, S7N 5A8, Canada; College of Water Resources and Architecture Engineering, Northwest Agriculture and Forestry University, Yangling 712100, China.

Phil Sollins College of Forestry, Oregon State University, Corvallis, OR 97331, USA.

Anastassi Stefanova Potsdam-Institute for Climate Impact Research, Post Box 601203, Potsdam 14412, Germany.

Jieqiong Su State Key Laboratory of Water Environment Simulation, School of Environment, Beijing Normal University, Beijing 100875, China; Chinese Academy for Environmental Planning, Ministry of Environmental Protection, Beijing 100012, China.

Kellie Vaché Biological and Ecological Engineering, Oregon State University, Corvallis, OR 97331, USA.

Tobias Vetter Potsdam Institute for Climate Impact Research (PIK), Telegraphenberg, 14473 Potsdam, Germany.

Marc Vis Department of Geography, University of Zurich, Winterthurerstrasse 190, 8057 Zurich, Switzerland.

Xuan Wang State Key Laboratory of Water Environment Simulation, School of Environment, Beijing Normal University, Beijing 100875, China; Key Laboratory for Water and Sediment Sciences of Ministry of Education, School of Environment, Beijing Normal University, Beijing 100875, China.

William Wolfe Geological Survey Lower Mississippi—Gulf Water Science Center, 640 Grassmere Park, Suite 100, Nashville, TN 37211, USA.

Zhifeng Yang State Key Laboratory of Water Environment Simulation, School of Environment, Beijing Normal University, Beijing 100875, China; Key Laboratory for Water and Sediment Sciences of Ministry of Education, School of Environment, Beijing Normal University, Beijing 100875, China.

Huaxia Yao Dorset Environmental Science Centre, Ontario Ministry of Environment and Climate Change, 1026 Bellwood Acres Road, Dorset, ON P0A 1E0, Canada.

Jun Yi Key Laboratory of Plant Nutrition and the Agri-Environment in Northwest China, Ministry of Agriculture, Northwest Agriculture and Forestry University, Yangling 712100, China.

Dejian Zhang College of Geographic Sciences, Fujian Normal University, Fuzhou 350007, China.

Jianguo Zhang Key Laboratory of Plant Nutrition and the Agri-Environment in Northwest China, Ministry of Agriculture, Northwest Agriculture and Forestry University, Yangling 712100, China.

Zhishan Zhang Shapotou Desert Research and Experimental Station, Cold and Arid Regions Environmental and Engineering Research Institute, Chinese Academy of Sciences, Lanzhou 730000, China; Key Laboratory of Stress Physiology and Ecology in Cold and Arid Regions Gansu Province, Lanzhou 730000, China.

Shouyan Zhao Management Office of Miyun Reservoir, Beijing 101512, China.

Ying Zhao Key Laboratory of Plant Nutrition and the Agri-Environment in Northwest China, Ministry of Agriculture, Northwest Agriculture and Forestry University, Yangling 712100, China; Department of Soil Science, University of Saskatchewan, Saskatoon, SK, S7N 5A8, Canada.

About the Guest Editors

Lutz Breuer holds the chair of Landscape, Water and Biogeochemical Cycles at Justus Liebig University Giessen (Germany) and is an expert in hydro-biogeochemistry. His activities focus on the intersection of experimental fieldwork and modelling. With a background in Physical Geography, he works on hydrological and nutrient turnover processes, primarily nitrogen. Apart from studies in German catchments of temperate climate, he has a number of ongoing projects in tropical ecosystems.

Philipp Kraft is a member of the chair of Landscape, Water and Biogeochemical Cycles at Justus Liebig University Giessen (Germany). He is a hydrological model developer and works on interfacing those with other models from a range of environmental disciplines. To obtain water quality data with high temporal resolution, he designs automatic, in-situ water quality measuring systems.

Preface to "Hydro-Ecological Modeling"

In recent decades, predictions of water balance, floods and droughts based on hydrological models improved drastically. However, the impact of ecological processes on water regimes and the ecological response to changing hydrological conditions has often been ignored, or understood as a static boundary condition and is scarcely included in the analyses of model performance. Simulating the complex interactions of water and ecosystem processes, however, requires the prediction of hydrological fluxes and states on the one side, and the coupling of the hydrological and ecosystem process models on the other.

This Special Issue presents twelve modeling studies exploring the interaction of ecological processes and water fluxes from Europe, Asia, North America, Africa and Australia. In these different climatic settings, half of the studies focus on the impact of ecological change on the water balance. Examples are the influence of bushfires, land use change driven by socio-economic processes or algae blooms that change the availability of water resources. The second set of studies focusses more on the problem, how water fluxes drive ecological processes. In these studies, water is understood as a habitat, a transport agent for nutrients, or as a crucial resource for plant development.

Lutz Breuer and Philipp Kraft
Guest Editors

Chapter 1:
Ecological Controls on Water Resources

Climate or Land Use?—Attribution of Changes in River Flooding in the Sahel Zone

Valentin Aich, Stefan Liersch, Tobias Vetter, Jafet C. M. Andersson,
Eva N. Müller and Fred F. Hattermann

Abstract: This study intends to contribute to the ongoing discussion on whether land use and land cover changes (LULC) or climate trends have the major influence on the observed increase of flood magnitudes in the Sahel. A simulation-based approach is used for attributing the observed trends to the postulated drivers. For this purpose, the ecohydrological model SWIM (Soil and Water Integrated Model) with a new, dynamic LULC module was set up for the Sahelian part of the Niger River until Niamey, including the main tributaries Sirba and Goroul. The model was driven with observed, reanalyzed climate and LULC data for the years 1950–2009. In order to quantify the shares of influence, one simulation was carried out with constant land cover as of 1950, and one including LULC. As quantitative measure, the gradients of the simulated trends were compared to the observed trend. The modeling studies showed that for the Sirba River only the simulation which included LULC was able to reproduce the observed trend. The simulation without LULC showed a positive trend for flood magnitudes, but underestimated the trend significantly. For the Goroul River and the local flood of the Niger River at Niamey, the simulations were only partly able to reproduce the observed trend. In conclusion, the new LULC module enabled some first quantitative insights into the relative influence of LULC and climatic changes. For the Sirba catchment, the results imply that LULC and climatic changes contribute in roughly equal shares to the observed increase in flooding. For the other parts of the subcatchment, the results are less clear but show, that climatic changes and LULC are drivers for the flood increase; however their shares cannot be quantified. Based on these modeling results, we argue for a two-pillar adaptation strategy to reduce current and future flood risk: Flood mitigation for reducing LULC-induced flood increase, and flood adaptation for a general reduction of flood vulnerability.

Reprinted from *Water*. Cite as: Aich, V.; Liersch, S.; Vetter, T.; Andersson, J.C.M.; Müller, E.N.; Hattermann, F.F. Climate or Land Use?—Attribution of Changes in River Flooding in the Sahel Zone. *Water* **2015**, *7*, 2796–2820.

1. Introduction

Catastrophic flooding in the Sahelian part of the Niger basin has become an increasing threat during the last decades, leading to more than ten million people affected since the year 2000 [1]. Tarhule *et al.* (2005) [2] were some of the first to bring the topic into academic research, referring to it as the "other Sahelian hazard". Aich *et al.* (2014) [1] recently published a comprehensive overview of flooding characteristics within the entire Niger basin, including a review of existing literature and damage statistics from different sources. They found that the Sahelian part of the Niger basin was particularly affected by catastrophic floods, with an almost exponential increase in people affected over the last decades. They also showed that the increasing flood risk was related to the extreme population growth, the increasing vulnerability of the population, and an increase in flood magnitude.

However, the reason for the increase in flood magnitude in the Sahel is still not fully understood. Descroix *et al.* (2012) [3] stated that climate is not the cause of the phenomenon, since the increasing discharges are accompanied by decreasing rainfall rates. This inconsistency is called "the Sahel Paradox" (SP) and is described in detail in Descroix *et al.* 2013 [4]. Based on their statistical analysis and field observations on infiltration, they argued that the effect of land use and land cover change (LULC), from the local to the meso-scale, caused the increased discharge in the region. The main processes were land clearing and the transformation of savannah into pasture, agricultural land or degraded savannah. This led to soil crusting and a decrease in infiltrability, which subsequently led to an increase in flood magnitude during the heavy rains of the Sahelian rainy season.

In contrast to their study, Aich *et al.* (2014) [1] identified climatic changes and a return to wet conditions as the major driver of increasing flood magnitudes in the Niger basin, including the Sahelian region. Aich *et al.* (2014) [1] used a data-based attribution approach and compared time series of maximum annual discharge (AMAX) with precipitation time series, as well as time series of flashiness of discharge, as a proxy for LULC. They showed that even though the LULC caused an increase in flashiness since at least the 1960s, the AMAX decreased until the 1980s and they concluded that LULC could not be the major driver of the increased flood regime in the Sahel. In addition, they demonstrated that the SP only existed during the 1970s and 1980s, after which the trends of precipitation and discharge again correlated.

This study intends to contribute to the discussion on the reasons for the increasing flood risk in the Sahelian part of the Niger basin. The specific research question is, to which share LULC and/or climatic changes cause the increase of river flooding in the area. To this end, a simulation-based attribution approach proposed by Merz *et al.* (2012) [5] is used. Merz *et al.* (2012) [5] introduced a hypothesis testing framework for attributing changes of flood regime, which is based on testing the consistency or inconsistency of plausible drivers with the observed flood trend and providing a confidence level for the attribution. They distinguished between a data-based and a simulation-based attribution approach. The data-based attribution compares flood time series or their statistics with those of the assumed driver, for example, by evaluating the correlation between the time series of the potential cause and effect variables. It is common and widely used in the literature (e.g., [1,6–11]). The simulation-based approach has been used in several studies with conceptual rainfall-runoff models [12–16]. Process-based hydrological models have not been widely used for attribution approaches. There are studies, which distinguish climate change and LULC impacts on historical trends in flood magnitude, but not systematically and within one modeling approach (e.g., [17]). To the best of our knowledge, the only study published which follows the protocol of Merz *et al.* (2012) [5] is that by Hundecha and Merz (2012) [18]. Hundecha and Merz (2012) [18] drove a hydrological model with a large number of stationary and non-stationary climate time series in order to study whether the observed flood trend was climate driven.

In this study, we analyze the effects of LULC on the flood trend using the process-based based ecohydrological model SWIM (Soil and Water Integrated Model) with integrated dynamic land use change. The ecohydrological model is applied to simulate flood discharges for the time period 1950–2009 with two different settings. The discharge is simulated with static land cover as of 1950 in order to show how the discharge would have developed over the last 60 years if there had not been any LULC. The second/control run implicitly includes past LULC. We hypothesize that the comparison of the modelled discharge of these two scenarios will give some initial quantitative insights into the relative share of LULC and climatic changes on changes in the Sahel flood regimes.

There is an ongoing debate over whether the observed "return to wet conditions" in West Africa itself can be attributed to global climate change or is still within the boundaries of natural climatic variability [19–21]; without taking sides, we refer to the recent changes of the precipitation patterns as "climatic" changes.

Since this study is the first to attribute LULC to flood trends via a process-based hydrological model following the proposed protocol of Merz *et al.* (2012) [5], it might additionally shed first light on the requirements of data quality/availability and the efficiency of the hydrological model in order to achieve robust attribution statements.

2. Materials and Methods

2.1. Regional Setting

The Sahelian part of the Niger River is located downstream of Diré in Mali and extends to around Niamey in the state of Niger (Figure 1). This part of the Niger basin until Niamey covers around 297,400 km^2. Its landscape is characterized by plateaus and smooth valleys with long slopes. The climate is semi-arid, with annual precipitation ranging from 267 mm in Ansongo to 540 mm in Niamey. The potential evapotranspiration in the area is 3500 mm per year [3]. The typically convective rainfalls occur within the rainy season between June and October. The vegetation in the region and its changes since the 1950s are described in detail in Descroix *et al.* (2012) [3]. The original bushy and woody savannah types have been replaced almost completely by crop fields, pasture and a patchwork of woody savannah vegetation called Tiger Bush. The northern part, in which the Sirba catchment is located, is too dry for effective rain fed agriculture and was therefore turned mainly into extensively or intensively used pastoral land (Figures 2 and 3). In contrast, the southern part in which the Sirba catchment is located has mainly been converted into cropland.

Figure 1. Map of the research area in West Africa including land use classes used in the model as base map in the year 2000. The orange, green, and red outlines mark the watershed of the gauging stations Alcongui (Goroul River), Garbe-Kourou (Sirba River) and the watershed of Niamey (Niger River). The grey dots show the grid of the PGFv2 climate reanalysis data set. The red dots show the grid of the climate data used for the analysis. The hatched area marks the region which is used for the quantification of land use and land cover changes.

6

Most of the discharge in the Sahelian part of the Niger originates in its upstream regions within the Guinean highlands. The Inner Niger Delta, a large wetland, limits the Upper Niger basin. The wetland smooths the river flow and protracts the peak. The outlet of the delta is close to Diré, and the flood, generated in the Guinean highlands, occurs between November and January. This flood, referred to as the "Guinean Flood", passes through Niamey between December and the beginning of March [1]. Transmission losses are high between Diré and Niamey, resulting in higher annual discharge at Diré compared to Niamey. Little additional runoff is generated locally in the Sahelian part [22]. This locally generated discharge comes mainly from the plateaus of the right-bank subbasins and results in another flood peak in the Sahelian Niger, previous to the Guinean Flood. This first peak occurs during the rainy season (July–November) and is called the "Red Flood" due to the red color of the sedimentary load of the local iron oxide rich soils. The two main tributaries are the Goroul (44,900 km^2) and Sirba Rivers (38,750 km^2), which are analyzed in this study (gauging stations Alcongui and Garbe-Kourou). Both are intermittent rivers, and the annual peaks vary substantially, with values between 35 and 300 m^3/s for Alcongui and 20 and 460 m^3/s for Garbe-Kourou (1950–2009). The vast subbasins to the East reach up into the central Sahara and contribute only a minor amount of inflow, and local tributaries are endorheic most of the year. The Guinean and Red Floods can usually be clearly distinguished. The peak of the Guinean Flood is already smoothed due to the large watershed and the dynamics of the Inner Niger Delta, whereas the peak of the Red Flood is more jagged and flashy. However, in years where the Red Flood is very low, a separate peak cannot be distinguished. This happened regularly during the 1950s and 1970s but has occurred significantly less often since the 1980s [1].

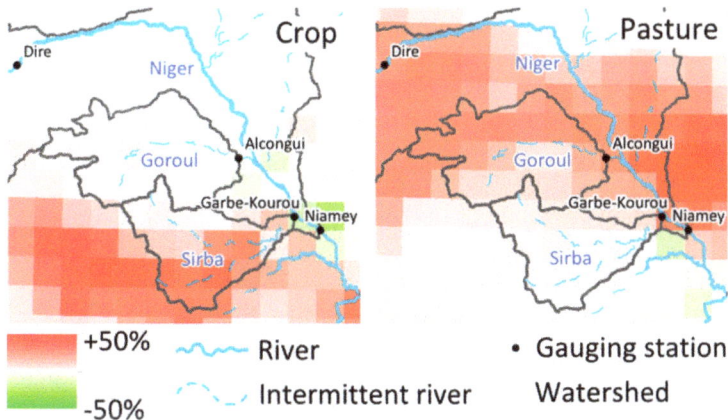

Figure 2. Land use and land cover changes between 1950 and 2005 for crop and pasture after Hurtt *et al.* (2011) [23].

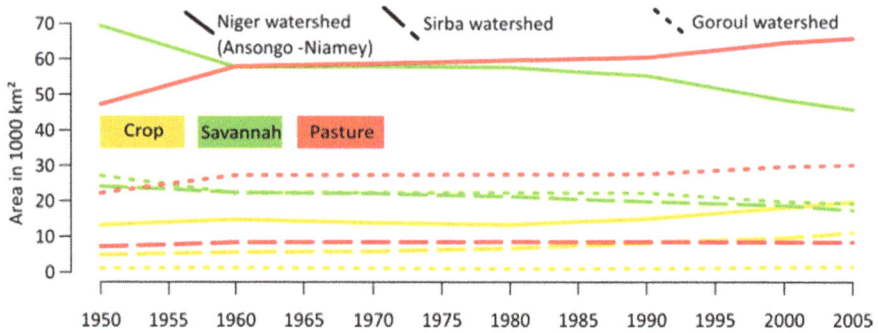

Figure 3. Changes in the main land use classes of crop, savannah, and pasture from 1950 until 2005 for the watershed of the Niger River between Ansongo and Niamey, and the catchments of the Sirba and Goroul Rivers (see area in Figure 1).

2.2. Ecohydrological Model and Model Set-Up

The ecohydrological model SWIM (Soil and Water Integrated Model) is a continuous-time (daily) and spatially semi-distributed model of intermediate complexity for river basins [24]. SWIM was developed based on two models: Soil and Water Assessment Tool (SWAT) [25] and MATSALU [26], with the aim to provide a tool for climate and LULC impact assessment in meso-scale and large river basins. It integrates hydrological processes, vegetation growth, nutrient cycling, erosion and sediment transport at the river basin scale [27]. The hydrological system of the model comprises four main segments: The soil surface, the root zone of the soil, the shallow aquifer, and the deep aquifer. On the soil surface, the surface runoff is estimated as a non-linear function of precipitation and a retention coefficient. It depends on soil water content, land use, and soil type (modified after the Soil Conservation Service curve number approach [28]). The soil column is subdivided into several layers. In these layers, the water balance is calculated, including precipitation, surface runoff, evapotranspiration, subsurface runoff, and percolation. Hydrological processes represented in the shallow aquifer are groundwater recharge, capillary rise to the soil profile, lateral flow, and percolation to the deep aquifer. Potential evapotranspiration is calculated using the method of Turc-Ivanov [29]. Actual evaporation from soil and transpiration by plants are simulated following the Ritchie concept [30].

A simplified Environmental Policy Integrated Climate (EPIC) approach [31] is integrated in the model for the simulation of arable crops and other general vegetation types (e.g., pasture, savannah, evergreen forest), using specific parameter values for each crop/vegetation type. The parameter settings of the newest version of the SWAT model are used for the aggregated vegetation types of the SWIM model [32]. These parameter settings have been widely used in the African context for LULC studies (e.g., [33–36]). The effects of the vegetation on the hydrological

8

processes include the cover-specific retention coefficient, impacting surface runoff and influencing the amount of transpiration. Transpiration is simulated as a function of potential evapotranspiration and leaf area index. A more detailed description of the representation of hydrological processes within SWIM is given in Huang *et al.* (2013) [27].

SWIM disaggregates a river basin into subbasins and hydrotopes. The subbasins are delineated on the basis of flow accumulation in a Digital Elevation Model. The hydrotopes are created by overlaying the subbasin map with maps of land use and soil. They represent the spatial units used to simulate all water flows and nutrient cycling in soil as well as vegetation growth based on the principle of similarity (*i.e.*, assuming that units within one subbasin that have the same land use and soil types behave similarly). The model was applied for impact studies in several basins in Africa and showed good efficiency for the whole Niger, the Blue Nile, the Limpopo and the Congo [36,37]. In addition, a multi-model intercomparison of hydrological models of Vetter *et al.* (2015) has shown that SWIM is quite capable of simulating flow in the Niger basin [38].

For this study, the model has been set up for the Sahelian part of the Niger River, from Diré in Mali to Niamey in the state of Niger. Since the model is only used for modeling river flows in the past, monitored discharges are routed into the model at the Diré gauge. The model includes 255 subbasins for the 297,000 km^2 area of the watershed. These subbasins are integrated to form three subcatchments which are the catchments Goroul (station Alcongui), Sirba (station Garbe-Kourou), and Niger (between the stations Diré and Niamey) (Figure 1). These subcatchments were calibrated individually in order to fit the model as closely as possible to the regional conditions (see Section 2.5).

2.3. Dynamic Land Use Change Module

The newly developed land use change module (LUCM) for the SWIM model is used for the first time in this study. It changes the land classes at any frequency or given point in time, while keeping the instantaneous balance of water and other modelled fluxes constant during the change, for example soil water content (SWC). This means that the number and the areas of hydrotopes within a subbasin can change and new hydrotopes can appear or others disappear.

The land-use status at given points in time (see Section 2.4.3) are read in by the LUCM (in this study every five years; Shorter frequencies up to a daily change are possible). The LUCM transforms the land classes and rearranges the hydrotopes on the basis of so-called stable units (SU). SU are areas within a subbasin and do not change their extent, like areas with uniform soil, for example. For these SU, fluxes like SWC remain constant during the change, even if the land class changes. Thereby, all given information on LULC on the subbasin level is used, and the transformation of the hydrotopes does not alter the balances of water or other relevant fluxes.

The following two examples shall illustrate the main processes when hydrotopes increase or decrease within a SU. If the shares of crop and pasture increase and the savannah deceases, the SWC of the old hydrotopes have to be newly distributed. The SWC of the savannah is reduced according to its areal share. The residual SWC is then added proportionally to crop and pasture, depending on their areal increase.

In another case, an existing SU consists of cropland and savannah. In the new land use, there is also pasture on the SU. Meanwhile, crop and savannah shrink. In this case, the SWC of crop and savannah is reduced proportionally and the residual SWC is completely added to the new pasture.

These examples show that the balance of the SWC remains constant during the change on the SU. This holds also for all other simulated fluxes. Only parameters connected to vegetation, such as biomass, for example, restart at zero for the additional area or are reduced relative to the areal changes.

2.4. Data

2.4.1. Climate Data

Climate data are used to drive the simulations and to analyze the total annual precipitation. For the modeling runs, a relatively dense spatial coverage of climate input is needed, since the climate forcing is interpolated for each subbasin (see Section 2.2). In the data-sparse research area, this can only be provided by reanalysis data sets. For this study, three different data sets have been analyzed and compared to data from six weather stations in the region, in order to check their performance in the face of the requirements for an attribution study with regard to accuracy. The WATCH Forcing Data 20th Century (WATCH) (1950–2001) [39], data from the Global Soil Wetness Project Phase 3 (GSWP3) [40] and the second version of the Global Meteorological Forcing Dataset for land surface modeling of Princeton University (PGFv2) [41] (Figures 4 and 5) have been selected as potential model input. The visual comparison of temperature and precipitation shows that all of the data sets generally have good correspondence with the measured data, yet also all of them have some deficits (PGFv2: Figures 4 and 5, WATCH: Supplementary Material Figures S1 and S2, GSWP3: Figures S3 and S4). Regarding precipitation, distinct deviations of single years or short periods do occur, as for example in all three data sets for the station Timbuktu during the early 1990s. However, the general trends in precipitation are represented in all three reanalysis products. Annual mean temperatures only rarely deviate more than 1 °C from the observed station data and reproduce the general trends also efficiently. Additional uncertainty derives from the limited evaluation. The comparison only takes six stations into account for the whole region. These points cannot represent the whole area. These uncertainties deriving from the climate data have to be taken into account when discussing the modeling results. The comparison implies that single years should not be compared between the modeling results and observations of discharge. However, since the major trends of the climate data are represented in the analysis, the climate data can be used for reproducing general trends with the model when results are carefully interpreted.

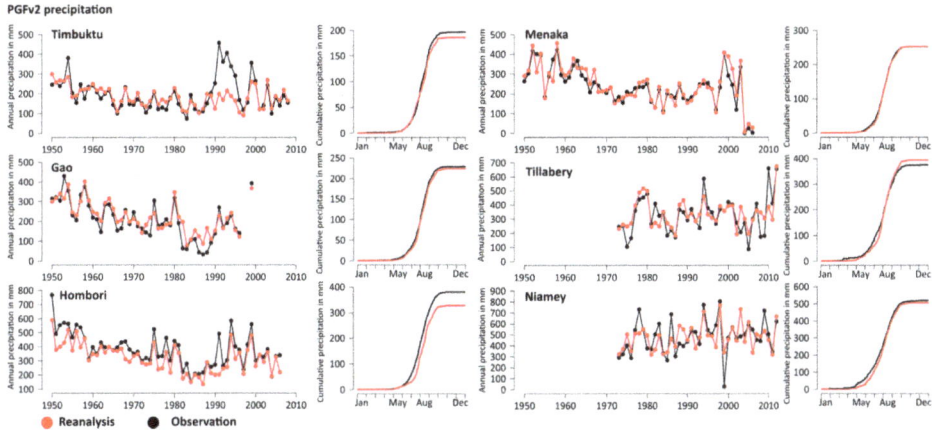

Figure 4. Comparison of precipitation from interpolated PGFv2 reanalysis data (red) with observations from six weather stations (black) in the research area. For each station, the annual precipitation (**left**) and the cumulative sum of the precipitation of the whole period (**right**) is depicted.

Since no systematic differences could be detected with regard to data quality, the PGFv2 data set was finally selected since it is the newest of the three data sets. It is a combination of a suite of global observation-based data sets with the National Centers for Environmental Prediction—National Center for Atmospheric Research reanalysis. The spatial resolution is at $0.5° \times 0.5°$ and for the modeling the 3-hourly data has been aggregated to daily data.

For the precipitation analysis, annual rainfall data is aggregated for each subcatchment (Niger, Sirba, Goroul) by building the mean over all data points in the subcatchment region (Figure 1). Since the Guinean Flood is completely generated in the Upper Niger basin, the related rainfall data was aggregated for this region outside the research area.

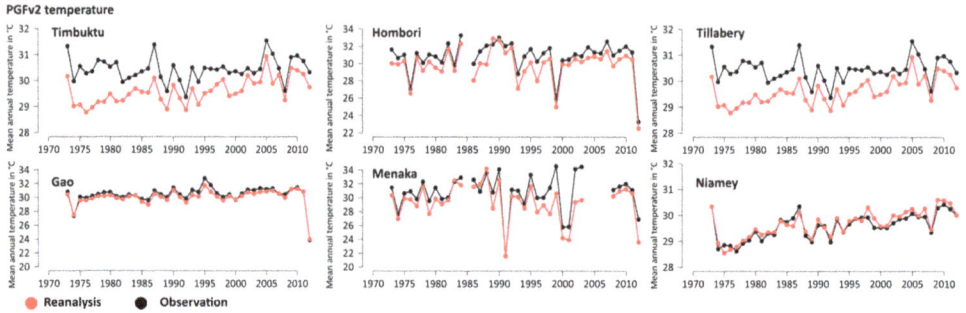

PGFv2 temperature

Figure 5. Comparison of mean annual temperature of interpolated PGFv2 reanalysis data (red) with observations from six weather stations (black) in the research area.

2.4.2. Discharge Data

Observed discharges from three monitoring stations (Figure 1) were used to calibrate and validate the model and for the analysis of the AMAX. The observed discharge at the Diré station was used as input for the model (see Section 2.2). The observations are part of the Niger-HYCOS monitoring network, managed by the Niger Basin Authority, which consists of daily water-level readings at more than 100 locations across the basin, as well as accompanying rating curves to compute discharge at these locations [42].

The AMAX for the time series of Alcongui on the Goroul River and Garbe-Kourou on the Sirba River are created by selecting the highest daily discharge per year, with gaps for years where the peak cannot be identified due to missing values. The time series for Niamey has two peaks per year (see Section 2.1). The second, the Guinean peak, occurs in most of the years after the 31st of December but is still assigned to the rainy season of the previous year. For the Red Flood, the procedure is not straightforward since the peak cannot be distinguished in dry years hidden from the Guinean Flood and thus it cannot be quantified for all years (see Section 2.1). For the simulation, this problem was solved by simulating the discharge without the inflow of Diré, leading to a clear AMAX value of the Red Flood, generated in the Sahelian part of the Niger basin after Diré. For the observation, it is not possible to filter only the discharge which is generated in the basin downstream of Diré. Therefore, identification of the AMAX depends on a clearly distinguishable first peak of the hydrograph between July and October. For years when the flood peak cannot be detected since it is too low and hidden under the Guinean flood peak, the AMAX time series contains gaps, which affects the statistical analysis (see Section 2.7).

12

2.4.3. Land Cover and Land Use Change Data

For the representation of the LULC in the model, maps of the land cover for different periods are needed. For this study, land use maps have been generated in five year steps from 1950 until 2005. There is no data available which provides this information to the degree of detail necessary for ecohydrological modeling at the meso-scale with regard to land classes. Therefore, the land cover maps have been derived using the information from two different data sets. In the first step, a base map was derived which has the necessary details concerning land cover at a fine spatial resolution. For the base map, land class information was derived from the GLC2000 data set [43]. It differentiates between 27 land classes and has a spatial resolution of $\frac{1}{112}$, which corresponds to 1 km at the equator. It is based on remote sensing data and includes a detailed legend. The GLC2000 classes occurring in the research area have been transformed to the classes of the ecohydrological model (Figure 1, see Section 2.2).

However, this map only represents the one point in time at which the data was collected (year 2000). Therefore, an additional data set was used which gives information about the change in land cover with regard to crop, pasture, and urban land but for different times in history. The information on areal changes in crop and pasture was obtained from the Land-Use Harmonization project [23]. The harmonized land use scenarios connect historical reconstructions with future scenarios and have been used as a basis for the Earth System Models of the fifth Assessment Report of the Intergovernmental Panel on Climate Change (IPCC). The historical data for the period 1500–2005 are based on the model HYDE [44,45], and contain information on changes to cropland, pasture and urban land at a 0.5° × 0.5° resolution on an annual basis. The reconstruction of the data set is based on satellite maps for the years they were available and for the more distant past by combining information on population density, soil suitability, distance to rivers or lakes, slopes, and specific biomes. Each grid point of the LULC data contained the percentage of crop, pasture and urban land (Figure 2).

With this information on changes, the base map was altered for each of the 5-year time steps in order to have land use maps which represent the LULC, as shown simplified in Figure 6. The LULC information was added to the base map on the subbasin level of the model, and existing land classes changed accordingly. The land classes of water, wetlands, sandy/stony desert and bare rock have been kept constant in the year 2000. When pasture, crop and/or urban land increase in a subbasin, other natural vegetation land classes like savannah or forest are proportionally reduced in the same subbasin. If crop, pasture and/or urban land decrease, the land classes of the natural vegetation increase proportionally.

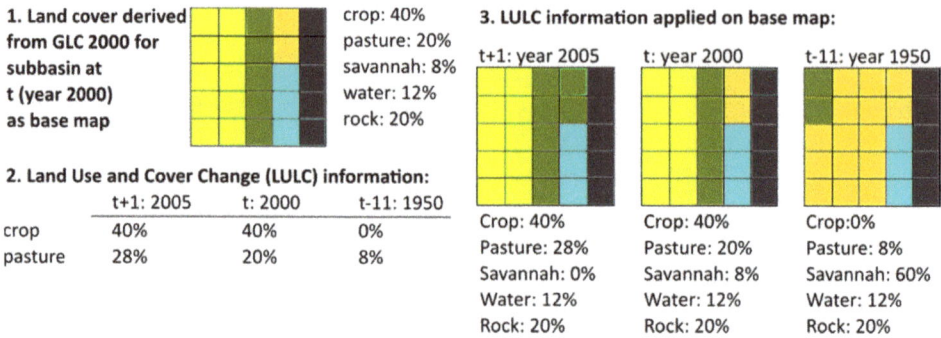

1. Land cover derived from GLC 2000 for subbasin at t (year 2000) as base map

crop: 40%
pasture: 20%
savannah: 8%
water: 12%
rock: 20%

3. LULC information applied on base map:

t+1: year 2005   t: year 2000   t-11: year 1950

2. Land Use and Cover Change (LULC) information:

	t+1: 2005	t: 2000	t-11: 1950
crop	40%	40%	0%
pasture	28%	20%	8%

Crop: 40%   Crop: 40%   Crop: 0%
Pasture: 28%   Pasture: 20%   Pasture: 8%
Savannah: 0%   Savannah: 8%   Savannah: 60%
Water: 12%   Water: 12%   Water: 12%
Rock: 20%   Rock: 20%   Rock: 20%

Figure 6. Exemplary process of how the information from the Land Use and Land Cover data of the Land-Use Harmonization project is used on the subbasin level to produce land cover maps for the different years which are then used by the land use change module. The whole square represents one exemplary subbasin, with small squares representing individual stable units (SU).

The emerging land class maps from 1950 until 2005 are used in the model as explained in Section 2.4. In Figure 3 the changes are quantified for the watershed of Niamey until Ansongo, corresponding to the area marked by climate grid points in Figure 1. The related spatial distribution is shown in Figure 2.

Crop types were derived from a data set for West African crops [46]. The four main types in the region are millet, sorghum, cowpea and rice. For every subbasin, the same crop type has been used for the whole period, according to the dominant crop in the data.

2.4.4. Soil and Topographic Data

Information on the soils in the research area for the modeling was derived from the Digital Soil Map of the World [47]. Relevant parameters for the model include depth, clay, silt and sand content, bulk density, porosity, available water capacity, field capacity, and saturated conductivity for each of the soil layers. For the delineation of the subbasins and necessary topographic information, a Digital Elevation Model derived from the Shuttle Radar Topography Missions at a 90 m resolution [48] was used.

2.5. Calibration of the Model

An accurate representation of the main hydrological processes and characteristics of the research is a crucial precondition for a robust modeling attribution. Therefore, the model was calibrated with a standard procedure for SWIM and SWAT [49] via an automatic calibration using the PEST software package [50]. This is commonly applied, as for example done in the studies of Vetter *et al.* [38]

on the rivers Rhine (Europe), Upper Niger (Africa), and Upper Yellow (Asia). The calibration was done individually for the Sirba (station Garbe-Kourou), the Goroul (station Alcongui) and the rest of the subcatchment between Diré and Niamey (station Niamey) to take into account the distinctly different geographic attributes. During the calibration, the LUCM was used in order to calibrate/validate the model with correct land use for the respective period and region. The calibrated parameters/factors are static and do not change within a subcatchment and over time. The focus of the calibration and model set-up for all subbasins was to achieve adequate efficiency for streamflow simulations on daily time steps, especially for high flows. Therefore, the main parameters/factors for the calibration were related to groundwater, river routing, saturated conductivity and potential evapotranspiration (Table 1). For the Sirba and Goroul catchments, the parameters related to groundwater and the factors for potential evapotranspiration and saturated conductivity showed the highest sensitivity. Routing parameters were somewhat less sensitive. For the Niger basin between Diré and Niamey without the subcatchments of Goroul and Sirba, the sensitivity for the routing and the potential evapotranspiration was higher and the sensitivity of other parameters lower. This is likely due to the fact that the model is fed monitored data from the Diré gauging station (see Section 2.2). Prior to the calibration, the available observed discharge data was checked visually and calibration periods were selected where distinct high and distinct low annual peaks are represented in order to cover a broad range of rainfall-runoff conditions. The period was finally selected when a period of eight years with a low amount of missing data was available (Goroul (Alcongui): 1963–1970, Sirba (Garbe-Kourou): 1986–1994, Niger (Niamey): 1988–1995). The validation periods are before or after the calibration period, depending on the availability of observations (Goroul (Alcongui): 1971–1978, Sirba (Garbe-Kourou): 1978–1985, Niger (Niamey): 1996–2003). Since there are not enough climatic observations available, reanalysis data is used to drive the model, also during the calibration. The results of the calibration are shown in Table 2.

Table 1. Calibrated parameters of the ecohydrological model SWIM (Soil and Water Integrated Model) with a short description.

Calibrated Parameter/Factors		Short Description
Groundwater related parameters	Groundwater recession	Rate at which groundwater flow is returned to the stream.
	Groundwater delay	The time it takes for water to leave the bottom of the root zone until it reaches the shallow aquifer where it can become groundwater flow (in days).
	Baseflow factor	The baseflow factor is used to calculate the return flow travel time. The return flow travel time is then used to calculate percolation in the soil from layer to layer.
Correction factor for saturated conductivity		The factor is applied for all soils
Correction factor for potential evapotranspiration		The factor is applied for all subbasins in the respective subbasin.
Routing coefficient		Routing coefficient to calculate the storage time constant of the flow from the initial estimation which is based on channel length and celerity.

Table 2. Calibration and validation results of the eco-hydrological model SWIM in Nash-Sutcliff efficiency (NSE) and percent bias (PBIAS) for three stations.

Gauging Stations	Calibration		Validation	
	NSE	PBIAS	NSE	PBIAS
Alcongui (Goroul)	0.58	−0.4	0.55	−16.9
Garbe-Kourou (Sirba)	0.66	22.1	0.49	34.5
Niamey (Niger)	0.86	12.2	0.87	6.9

2.6. Sensitivity Analysis of the Effects of LULC on the Hydrological Regime

The representation of the LULC and their effect on the hydrological processes are important for an understanding of the discharge changes. In Figure 7, the effects of different land cover on the modelled hydrological regime are shown for the Sirba and Goroul watersheds. The watersheds have been modelled with three set-ups, with either crop, pasture or savannah vegetation covering the entire subcatchments. The model is run for a period of 10 years (1985–1995). Regarding precipitation, the 10-year period was selected with a rather dry beginning and above-average wet conditions at the end in order to allow for different rainfall-runoff conditions.

The subcatchments covered solely with crops show a strong increase in peak discharges. In the wetter Goroul catchment, this holds true even for the dry years at the beginning of the 10-year period. The results for crop cover correspond to the changes as described in detail in Amogu et al. (2010) [22], Descroix et al. 2012 [3] and Descroix et al. 2013 [4] for the research area. The LULC processes lead to a decrease in infiltrability and more direct runoff and regeneration of groundwater. They attribute the regeneration of the groundwater to the fast infiltration in the rivers, which carry more water due to the increase in runoff. This effects lead to a higher frequency of flow changes and flashiness. For pasture, this process seems to be much weaker in the modeling results.

These results reflect the sensitivity analysis of different studies of LULC in Africa, using the same land class parameters of the SWAT model. Awotowi *et al.* (2014) [33] show for West Africa that land use classes of SWAT are generally suitable for West African conditions. They found similar effects of LULC on the hydrology for the Volta basin, where cropland replaced savannah and grassland. Other studies undertaken with SWAT in East Africa studying the effects of LULC focus more on the transformation of forest to cropland [34,35,51]. Concerning the effect of the transformation from savannah into pasture as taken place mainly in the Goroul catchment, it was not possible to test whether the parameterization of the model reflects the hydrological processes. There is no quantitative data or literature available to the authors which could be used to verify the modelled effects. The potential misparametrization of pasture is discussed in more detail in Sections 3.2 and 3.3.

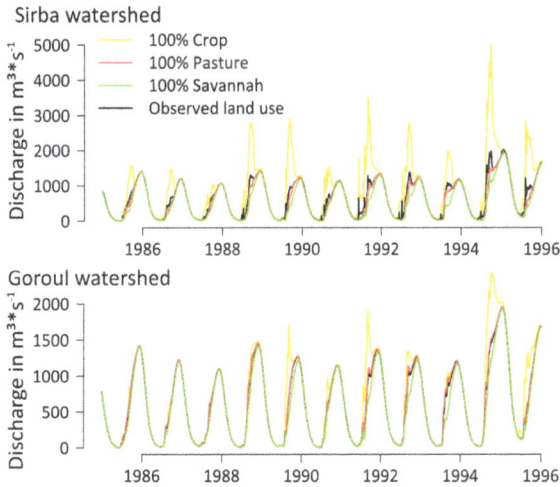

Figure 7. Comparison of discharges with four different land use coverages (100% crop, 100% pasture, 100% Savanna and land use as observed) for the Sirba and Goroul watersheds.

2.7. Statistical Methods

In order to make general trends of the time series more clearly visible, the local regression fitting technique LOESS was used [52]. This is a nonparametric regression method that combines multiple regression models in a k-nearest-neighbor-based meta-model [53]. When plotted, it generates a smooth curve through a set of data points. It is used to depict nonlinear trends in time series.

Since the absolute values of discharge differ distinctively between the different parts of the catchment, AMAX anomalies are used in order to make the results more easily comparable. The time series of AMAX anomalies is given by the time series of AMAX of the individual years divided by the mean AMAX of the entire AMAX time series of 1950–2009, given as a percentage. The annual rainfall time series are transformed to anomalies accordingly.

Linear trends in the observed and simulated AMAX time series are estimated using the Theil-Sen estimator [54,55]. It estimates the slope of a trend and is widely used, since it is insensitive to outliers. Since serial independence is a requirement of the test, the data was checked for autocorrelations using the Durbin-Watson statistic test [56,57]. It tests the null hypothesis that the residuals from an ordinary least-squares regression are not autocorrelated against the alternative that the residuals are autocorrelated. If an autocorrelation of the first order was found, trend-free pre-whitening was applied according to the method proposed by Yue *et al.* (2002) [58]. It is a procedure to remove serial correlation from time series, and hence to eliminate the effect of serial correlation on the Theil-Sen estimator.

As a result of the gaps in the time series for the Red Flood (see Section 2.4.2), it is not possible to generate a local regression curve and to identify the minimum of the AMAX time series. The linear trends are therefore only calculated from 1984 to 2009 with five missing years (1986, 1987, 1993, 1995, 1996). This means that the AMAX trends are calculated consistently for the simulations even with these gaps and can therefore be included in the interpretation.

For the calibration and validation, the methods of Nash and Sutcliff (1970) (NSE) [59], and percent bias (PBIAS) have been used. The NSE is calculated using the formula.

$$NSE = 1 - \frac{\sum_{i=1}^{n} (Q_{obs} - Q_{sim})^2}{\sum_{i=1}^{n} (Q_{obs} - \overline{Q_{obs}})^2}$$

Values for the NSE range from 1 to negative ∞ values. An NSE of 0 means that the model is no better than using average annual discharge as a predictor. If NSE = 1, it means that the model is perfectly aligned with the observations. The PBIAS indicates the over- or underestimation of discharge during the calibration or validation period as a percentage. For the evaluation of the NSE and PBIAS the terminology of Moriasi *et al.* [60] is used (for NSE: very good: 0.75–1.0, good: 0.65–0.75, satisfactory: 0.5–0.65, unsatisfactory: <0.5; for PBIAS: very good: $<|10|$, good: $|10|-|15|$, satisfactory: $|15|-|25|$, unsatisfactory: $\geq|25|$).

2.8. Hypothesis Testing Framework

In order to attribute the increase of discharge observed in the research area to LULC and/or climatic changes, a simulation-based approach within a hypothesis testing framework as proposed by Merz *et al.* [5] is applied. The framework consists

18

of the evidence of consistency, the evidence of inconsistency, and the provision of confidence level. In order to show the consistency or inconsistency, the observed trend is compared to the change simulated by the hydrological model, either with or without considering the potential driver LULC. The model which includes all postulated drivers, as in our case both changes in LULC and climatic regime, should be able to reproduce the observed trends and are therefore control runs. The simulations without one of the postulated drivers, might result in no trends or differing trends. This difference can then be used for the quantification of the influence. In other words, if the model is capable of simulating the observed trend without LULC, it would mean that LULC has no or little influence on the discharge trend and vice versa (within the uncertainty limits of model runs).

In this study, the observed trend is an increase of flood peak magnitude since the 1970s/1980s. Therefore, the AMAX is derived from the observed and simulated daily time series of discharge. The AMAX time series have been transformed from absolute values to anomalies (see Section 2.7) in order to be able to compare the gradients amongst the subcatchments. The gradients of the trends are used as measure for the comparison. Ideally, the second/control run including LULC shows a similar gradient like the observed trend. The difference between the gradients of the run without LULC and the observed then consequently indicates the share of the climatic variability.

Annual rainfall is used as general indicator of the wetness trend in the respective watershed. The rainfall trend mainly helps to illustrate the SP (see Section 2.1.), and whether the simulations are able to reproduce the phenomenon when including LULC.

3. Results and Discussion

3.1. Validation of the Model

To quantify the efficiency of the model, the NSE and PBIAS were used (see Section 2.7). The model showed very high efficiency at the gauging station Niamey for the validation period (NSE: 0.87/PBIAS: 6.9) (Figure 8). For the Goroul basin, the results are satisfactory for the NSE (0.55) and slightly unsatisfactory for the Sirba basin (0.49). For the PBIAS, only the validation of the station Alcongui shows satisfactory efficiency (−16.9). At the station Garbe-Kourou, the PBIAS is over 25 and therefore unsatisfactory. The very good performance of the model for the station Niger can be explained by the fact that monitored data was fed into the model at the station Diré, while the reason for the weaker performance of the model in the watersheds of the Goroul and Sirba Rivers after the intensive calibration efforts is unclear. We assume that the climatic reanalysis data for these subcatchments are at above-average deficiency and/or that the land use data for these regions are not accurate. In fact, some years with similar observed hydrographs, like 1974 and 1975 at the Goroul river, for example, are simulated once very accurately and yet in the other year are deficient. An additional reason for the difficulties could be that the discharge of rivers in dry environments is generally more sensitive to climate input due to the low runoff-coefficients [37], which is especially the case for the intermittent rivers Goroul and Sirba. Inaccurate climate forcing is therefore more likely to affect the model performance in drier regions. The reasons, therefore, are the proportionally higher losses in smaller streams through evapotranspiration, transmission losses, *etc.* compared to large rivers. A certain decrease in rainfall thus leads to a proportionally larger decrease in discharge in dry areas compared to wetter areas [37]. Other explanations for the lower performance might be data quality of the streamflow, the parametrization of the land use or other deficits in the model structure, e.g., for the representation of the groundwater. These input data related problems are taken into account in the discussion (see Sections 3.2 and 3.3).

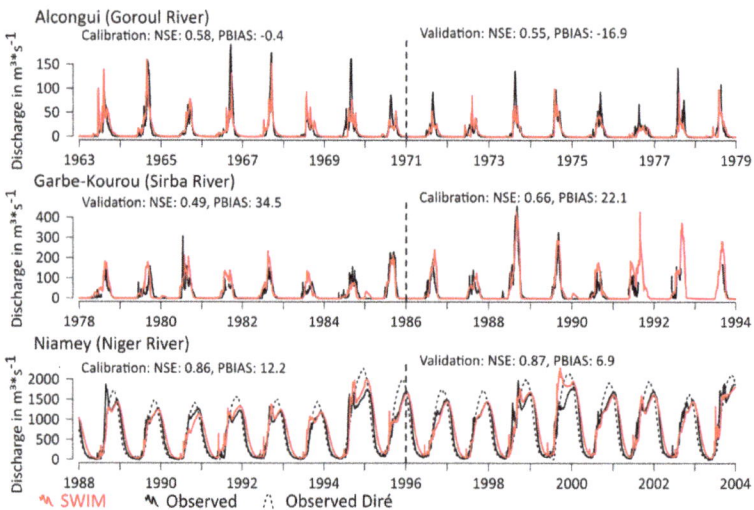

Figure 8. Validation and calibration of the SWIM model for the watersheds of Alcongui, Garbe-Kourou and Niamey for eight-year periods using PGFv2 reanalysis climate forcing. For Niamey, the measured discharge at the Diré gauge is additionally plotted, which is fed into the model.

Since in this study absolute numbers of discharge are not analyzed but only anomalies, the NSE is more meaningful than the PBIAS. Therefore, the output of the model is used in the study for the Niamey gauge but also for the smaller subcatchments, with lower but still satisfying model efficiency, when looking at the calibration and validation phases.

3.2. Attribution of Trends in Annual Maximum Discharges

3.2.1. Simulation Results

The gradients of the trend lines for all AMAX anomaly time series are used as a quantitative attribution measure as specified in Section 2.8. (see trend lines in the lower part of Figure 9 and gradients in Table 3) All estimated observed and simulated trends are positive and statistically significant ($\alpha = 0.05$).

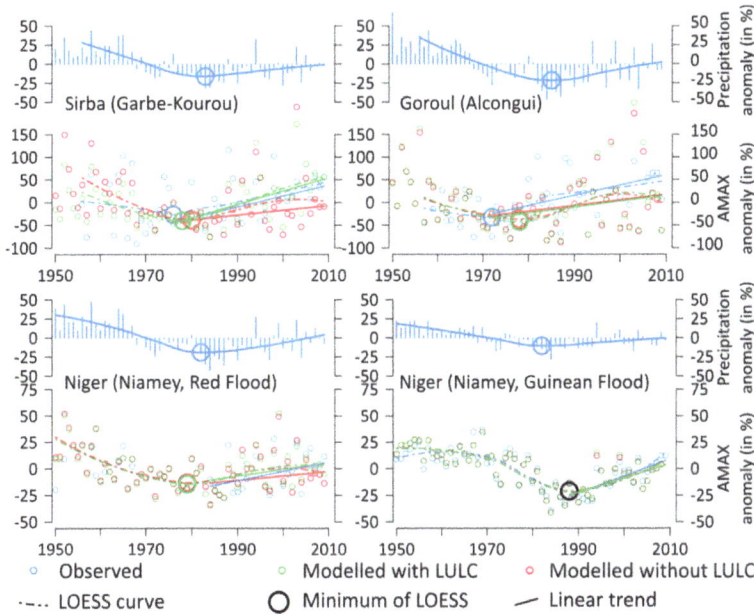

Figure 9. Anomalies of annual maximum discharges for the gauging stations Alcongui (Goroul River), Garbe-Kourou (Sirba River) and differentiated between the Red and the Guinean Flood for Niamey (Niger River). On the top of each region, rainfall anomalies over the respective catchment are plotted. A LOESS curve with a minimum point is added as a dashed line and the Theil-Sen estimators for the discharge trends are plotted as bold lines, beginning at the minimum of the observed discharge points. Please note that, for the observed values of the Red Flood at Niamey, the time series is incomplete (see Section 2.4.2). Therefore, the LOESS curve is not plotted and the trends start at 1984 (see Section 2.6). For the Guinean Flood at Niamey, all minima are on the same point and the circle is plotted in black.

In the Sirba catchment, the simulation run including LULC reproduces the trend of AMAX anomaly adequately (sim$_{with\ LULC}$: 2.94, obs.: 2.42, Table 3), whereas the

21

simulation run that assumed no LULC since 1950s does not reproduce the observed trend of AMAX anomaly adequately ($sim_{without\ LULC}$: 1.05).

Table 3. Gradient of the trends of AMAX anomalies (see Figure 4) as estimated with the Theil-Sen approach.

Gauging Stations	Observed	Simulation w. LULC		Simulation w/o LULC	
		Gradient	% of Observed Gradient	Gradient	% of Observed Gradient
Sirba River (Garbe-Kourou)	2.42	2.94	121%	1.05	43%
Goroul River (Alcongui)	2.37	1.38	58%	1.29	54%
Niger River (Niamey, Red Flood)	0.87	0.63	72%	0.56	64%
Niger River (Niamey, Guinean Flood)	1.61	1.37	85%	1.36	84%

In contrast, for the Guinean flood of the Niger River at Niamey, the AMAX anomaly gradients of both simulations are similar to the observed trends of the AMAX anomaly ($sim_{with\ LULC}$: 1.37, $sim_{without\ LULC}$: 1.36, obs: 1.61). For the so-called Red Flood of the Niger River at Niamey, the simulation with LULC is closer to the observed AMAX trend than the run without considering LULC ($sim_{with\ LULC}$: 0.63, $sim_{without\ LULC}$: 0.56, obs.: 0.87), but both are not able to reproduce the trend adequately.

In the Goroul catchment, the gradients of both simulation runs ($sim_{with\ LULC}$: 1.38, $sim_{without\ LULC}$: 1.29, obs.: 2.37) significantly underestimate the observed gradient of AMAX anomalies.

The Sahel Paradox (SP) of the Sirba catchment is distinct with an offset between the minima of AMAX and annual rainfall of approximately 10 years. Both simulations show this effect while the SP of the control run is more distinct, with an AMAX minimum close to the observed AMAX minimum.

In the Goroul catchment the SP is very pronounced, too, with an offset of approximately 14 years. Both simulations show the effect while they underestimate the offset by approximately seven years. For the Red Flood at Niamey, it is not possible to detect the magnitude of the SP shift of the observed AMAX due to the missing LOESS curve and minimum (see Section 2.7). However, both simulations show an offset of around five years, which is in accordance with the findings from the Sirba River. For the Guinean Flood the effect of the SP does not occur and rainfall and AMAX minima do not show the paradoxical offset.

3.2.2. Discussion of Simulation Results

The effects of LULC are most pronounced in the model runs for the Sirba subcatchment. The main LULC in this watershed is a distinct increase in cropland and a reduction of natural savannah (Figure 3). This leads to an increase of surface flow and less evapotranspiration, both represented in the model parametrization. Because the run without LULC generates a trend gradient of 43% of the gradient of the observed trend, this share can be attributed to the climatic changes and, consequently, the other half to the LULC.

The Guinean Flood is almost exclusively controlled by the hydrological setting upstream of the Sahelian Niger. Since this study only simulates the basin downstream of Diré, the LULC effects upstream are not simulated. If there were effects, they are inherent in the observed discharge which was fed into the model at Diré. Still, the absence of any LULC effect when flowing through the Sahelian part, as well as the correlation between the rainfall and the AMAX trends, supports the finding of Aich *et al.* [1]. They identified the climate as the main driver of the Guinean Flood. The differences between the trend magnitudes of rain and AMAX can be explained by the sensitivity of the Niger basin, which causes a higher increase in discharge compared to the magnitude of rainfall [37].

For the Red Flood of the Niger River at Niamey, there is almost no difference in the gradients of the simulations with and without LULC. The runs without LULC and hence with only the climatic forcing causing the trend can explain 64% of the observed AMAX anomaly trend. The modelled effect of LULC is small and reproduces 72% of the observed trend. These results of the Red Flood correspond to the results of the Goroul catchment, which is the beside the Sirba catchment the second major tributary catchment generating the Red Flood (see Section 2.1). Therefore they are discussed together.

In the Goroul catchment, the poor performance of the model runs with and without LULC only allows a partial quantification of the relative effect of the drivers. Following the simulation results without LULC, the climatic part of the simulations explains 54% of the observed trend gradient. The cause for the remaining 46% is unclear, since the runs with LULC do not reproduce the trend significantly better. Therefore, it is not possible to give robust statements for the Goroul catchment and for the Red Flood as to the influence of the assumed drivers and quantification is a fortiori uncertain. The results might even indicate that there is an additional driver forcing the AMAX trends besides climatic changes and LULC, which is not represented in the model and has not been postulated. Another general point is the quality of the observed discharge data and the reanalyzed climate data. Especially in the Goroul catchment, the validation shows that some peaks are missed completely by the model (e.g., Alcongui 1977), or that peaks which are modelled do not occur at all in the discharge data (e.g., Alcongui 1963). This indicates the general deficit of the employed observed reanalyzed data and needs to be taken into account for the interpretation as well. However, a more probable explanation is the combination of inadequate information on LULC and/or a deficient representation of LULC in the model. The dominant change in the reanalysis data set of LULC in the watershed is a transformation of savannah to pasture (Figure 2). These two land cover types do not differ substantially in their modelled effects on the hydrological regime (see

Section 2.6, Figure 7). Therefore, the similar and poor results for the runs with and without the LUCM are a logical consequence. The parameters for pasture of the SWAT land class parameters are most likely optimized for pastures in the temperate zone. The hydrological effects of pasture in the Sahel differ probably from these temperate pastures. However, since no other data was available to the authors in order to check or even change the effects and the related parameters of pasture in the Sahel, this deficit could neither be verified nor corrected. We assume that especially the infiltration capacities of Sahelian pastures are lower than of pastures in the temperate zones, e.g., due to soil-crusting effects and less dense vegetation. An additional source of uncertainty is the undifferentiated information on LULC with regard to savannah. Descroix *et al.* (2012) [3] describe the LULC of this region with a change from the original bushy and woody savannah to a degraded savannah (Tiger Bush, see Section 2.1) and sandy slopes due to land clearing. These changes are not represented in the LULC data (see Section 2.4.3).

The SP is a special aspect of the increase of flood magnitudes. Following our results, the same statements and explanations, which can explain the increase in flood magnitude, hold also for the SP. For all time series (except the ones from the Guinean Flood of the Niger River at Niamey, where the SP did not occur), the SP phenomenon could be reproduced without LULC, but to a lower degree than when accounting for LULC. We assume that this effect is based on changes in the frequency of precipitation in the climatic forcing. The findings of Panthou *et al.* [61,62] of increasing convective precipitation supports this theory that heavy precipitation leads to an increase of the run-off coefficient, and finally to an increase in discharge. This effect is also shown by means of measurements in Amogu *et al.* [22]. Eventually, the combined actions of LULC and changes in precipitation patterns seem responsible for the observed SP in the 1970s/1980s.

Another finding is the non-stationary influence of climate. The model runs without LULC, and hence with only climate as the driver show upward trends in AMAX for both tributaries and the Red Flood of the Niger River at Niamey. These increases are, however, stronger than the respective annual rainfall anomalies would suggest. Therefore, we assume that changes in the rainfall patterns, most likely an increase in the frequency of heavy precipitation, contribute to the trend. This is supported by state-of-the-art analysis of climatic changes in the Sahel, for example by Panthou *et al.* [61,62].

3.3. Discussion of the Methodological Framework Employed and Related Uncertainties

We acknowledge that simulation runs and the calibration procedure are affected by a wide range of uncertainties regarding the processes in the model, their parametrization and the driver data. Relevant processes like evapotranspiration and surface runoff are not entirely physically based in the model but simplified and calibrated with a factor or empirical (e.g., the factor for potential evapotranspiration and the curve number approach). Therefore, the effects of LULC on these processes could be represented inadequately. However, the SWIM model showed in this study, like SWAT model in other LULC studies in the region [33], that the process-based models are generally able to reproduce the effects of LULC.

The data on LULC as the key driver of the attribution study cannot be easily improved regarding temporal and spatial resolution since no other data products or observation on LULC are available for the region. There is only a qualitative confirmation of the used LULC data by studies which are based on observations [3,22]. The global LULC data set seemed therefore partly reliable, but, for example degradation of savannah (Tiger Bush, see Section 2.1), as reported in the studies are not represented in such a detail. We therefore recommended for similar studies the use of regional and more detailed data sets, if available.

Regarding the uncertainty coming from the parametrization, especially the representation of land cover types in the model has to be adequately included in terms of their hydrological attributes. The robustness of a simulation-based attribution study is therefore not only dependent on the general validation results of the model, but rather on the ability of the model to reflect well-known changes of the postulated driver(s) adequately. Especially the parametrization of pasture in the model is questionable, but could, due to the lack of related field data, not be improved. A way to avoid this uncertainty might be the calibration of each of the main changing land cover types separately by looking at small catchments which are dominated by one of these land uses [63]. In doing so, the sensitivity of the model against the impact of each change can be verified and tested whether the representation is adequate or not. Unfortunately, discharge data for the validation of the model performance on such homogeneous areas were not available for the research area of this study.

Additional uncertainty is related to the general methodological framework. A modeling attribution study only can provide robust results, if all potential drivers of the observed change were adequately represented in the model. Two drivers, LULC and climatic changes, which are mentioned by qualitative studies as potential sources for the increasing flood trend [3,22] are tested in this study. However, we cannot exclude the influence of a third or more drivers for the observed changes. For example the effects of the extreme population growth in the region, might influence the hydrological regime, possibly via sealing of soils near settlements. Still, the two considered drivers LULC and climate are the only potential drivers in the current literature of the Niger, which makes the assumption more reliable.

Also the assumption of linear trends in the AMAX anomaly time series and accordingly in the LULC and climatic changes is probably not realistic.

25

However, comparing non-linear trends would request an even higher level of data quality/density, which is not available in the research area.

Another critical issue is model performance with respect to other important state variables of the model like groundwater flows and evapotranspiration. In this study, the performance of the model was only tested with regard to discharge, since no quantitative data was available on groundwater dynamics. However, in the Sahelian part of the Niger, groundwater is known to play a crucial role [64] and it cannot be excluded that the model underestimates the effects of a potential relevant process related to LULC.

However, keeping model uncertainty in mind—Even under conditions of comparatively poor data quality and availability, the simulation-based approach using a process-based based model, shed new light on attribution of increasing floods in the Sahelian part of the Niger by developing some first quantitative measure of comparing the linear trend estimates of AMAX anomalies. In addition, even without adequately representing the LULC in the Goroul catchment and for the Red Flood at the station Niamey, the method could show that at least climatic changes contribute substantially to the flood increase in the region.

4. Conclusions

The research question, to which share LULC and/or climatic changes cause the increase of flooding in the Sahelian part of the Niger basin, can only be answered partly by this study. The most reliable conclusions can be drawn from the results of the Sirba catchment, since the simulation including LULC is able to reproduce the observed trend adequately. The influences of both drivers on the observed trend of AMAX seem to be roughly equal with a share of 43% which can be—within the known limitations—attributed to the climatic changes.

For the Goroul subcatchment and the Red Flood of the Niger River, the results are in the same order of magnitude with shares of the climatic forcing of 54% (Goroul) and 64% (Niamey Red Flood). Since the model was not able to reproduce the influence of LULC adequately, the results for these two stations are uncertain and only partly reliable. For Goroul and the Red Flood, we conclude that climatic changes have A major influence on the observed trend of AMAX and LULC also contributes, but to an unknown amount.

Using a process-based eco-hydrological model seems to be a valid method for attributing an increase of flooding. Even though main processes related to LULC, like e.g., evapotranspiration, are simplified and calibrated in the model or empirical like the curve number approach for surface runoff, the effects of LULC on the hydrology are assumed to be generally well represented since the modelled effects reflect observed effects and other studies could show a general suitability of process-based models for LULC studies. Still we see a general need for hydrological science to overcome the calibration of land use sensitive parameters in favor of more physically-based models, especially for LULC studies. These efforts are, however, still limited by data availability, computing power and partly also detailed process understanding.

In regard of the specific methodology applied in this study, the gradient prooved to be an uncomplicated and intuitive measure for comparing simulated and observed trends and estimating the influence of different drivers. However, the demands for a robust attribution on the model and data quality/density are high. Testing the method in a data-richer environment, where especially more information on LULC are available, would help to get a better understanding of its robustness and reliability.

Regarding flood mitigation and adaptation strategies, the modeling framework could be employed to assess different land management options. If the flood risk is to a significant extent due to LULC (rather than due to a climatic change), it means that it can also be counteracted. State-of-the-art options implemented locally can reduce surface runoff, for example, by reforestation, smart planting techniques which also reduce erosion, shifting cultivation *etc.*, such as those published in the World Overview of Conservation Approaches and Technologies (WOCAT) [65] or weADAPT [66]. However, in the face of the existing flood risk in the region and the return to wet conditions, mitigation is not enough. There is a strong need for immediate adaptation measures and we argue for early-warning systems, investments in flood protection infrastructure and flood-smart settlement policies in the riverine nations. To ensure cost-efficient implementation, a simulation-based approach can be further used to assess the relative merits of both mitigation and adaptation measures [67].

Supplementary Materials: Supplementary materials can be found at http://www.mdpi.com/2073-4441/7/6/2796/s1.

Acknowledgments: We thank the IMPACT2C project for financing this study and the Niger Basin Authority (NBA) for providing data.

Conflicts of Interest: The authors declare no conflict of interest.

References

1. Aich, V.; Koné, B.; Hattermann, F.F.; Müller, E.N. Floods in the Niger basin—Analysis and attribution. *Nat. Hazards Earth Syst. Sci. Discuss.* **2014**, *2*, 5171–5212.

2. Tarhule, A. Damaging rainfall and flooding: The other sahel hazards. *Clim. Chang.* **2005**, *72*, 355–377.

3. Descroix, L.; Genthon, P.; Amogu, O.; Rajot, J.-L.; Sighomnou, D.; Vauclin, M. Change in Sahelian Rivers hydrograph: The case of recent red floods of the Niger River in the Niamey region. *Glob. Planet. Chang.* **2012**, *98–99*, 18–30.

4.   Descroix, L.; Bouzou, I.; Genthon, P.; Sighomnou, D.; Mahe, G.; Mamadou, I.; Vandervaere, J.-P.; Gautier, E.; Faran, O.; Rajot, J.-L.; *et al.* Impact of drought and land—Use changes on surface—Water quality and quantity: The sahelian paradox. In *Current Perspectives in Contaminant Hydrology and Water Resources Sustainability*; Bradley, P., Ed.; InTech: Rijeka, Croatia, 2013; pp. 243–271.

5.   Merz, B.; Vorogushyn, S.; Uhlemann, S.; Delgado, J.; Hundecha, Y. HESS Opinions "More efforts and scientific rigour are needed to attribute trends in flood time series.". *Hydrol. Earth Syst. Sci. Discuss.* **2012**, *16*, 1379–1387.

6.   Giuntoli, I.; Renard, B.; Vidal, J.-P.; Bard, A. Low flows in France and their relationship to large-scale climate indices. *J. Hydrol.* **2013**, *482*, 105–118.

7.   Murphy, C.; Harrigan, S.; Hall, J.; Wilby, R.L. Climate-Driven trends in mean and high flows from a network of reference stations in Ireland. *Hydrol. Sci. J.* **2013**, *58*, 755–772.

8.   Vorogushyn, S.; Merz, B. What drives flood trends along the Rhine River: Climate or river training? *Hydrol. Earth Syst. Sci. Discuss.* **2012**, *9*, 13537–13567.

9.   Mediero, L.; Santillán, D.; Garrote, L.; Granados, A. Detection and attribution of trends in magnitude, frequency and timing of floods in Spain. *J. Hydrol.* **2014**, *517*, 1072–1088.

10. Villarini, G.; Strong, A. Roles of climate and agricultural practices in discharge changes in an agricultural watershed in Iowa. *Agric. Ecosyst. Environ.* **2014**, *188*, 204–211.

11. Prosdocimi, I.; Kjeldsen, T.R.; Svensson, C. Non-Stationarity in annual and seasonal series of peak flow and precipitation in the UK. *Nat. Hazards Earth Syst. Sci.* **2014**, *14*, 1125–1144.

12. Harrigan, S.; Murphy, C.; Hall, J.; Wilby, R.L.; Sweeney, J. Attribution of detected changes in streamflow using multiple working hypotheses. *Hydrol. Earth Syst. Sci.* **2014**, *18*, 1935–1952.

13. Gebrehiwot, S.G.; Seibert, J.; Gärdenäs, A.I.; Mellander, P.-E.; Bishop, K. Hydrological change detection using modeling: Half a century of runoff from four rivers in the Blue Nile Basin. *Water Resour. Res.* **2013**, *49*, 3842–3851.

14. Schreider, S.Y.; Jakeman, A.J.; Letcher, R.A.; Nathan, R.J.; Neal, B.P.; Beavis, S.G. Detecting changes in streamflow response to changes in non-climatic catchment conditions: Farm dam development in the Murray-Darling basin, Australia. *J. Hydrol.* **2002**, *262*, 84–98.

15. Seibert, J.; McDonnell, J.J. Land-Cover impacts on streamflow: A change-detection modelling approach that incorporates parameter uncertainty. *Hydrol. Sci. J.* **2010**, *55*, 316–332.

16. Andréassian, V.; Parent, E.; Michel, C. A distribution-free test to detect gradual changes in watershed behavior. *Water Resour. Res.* **2003**.

17. Hattermann, F.; Kundzewicz, Z.W.; Huang, S.; Vetter, T.; Kron, W.; Burghoff, O.; Merz, B.; Bronstert, A.; Krysanova, V.; Gerstengarbe, F.-W.; *et al.* Flood risk from a holistic perspective-observed changes in Germany. In *Changes of Flood Risk in Europe*; Kundzewicz, Z.W., Ed.; IAHS Press: Wallingford, UK, 2012; pp. 212–237.

18. Hundecha, Y.; Merz, B. Exploring the relationship between changes in climate and floods using a model-based analysis. *Water Resour. Res.* **2012**.

19. Kandji, S.T.; Verchot, L.; Mackensen, J. *Climate Change and Variability in the Sahel Region: Impacts and Adaptation Strategies in the Agricultural Sector*; World Agroforestry Centre: Nairobi, Kenya, 2006.

20. Jury, M.R. A return to wet conditions over Africa: 1995–2010. *Theor. Appl. Climatol.* **2013**, *111*, 471–481.

21. Mahe, G.; Lienou, G.; Descroix, L.; Bamba, F.; Paturel, J.E.; Laraque, A.; Meddi, M.; Habaieb, H.; Adeaga, O.; Dieulin, C.; *et al.* The rivers of Africa: Witness of climate change and human impact on the environment. *Hydrol. Process.* **2013**, *27*, 2105–2114.

22. Amogu, O.; Descroix, L.; Yéro, K. Increasing river flows in the Sahel? *Water* **2010**, *2*, 170–199.

23. Hurtt, G.C.; Chini, L.P.; Frolking, S.; Betts, R.A.; Feddema, J.; Fischer, G.; Fisk, J.P.; Hibbard, K.; Houghton, R.A.; Janetos, A.; *et al.* Harmonization of land-use scenarios for the period 1500–2100: 600 years of global gridded annual land-use transitions, wood harvest, and resulting secondary lands. *Clim. Chang.* **2011**, *109*, 117–161.

24. Krysanova, V.; Müller-Wohlfeil, D.-I.; Becker, A. Development and test of a spatially distributed hydrological/water quality model for mesoscale watersheds. *Ecol. Model.* **1998**, *106*, 261–289.

25. Arnold, J.G.; Allen, P.M.; Bernhardt, G. A comprehensive surface-groundwater flow model. *J. Hydrol.* **1993**, *142*, 47–69.

26. Krysanova, V.; Meiner, A.; Roosaare, J.; Vasilyev, A. Simulation modelling of the coastal waters pollution from agricultural watershed. *Ecol. Model.* **1989**, *49*, 7–29.

27. Huang, S.; Krysanova, V.; Hattermann, F. Projection of low flow conditions in Germany under climate change by combining three RCMs and a regional hydrological model. *Acta Geophys.* **2013**, *61*, 151–193.

28. Arnold, J.; Williams, J.; Nicks, A.; Sammons, N. *SWRRB: A Basin Scale Simulation Model for Soil and Water Resources Management*; Texas A&M University Press: College Station, TX, USA, 1990.

29. Wendling, U.; Schellin, H. Neue Ergebnisse zur Berechnung der potentiellen Evapotranspiration. *Z. Meteorol.* **1986**, *36*, 214–217.

30. Ritchie, J. Model for predicting evaporation from a row crop with incomplete cover. *Water Resour. Res.* **1972**, *8*, 1204–1213.

31. Williams, J.R.; Renard, K.G.; Dyke, P.T. EPIC: A new method for assessing erosion's effect on soil productivity. *J. Soil Water Conserv.* **1983**, *38*, 381–383.

32. Arnold, J.G.; Kiniry, J.R.; Srinivasan, R.; Williams, J.R.; Haney, E.B.; Neitsch, S.L. *SWAT 2012 Input/Output Documentation*; Texas Water Resources Institute: Forney, TX, USA, 2013.

33. Awotwi, A.; Yeboah, F.; Kumi, M. Assessing the impact of land cover changes on water balance components of White Volta Basin in West Africa. *Water Environ. J.* **2015**, *29*, 259–267.

34. Baker, T.J.; Miller, S.N. Using the Soil and Water Assessment Tool (SWAT) to assess land use impact on water resources in an East African watershed. *J. Hydrol.* **2013**, *486*, 100–111.

35. Nyeko, M.; D'Urso, G.; Immerzeel, W.W. Adaptive simulation of the impact of changes in land use on water resources in the lower Aswa basin. *J. Agric. Eng.* **2013**, *43*, 24.

36. Liersch, S.; Cools, J.; Kone, B.; Koch, H.; Diallo, M.; Reinhardt, J.; Fournet, S.; Aich, V.; Hattermann, F.F. Vulnerability of rice production in the Inner Niger Delta to water resources management under climate variability and change. *Environ. Sci. Policy* **2013**, *34*, 8–33.

37. Aich, V.; Liersch, S.; Vetter, T.; Huang, S.; Tecklenburg, J.; Hoffmann, P.; Koch, H.; Fournet, S.; Krysanova, V.; Müller, E.N.; *et al.* Comparing impacts of climate change on streamflow in four large African river basins. *Hydrol. Earth Syst. Sci.* **2014**, *18*, 1305–1321.

38. Vetter, T.; Huang, S.; Aich, V.; Yang, T.; Wang, X.; Krysanova, V.; Hattermann, F. Multi-Model climate impact assessment and intercomparison for three large-scale river basins on three continents. *Earth Syst. Dyn. Discuss.* **2014**, *5*, 849–900.

39. Weedon, G.P.; Gomes, S.; Viterbo, P.; Shuttleworth, W.J.; Blyth, E.; Österle, H.; Adam, J.C.; Bellouin, N.; Boucher, O.; Best, M. Creation of the WATCH forcing data and its use to assess global and regional reference crop evaporation over land during the twentieth century. *J. Hydrometeorol.* **2011**, *12*, 823–848.

40. Kim, H. GSWP3. Available online: http://hydro.iis.u-tokyo.ac.jp/GSWP3/exp1.html (accessed on 24 April 2015).

41. Sheffield, J.; Goteti, G.; Wood, E.F. Development of a 50-year high-resolution global dataset of meteorological forcings for land surface modeling. *J. Clim.* **2006**, *19*, 3088–3111.

42. Niger Basin Authority NIGER-HYCOS. Available online: http://nigerhycos.abn.ne/ (accessed on 1 March 2012).

43. Bartholomé, E.; Belward, A.S. GLC2000: A new approach to global land cover mapping from Earth observation data. *Int. J. Remote Sens.* **2005**, *26*, 1959–1977.

44. Klein Goldewijk, K.; Beusen, A.; Janssen, P. Long-Term dynamic modeling of global population and built-up area in a spatially explicit way: HYDE 3.1. *Holocene* **2010**, *20*, 565–573.

45. Klein Goldewijk, K.; Beusen, A.; van Drecht, G.; de Vos, M. The HYDE 3.1 spatially explicit database of human-induced global land-use change over the past 12,000 years. *Glob. Ecol. Biogeogr.* **2011**, *20*, 73–86.

46. Ramankutty, N. Croplands in West Africa: A geographically explicit dataset for use in models. *Earth Interact.* **2004**, *8*, 1–22.

47. FAO; IIASA; ISRIC; ISSC. JRC Harmonized World Soil Database v 1.2. Available online: http://webarchive.iiasa.ac.at/Research/LUC/External-World-soil-database/ HTML/ (accessed on 24 April 2015).

48. Jarvis, A.; Reuter, H.I.; Nelson, A.; Guevara, E. Hole-Filled Seamless SRTM Data V4. Available online: http://srtm.csi.cgiar.org/ (accessed on 13 January 2015).

49. Arnold, J.G.; Moriasi, D.N.; Gassman, P.W.; Abbaspour, K.C.; White, M.J.; Srinivasan, R.; Santhi, C.; Harmel, R.D.; van Griensven, A.; van Liew, M.W.; *et al.* SWAT: Model use, calibration, and validation. *Trans. Am. Soc. Agric. Biol. Eng.* **2012**, *55*, 1491–1508.

50. Doherty, J. *PEST—Model-Independent Parameter Estimation, User Manua*, 5th ed.; Computing, W.N., Ed.; Watermark Numerical Computing: Brisbane Area, Australia, 2005.

51. Nobert, J.; Jeremiah, J. Hydrological response of watershed systems to land use/cover change: A case of wami river basin. *Open Hydrol. J.* **2012**, *6*, 78–87.

52. Cleveland, W.S.; Devlin, S.J. Locally weighted regression: An approach to regression analysis by local fitting. *J. Am. Stat. Assoc.* **1988**, *83*, 596–610.

53. Altman, N.S. An introduction to kernel and nearest-neighbor nonparametric regression. *Am. Stat.* **2012**, *46*, 175–185.

54. Sen, P. Estimates of the regression coefficient based on Kendall's tau. *J. Am. Stat. Assoc.* **1968**, *63*, 1379–1389.

55. Theil, H. A rank-invariant method of linear and polynomial regression analysis. *Ned. Acad. Wetensch. Proc.* **1950**, *53*, 386–392.

56. Durbin, J.; Watson, G. Testing for serial correlation in least squares regression. I. *Biometrika* **1950**, *37*, 409–428.

57. Durbin, J.; Watson, G. Testing for serial correlation in least squares regression. II. *Biometrika* **1951**, *38*, 159–178.

58. Yue, S.; Pilon, P.; Phinney, B.; Cavadias, G. The influence of autocorrelation on the ability to detect trend in hydrological series. *Hydrol. Process.* **2002**, *16*, 1807–1829.

59. Nash, J.E.; Sutcliffe, J.V. River flow forecasting through conceptual models part I—A discussion of principles. *J. Hydrol.* **1970**, *10*, 282–290.

60. Moriasi, D.; Arnold, J. Model evaluation guidelines for systematic quantification of accuracy in watershed simulations. *Trans. ASABE* **2007**, *50*, 885–900.

61. Panthou, G.; Vischel, T.; Lebel, T. Recent trends in the regime of extreme rainfall in the Central Sahel. *Int. J. Climatol.* **2014**, *34*, 3998–4006.

62. Panthou, G.; Vischel, T.; Lebel, T.; Blanchet, J.; Quantin, G.; Ali, A. Extreme rainfall in West Africa: A regional modeling. *Water Resour. Res.* **2012**.

63. Strömqvist, J.; Arheimer, B.; Dahné, J.; Donnelly, C.; Lindström, G. Water and nutrient predictions in ungauged basins: Set-up and evaluation of a model at the national scale. *Hydrol. Sci. J.* **2012**, *57*, 229–247.

64. Mahé, G. Surface/groundwater interactions in the Bani and Nakambe rivers, tributaries of the Niger and Volta basins, West Africa. *Hydrol. Sci. J.* **2009**, *54*, 704–712.

65. WOCAT. Available online: https://www.wocat.net/ (accessed on 21 January 2015).

66. weADAPT 4.0. Available online: https://weadapt.org/ (accessed on 21 January 2015).

67. Andersson, J.C.M.; Zehnder, A.J.B.; Wehrli, B.; Jewitt, G.P.W.; Abbaspour, K.C.; Yang, H. Improving crop yield and water productivity by ecological sanitation and water harvesting in South Africa. *Environ. Sci. Technol.* **2013**, *47*, 4341–4348.

Combined Impacts of Medium Term Socio-Economic Changes and Climate Change on Water Resources in a Managed Mediterranean Catchment

Anastassi Stefanova, Cornelia Hesse and Valentina Krysanova

Abstract: Climate projections agree on a dryer and warmer future for the Mediterranean. Consequently, the region is likely to face serious problems regarding water availability and quality in the future. We investigated potential climate change impacts, alone (for three scenario periods) and in combination with four socio-economic scenarios (for the near future) on water resources in a Mediterranean catchment, whose economy relies on irrigated agriculture and tourism. For that, the Soil and Water Integrated Model (SWIM) was applied to the drainage area of the Mar Menor coastal lagoon, using a set of 15 climate scenarios and different land use maps and management settings. We assessed the long-term average seasonal and annual changes in generated runoff, groundwater recharge and actual evapotranspiration in the catchment, as well as on water inflow and nutrients input to the lagoon. The projected average annual changes in precipitation are small for the first scenario period, and so are the simulated impacts on all investigated components, on average. The negative trend of potential climate change impacts on water resources (*i.e.*, decrease in all analyzed components) becomes pronounced in the second and third scenario periods. The applied socio-economic scenarios intensify, reduce or even reverse the climate-induced impacts, depending on the assumed land use and management changes.

Reprinted from *Water*. Cite as: Stefanova, A.; Hesse, C.; Krysanova, V. Combined Impacts of Medium Term Socio-Economic Changes and Climate Change on Water Resources in a Managed Mediterranean Catchment. *Water* **2015**, *7*, 1538–1567.

1. Introduction

The climate of the Mediterranean region is especially vulnerable to potential changes in the global circulation processes [1]. It is therefore not surprisingly, that Giorgi [2] identified the region of the Mediterranean as a primary climate change "Hot-Spot" based on projections of Phase 3 of the Coupled Model Intercomparison Project (CMIP3) for the late 21st century. More recently, Diffenbaugh and Giorgi [3] confirmed this result using a similar but more comprehensive approach and an ensemble of CMIP5 Representative Concentration Pathway (RCP) 8.5 and RCP4.5 simulations.

The agreement among climate models for the Mediterranean region is considerably strong compared to other regions in the world [4] and so are the trends for temperature and precipitation changes. Several studies state that the Mediterranean region is very likely to become dryer and warmer in future, as global warming continues (e.g., [5–9]). In consequence, the area is expected to face serious problems, such as agricultural production losses, land degradation and habitat losses [10]. Moreover, a decrease in water availability will in addition enhance the competition for water between economic sectors and by that the vulnerability of the Mediterranean countries to changes in climate [11].

Our study area, the catchment of the Mar Menor, a hyper saline lagoon in southeast Spain, is one of the driest and hottest regions in Spain and on the Iberian Peninsula [12]. The region is exposed to a severe water stress under the current climate [13] and still, due to intensive irrigation, is one of Europe's major horticultural producers and exporters [14,15]. Apart from agriculture, mass tourism along the shoreline of the lagoon is also highly important for the local economy. The rapid development of both sectors during the last decades was possible due to the Tagus-Segura Inter Basin Transfer (IBT) that has been delivering water from the Tagus River to the Mar Menor catchment since 1978 [16].

Since then, the major watercourse in the catchment, the Albujon wadi, inputs regularly and especially during the wet period high amounts of nutrients from the adjustment agricultural fields to the lagoon [17]. Moreover, insufficiently treated effluents coming mainly from the touristic areas and reaching a maximum during the touristic peak in summer have been discharged into the same wadi over a long period [18]. These inputs have resulted in a rapid increase of the pollution in the lagoon [19–22], and need to be addressed urgently in order to prevent the ecosystem from further degradation.

In addition to current anthropogenic pressures, climate change is expected to pose further stress to the Mar Menor and its drainage area. An increase in sea water temperature, for example, is very likely to lead to an intensification of the eutrophication processes in the lagoon, and finally to a collapse of the water body [23]. Climate change will also affect the water resources in the donor basin of the IBT, the Tagus River Basin. According to Killsby et al. [23] the discharge of the Tagus River is likely to get reduced by almost a half (49%) until the end of the century, due to climate change. Consequently, the water resources transferred to the Mar Menor catchment could also decrease, which will certainly affect the agricultural and touristic sectors, and by that also the water and nutrient inputs to the lagoon. The natural, non-managed water resources in the catchment will be affected by climate change too and will most probably further decline, as climate becomes dryer. CMIP3 model simulations for the Mediterranean region project a long-term decrease in precipitation of about 15% for the end of the century (2070–2099) and of 8% for the near future (2020–2049) compared to 1950–2000 [24]. Based on their results, Mariotti *et al.* [24] expect a decrease in runoff and river discharge as well, which would certainly reduce the water available for irrigation and other uses, as climate change continues. A study on the impacts of climate change on water resources in Spain in particular [25] identified the Segura region, in which the catchment of the Albujon wadi is located, as highly critical regarding the hydrological implications of future climate. More recently Mariotti *et al.* [26] investigated the long-term climate changes in the Mediterranean region using the newly available CMIP5 model simulations, and obtained similar results as in their previous study. The arid and semi-arid regions in the Mediterranean are projected to become dryer, especially during the wet season (December–February), whereas there is no significant decrease in precipitation projected for the already dry summer (June–August) season [26]. Another study comparing the A1B emission scenario (balanced emphasis on all energy sources) simulations of CMIP3 with the RCP4.5 and 8.5 simulations of CMIP5 for the Mediterranean region on the seasonal basis [27] also identified a consistency between both types of scenarios. This indicates a robustness of future climate trends simulated for the region [27], and thus a warmer and dryer future for the Mar Menor catchment becomes even more plausible.

We therefore decided to assess the response of the Mar Menor catchment to potential changes in climate because of its importance for the ecological status of the lagoon and the region's economy. Moreover, as the water resources in the catchment are strongly human influenced, the vulnerability to climate change was studied in combination with potential changes in land use and water management.

For that, we firstly quantified the total amount of water and nutrient inputs to the Mar Menor using the eco-hydrological Soil and Water Integrated Model (SWIM, [28]). The model was then driven by a set of 15 regional climate scenarios from the ENSEMBLES project [29] for one reference and three future scenario periods, of 30 years each. Next, four different socio-economic scenarios, including new land use maps and water management settings were run in combination with the 15 climate scenarios for the near future scenario period.

The results of this study were used for further analysis. They were used as input data to a specific lagoon model that can investigate the lagoons response to changes in its drainage basin, and also as scientifically founded information for stakeholders and policy makers and their future planning and decisions.

2. Case Study Area Description

The catchment of the Mar Menor is situated on the Mediterranean coast in southeast Spain, in the region of Murcia. It covers an area of about 1380 km^2 and overlaps almost entirely with the basin of the Campo de Cartagena aquifer (Figure 1). The Albujon wadi is the major watercourse in the catchment and drains, together with several smaller, so called ramblas (ephemeral water courses with uncontinuous flow), into the Mar Menor. However, there are no gauging stations in the catchment, not even on the Albujon wadi. The main soils are deep Cambisols with low permeability (76%), and agricultural land occupies about 82% of the area.

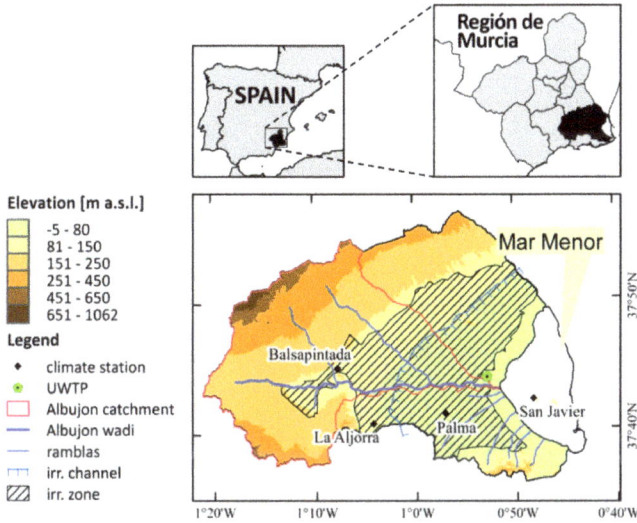

Figure 1. Geographical location of the Mar Menor catchment and major characteristics of the case study area.

The drainage area has a gentle slope and elevations ranging from 1062 m.a.s.l at the low mountainous range Sierra de Carrascoy in the West to −5 m.a.s.l. at the Mediterranean coast in the East. The climate is semi-arid Mediterranean, characterized by dry and hot summers and mild winters. The mean annual temperature is about 18 °C, and the mean annual precipitation about 300 mm, mostly occurring during short episodic storm events in autumn and spring. The estimated potential evapotranspiration is about three times higher than the mean annual precipitation and ranges between 800 mm· y^{-1} and 1200 mm· y^{-1} [30].

Since 1978, the catchment receives water for irrigation and public supply through the Tagus Segura Inter Basin Transfer. The water is transported from the Entrepenas and Buendia reservoirs in the Upper Tagus to the Talave reservoir in the Segura catchment and then redistributed among different sectors and regions, one of them being the Mar Menor catchment. The availability of additional water in the catchment has led to drastic economic and environmental changes over the last decades.

The so called Campo de Cartagena Irrigation District was established, and with it the agricultural practices in the catchment have changed from dry crop farming to intensively irrigated and fertilized fruits and vegetables (mainly lettuce and melon). This resulted in an increased nutrient input from the agricultural fields to water flows and to serious pollution problems in the Mar Menor [17].

Moreover, the former overexploitation of groundwater resources decreased, and surplus of irrigation water is infiltrating into the soil and recharging the aquifer. This has led to a rise in the phreatic levels of the aquifer and in consequence to a continuous flow of the Albujon wadi [22], which in addition also contributes to the pollution of the lagoon.

In parallel to agriculture, tourism was also developing rapidly. The permanent population in the catchment (about 10,000 inhabitants) rises by a factor of ten during the touristic peak in summer. In consequence the sewage water release also drastically increases. As a consequence, large amounts of untreated and insufficiently treated waste water are discharged regularly into the Mar Menor, which has led to planktonic changes [20] and the proliferation of jellyfish over the last decades [22]. Nowadays, all of the treated effluent from one of the smaller treatment plants, Torre Pachecco, and about half of the effluents (56%) from the enlarged and modernized major treatment plant, Los Alcazares (in operation since 2008), is reused for irrigation. By that, the nutrient input from point sources could be, at least partly, reduced.

3. Methods and Materials

3.1. The Eco-Hydrological Model SWIM

SWIM is a semi-distributed, process-based model simulating hydrological processes, vegetation growth and nutrient cycling at the river basin scale. It is driven by daily temperature (minimum, maximum and average), precipitation, solar radiation and air humidity, and requires a subbasin map (could be derived from a digital elevation model), a land cover map and a soil map with associated soil profile characteristics as spatial input data. All relevant processes related to water, plant and nutrient dynamics are calculated on the highest level of disaggregation, the hydrotope level. Hydrotopes are units within one subbasin that have a unique combination of land use and soil type. Next, the model outputs are aggregated at the subbasin level, and the lateral flows of water and nutrients are routed via river network to the outlet. A full description of the model structure and the simulated processes is given in the SWIM manual [28].

Before being suitable for any kind of impact assessment, the model should be set up for the specific area and calibrated and validated towards observed data. For that, apart from the above-mentioned input data, additional information on water and land management can be implemented, in order to better represent the hydrological situation and nutrient cycling in the catchment. These data are related to point sources of pollution (e.g., effluents), diffuse pollution (e.g., fertilizer rates, types and application dates), water abstraction (e.g., wells), water transfers, agricultural practices (e.g., irrigation scheme) and others. For model calibration, continuously long time series of measured daily river discharges and observed nutrient concentrations or loads (in case water quality is also subject to further assessment) are needed.

3.2. Model Setup, Calibration and Validation

The spatial and temporal input data used to set up the SWIM model for the Mar Menor catchment are listed in Table 1. The catchment was discretized into 215 subbasins and 1068 hydrotopes. By intersecting the hydrotopes map with the shape file of the Campo de Cartagena Irrigation Zone a total area of 500 km^2 irrigated agricultural hydrotopes was identified. These hydrotopes were assigned a constant amount of water, added as daily precipitation during the irrigation period from March to September.

The amount of irrigation water supplied to the Mar Menor catchment depends on several factors and can vary from year to year [16]. According to the Irrigation Agency of Campo de Cartagena (Comunidad de Regantes del Campo de Cartagena, CRCC [16]) the Tagus-Segura Inter-Basin Transfer delivers on average about 122 hm^3 of water per year to the Mar Menor catchment. Another 13.2 hm^3 are reused effluents from the Urban Waste Water Treatment Plants (UWWTPs), and about 4.2 hm^3 are diverted directly from the Segura basin. Desalination plants have the smallest share in the total water used for irrigation (2.2 hm^3) [16]. In addition to that, about 1300 ponds with a total capacity of more than 21 hm^3 [16] exist in the area, but there is no further information about their operation.

Table 1. Overview of spatial and temporal input data used to set up the SWIM model.

Type of Input Data	Data and Source
Observed climate	5 stations (4 in the basin), period: 2000–2011, Source: Sistema de Informacíon Agraria de Murcia
DEM	20 m × 20 m SRTM (Shuttle Radar Topography Mission), Source: CGIAR Consortium for Spatial Information
Land use	CORINE Land Cover 2006 vector product, Version 13, Source: European Environmental Agency main crops: melons, lettuce, Source: [25] fertilization: 270 kg·N/ha, 110 kg·P/ha, Source: University of Murcia
Soil map and soil parameterization	1 km × 1 km Raster map, Source: Harmonized World Soil Database (HWSD) Soil parameters: HWSD and estimated using the German soil mapping guidelines [31]

Therefore, as the actual amount of irrigation water used in the catchment every year is uncertain, we decided to apply an average of 150 hm^3 per year (as sum of all sources plus half of the potential water stored in ponds) to the whole irrigated area, or in other words, of 1.51 mm per day to the irrigated hydrotopes. This value was assumed constant in all simulations, including the socio-economic scenarios, and only the size of the irrigated area was increased or decreased accordingly.

The effluents from the major UWWTP, collecting the sewages of Los Alcazares, were implemented in three different ways. For the calibration and validation periods, the average monthly values of discharged water, NO_3-N, NH_4-N and PO_4-P loads to the Albujon wadi, estimated from available data (provided by the University of Murcia), were implemented as constant daily inputs. For the climate change scenario runs, an average year using the available monthly data from the period after construction of the new treatment plant was used. For the socio-economic scenarios, we firstly estimated the share of sewage water from the permanent population and of that from the touristic activities. Next, the assumed changes in population were applied to one part of the effluent constantly over the whole year, while the assumed changes in tourism were applied to the second part of the effluents during the touristic peaks only.

Further water management practices, that are less important for the water flow in the catchment, such as the reuse of water from the desalination plant or the discharge of agricultural water surplus drainages into the Mar Menor, could not be implemented, due to the lack of data.

After setup, the model should be calibrated towards discharge (Q) and nutrient loads (NO_3-N, NH_4-N and PO_4-P). For that, only data from a measuring campaign performed between September 2002 and July 2006 with a total of 25 measurements for the mouth of the Albujon wadi were available from the University of Murcia. As this number was too little for a usual calibration based on performance criteria (e.g., Nash-Sutcliffe efficiency or percent bias), we decided to compare the model outputs with literature values. So, the simulated daily discharges were averaged to biweekly means and compared to the fortnightly discharges measured by Garcia-Pintado et al. [18] in the period between October 2002 and February 2004 by graphical fitting. Furthermore, the simulated and estimated by Garcia-Pintado et al. [18] average annual nitrate nitrogen, ammonium nitrogen and phosphate phosphorus loads for the same period were also compared.

3.3. Estimation of Water Inflow and Nutrients Input to the Lagoon

Previous studies estimated the water and nutrient input to the Mar Menor considering (a) the flow of the Albujon wadi and the Rambla de la Maraña, which is artificially connected to it; and (b) the drainages from some channels/pipelines at the mouth of the Albujon wadi (e.g., [17,18]). The above mentioned channels bring agricultural water surplus to a desalination plant close to the mouth of the Albujon wadi, which is then treated and reused for irrigation or in case the plant's capacity is reached, discharged directly into the lagoon (e.g., [17,18]). There is however, a number of small ramblas (ephemeral watercourses) with uncontinuous flows that bring additional water and nutrients to the lagoon during storm events.

We included these ramblas and their catchments in our model setup by using the same calibration parameters as for the Albujon wadi and simulated their discharges and nutrient loads. By adding up all watercourses flowing into the Mar Menor, we then estimated the average total annual inflow and nutrient inputs to the lagoon from its drainage basin. A similar method has been also successfully applied to the drainage areas of the Ria de Aveiro [32] and the Vistula Lagoon [33]. Furthermore, in order to estimate the share of the infiltrated irrigation water on the total inflow an additional model run without any irrigation of the agricultural land was done.

The above mentioned drainages of agricultural water surplus as well as their partial reuse for irrigation after treatment in the desalination plant were not considered in our simulations, as no information about the volumes or the operation schemes of the plant were available.

3.4. Climate Change Scenarios

As the resolution of General Circulation Models (GCMs) is too coarse ($\geqslant$100 km) for regional studies, it is preferable to use climate scenarios from the Regional Climate Models (RCMs) with a higher resolution for an adequate impact assessment [34]. Moreover, as climate projections (e.g., temperature and precipitation trends) of different RCMs for the same region can vary significantly [35], a multi-model approach using scenarios from several RCMs that allows estimating the range of uncertainty is recommended [36].

The climate scenario data used in our study was obtained from the ENSEMBLES project [34]. It consists of a set of 15 climate scenarios, each being the output of a unique combination of different RCMs and GCMs (providing initial and boundary conditions). All models in the ENSEMBLES project have been driven by the A1B emission scenario, which can be described as moderate regarding future projections of atmospheric CO_2 concentrations. The resolution of the applied scenarios is 25 km, and the available time frames are 1951–2098 or 1951–2100, depending on the scenario. For consistency, the end year of all scenario time periods was set to the year 2098.

Before applying the climate scenarios in SWIM simulations for impact assessment, we evaluated their ability in simulating the present climate as well as their climate change signals, considering temperature and precipitation.

For the first task, we used the longest available time period (2000–2011) of daily mean temperature and precipitation records from a station inside the catchment (Balsapintada) to calculate the average monthly and average annual means and compare these with the average climate scenario data of the nearest grid cell for the same period (Figure 2).

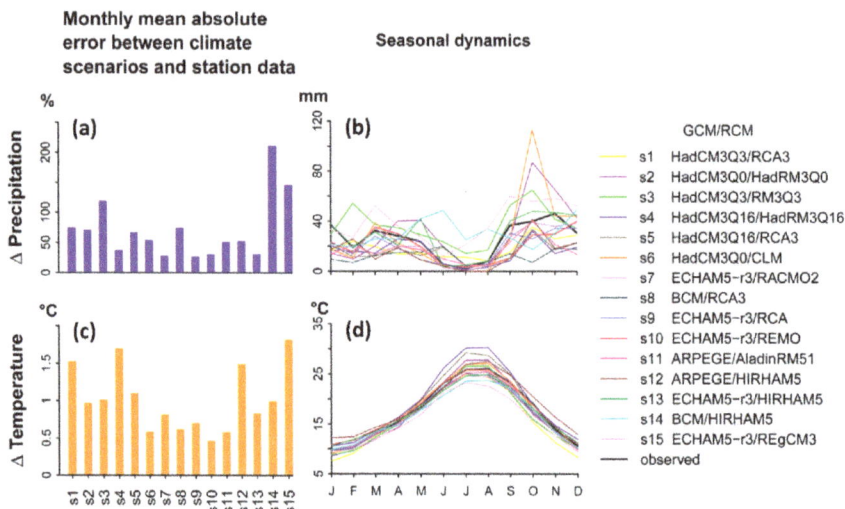

Figure 2. Evaluation of climate scenario data for the historical period by comparing the modeled and observed station data in the catchment for the years 2000–2011: Mean monthly absolute error between modeled and observed station data for precipitation (**a**) and temperature (**c**); and average seasonal dynamics of modeled and observed station data for precipitation (**b**) and temperature (**d**).

As visible in the graph, some scenarios reproduce well the annual dynamics of precipitation (e.g., s7, s9, s10), while others clearly overestimate rainfall in autumn (e.g., s6) or underestimate it in spring (e.g., s12). The smallest mean annual absolute error for precipitation between scenario data and station data was calculated for the s9 scenario (27%), and the biggest, reaching a bias of 211%, was found for the s14 scenario.

The agreement between the modeled and the observed temperatures is clearly higher. Still, some scenarios underestimate average monthly temperatures throughout the whole year (e.g., s15), while others overestimate the annual dynamics (e.g., s4). The s10 scenario produces the smallest annual absolute error (0.5 °C) in this case and the s15 the biggest (1.5 °C).

The climate change signals for temperature and precipitation were obtained for three chosen scenario periods (Section 3.6). We calculated the long-term average monthly and annual differences between each of the three future periods (p1, p2 and p3) and the reference period (p0). The results are shown in Figure 3.

The average annual precipitation in the Mar Menor catchment is projected to decrease by −1.6% in the first, −10.7% in the second, and −18.3% in the third scenario period.

The calculated average monthly changes for p1 do not show any shifts in the seasonality of future precipitation. The projected changes range between −51% in March and 144% in May for different scenarios. In January, March–June, August, and October, about one half of the climate scenarios project higher precipitation rates compared to the reference period, while there is a negative trend for the remaining months. The median projected changes for all months in p1 are in the range of ±20%. In the second and third scenario periods the decreasing trend becomes more clear, and seasonal dependency of the projected changes can be observed. In p2, the mean precipitation decreases during the wet period (between −10% and −30%), except for January and slightly increases (about 10%) during the dry period in summer. In p3 it decreases for all months (between −20% and −40%), except for February and August. It is also visible from the graphs that the disagreement between scenarios decreases from period p2 to period p3, as the ranges of future projections become smaller.

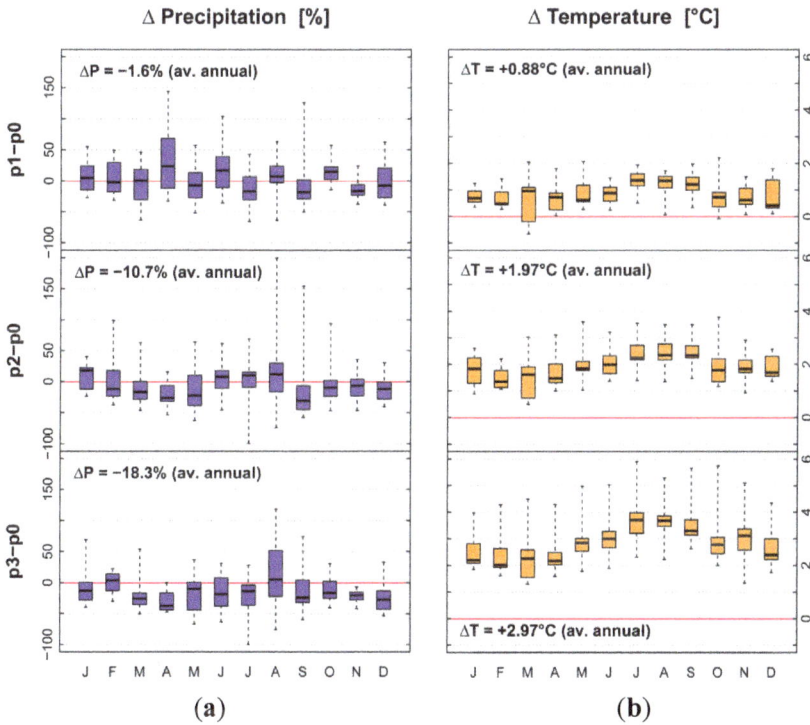

Figure 3. Average monthly climate change signals for the three future scenario periods (p1, p2 and p3) compared to the reference period (p0) for precipitation (**a**) and temperature (**b**) shown as boxplots, where the whiskers represent the min/max values, the boxes the 25th/75th percentiles and the thick lines the median values of changes for 15 climate scenarios.

Regarding air temperature in the catchment, the climate scenarios project an average annual increase of 0.88 °C in p1, 1.97 °C in p2 and 2.97 °C in p3. Some of the climate scenarios (s9, s1 and s15) in p1 have negative signals in April and October but still, in general, we can observe a clear increase in temperature among scenarios in all months and periods. The increase is slightly higher (about 1 °C) during the summer months, and unlike precipitation, the range of future projections (whiskers in boxplots) increases with time and is the biggest for the last scenario period.

3.5. Socio-Economic Scenarios

The socio-economic scenarios used in this study are the product of a complex multi-stage process, including discussions focus groups, citizen juries and scenario workshops. Firstly, narrative storylines representing four different directions of the economic and environmental development for the near future (around year 2030) in the catchment were developed. Their main aspects are shortly described in the following.

The "business as usual" (BAU) scenario represents a possible future of the catchment based on the continuation of current trends of the economic development. In particular, this means a strong increase in tourism along with slight decrease in the agricultural sector and their observed negative effects on the environment and the lagoon (high nutrients input).

The "crisis" (CRI) scenario assumes a negative development of the local economy (a decrease in tourism and a strong decrease of agricultural activities), which leaves no space for environmental protection measures, and hence is likely to lead to further environmental and ecological degradation.

The "managed horizons" (MH) scenario considers a possible future based on economic growth (*i.e.*, increase in touristic development and agricultural area) along with an improvement of the environmental situation through the introduction of appropriate measures (e.g., decrease of mineral fertilization).

And finally, the "set-aside" (SET) scenario describes a possible future of a shrinking economy (*i.e.*, a decrease in tourism and agriculture), which is meant to improve the environmental situation in the catchment and the ecological status of the lagoon.

A detailed description of the scenario development can be found in the Lagoons deliverables [37,38]. The narrative storylines of each socio-economic scenario are presented in Deliverable 4.2 [38].

For the purpose of modeling and the assessment of potential changes the qualitative scenarios were translated into quantitative ones, using statistical data from the Statistical Office of the European Communities (EUROSTAT) and expert knowledge. The assumed relative changes were then transferred into new land use maps and a modified set of SWIM input parameters (effluent characteristics and fertilizer amounts) for each scenario. The effluents from the UWWTP (discharge and nutrient loads) were modified according to the assumed changes in permanent population and touristic activities. The changes of SWIM input data for each scenario are shown in Table 2.

Table 2. Assumed management related relative changes in population, tourism and agricultural practices for the four socio-economic scenarios (BAU, CRI, MH and SET).

Parameter	BAU	CRI	MH	SET
Population (%)	28	−20	10	−10
Tourism (%)	2	−10	4	−5
Min. fertilization (%)	-	−20	−15	−20
Org. fertilization (%)	-	−20	+15	+20
Irrigation (%)	−22	−45	+5	−25

The new land use maps were created by implementing the assumed changes in agricultural land and land cover type. For that, different criteria, such as the soil quality (in terms of water holding capacity), distance to the lagoon/urban areas/major irrigation channel, catchment morphology, rainfall distribution and others were used. The assumed changes in irrigation were implemented through changes in the size of the irrigated area, whereas the amount of irrigation water per hectare remained constant. The implementation of the assumed changes for each scenario is briefly described in the following. The reference land use map and the four new land use maps are presented in Figure 4.

For the BAU scenario all irrigation units from the "Zona Regable Occidental" and some irrigation units from the "Zona Regable Oriental" (located most far away from the major supply channel) were excluded from the irrigated area. Besides, 14% of the agricultural land was converted to fallow. For that, areas outside the new irrigation zone, having low precipitation rates and low quality soils (low water holding capacity) were chosen.

In the CRI scenario, some more irrigation units from the "Zona Regable Oriental" were excluded from the irrigated area, resulting in a narrow irrigation stripe along the major supply channel. Agriculture land was reduced by 30% (converted to fallow), using the same criteria as for the BAU scenario. In addition, forest was reduced by 20%. Deforestation was implemented preferably close to urban areas (which is still mostly in the mountainous areas of the catchment) and on soils with lower water holding capacity.

In the MH scenario the irrigated area was slightly extended. For that, land units outside the reference irrigation zone and close to the major supply channel were chosen. All of the abandoned land (fallow) and 5% of the land cover "heather" were converted into new agricultural land. The conversion was made preferably within the new irrigation zone but also outside the zone on good quality soils (high water holding capacity).

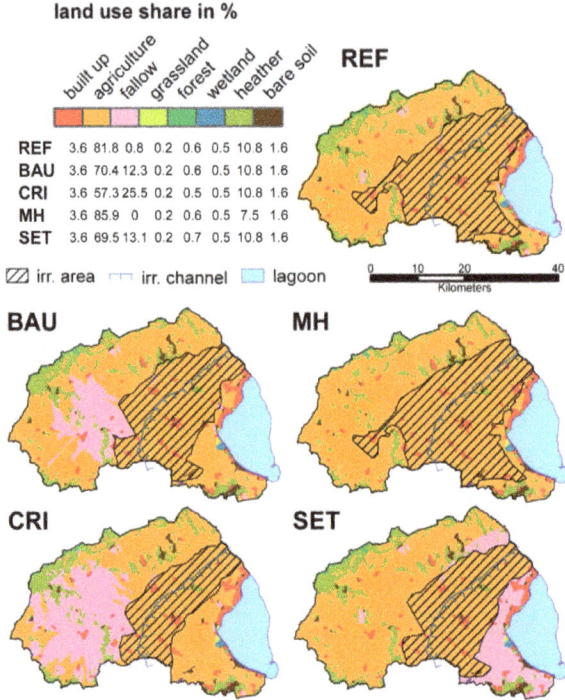

Figure 4. Land use maps for the reference (REF) and scenario (BAU, CRI, MH and SET) conditions for the Mar Menor catchment, as well as shares of land use classes for the reference conditions and the four socio-economic scenarios.

In the SET scenario the irrigated area was reduced similarly as in the BAU scenario (exclusion of irrigation units located most far away from the main supply channel). The abandonment of agricultural land was on purpose and had an intension to improve the environmental situation in the catchment. Therefore the changes were implemented as close as possible to the lagoon, in order to create a kind of buffer strip along the water body. Some less suitable agricultural units (with lower water holding capacity) outside the irrigation zone were converted to fallow as well as some areas at high elevations were even afforested. In total, the agricultural area was reduced by 15%.

3.6. Approach for Impact Assessment

For climate change impact assessment, we ran the calibrated model with each of the 15 ENSEMBLES scenarios for one reference (p0: 1971–2000) and three

44

future (p1: 2011–2040, p2: 2041–2070 and p3: 2071–2098) scenario periods using the reference model setup (reference land use map and reference management settings). This approach has been already successfully used in Stefanova *et al.* [32] and Hesse *et al.* [33] for climate change impact assessment in the catchments of the Ria de Aveiro in Portugal [32] and the Vistula Lagoon in Poland and Kaliningrad [33].

Next, each of the four socio-economic model setups (four different land use maps in combination with four different management settings) was run with the ENSEMBLES climate data for the first scenario period (p1), assuming that the socio-economic changes around the year 2030 correspond to climate in 2011–2040. In total, the outputs of 120 SWIM simulations (15 climate scenarios × 4 periods + 15 climate scenarios × 1 period × 4 socio-economic scenarios) were used for impact assessment.

The responses of water resources to climate change only, combined climate and socio-economic changes and socio-economic changes only were estimated by calculating differences in model outputs between periods and scenarios in the following ways:

(a)   Climate change impacts:

p1(reference), p2(reference), p3(reference) − p0(reference)

(b)   Combined impacts:

p1(BAU), p1(CRI), p1(MH), p1(SET) − p0 (reference)

(c)   Socio-economic impacts:

p1(BAU), p1(CRI), p1(MH), p1(SET) − p1 (reference)

We calculated the long-term mean annual and monthly relative changes in average daily total water inflow (Q), nitrate nitrogen (NO_3-N), ammonium nitrogen (NH_4-N) and phosphate phosphorus (PO_4-P) input to the lagoon, as well as differences in total annual and monthly groundwater recharge (GWR) and actual evapotranspiration (ETa) in the catchment.

4. Results and Discussion

4.1. Model Performance

Despite all uncertainties related to measured data (see Section 3.2), the simulation results for the Albujon wadi show a satisfactory model performance (Figure 5). With regard to river discharge (Q) we can observe that SWIM is able to reproduce adequately most of the winter peaks (e.g., November 2003 and April 2004) and summer low flows (e.g., Septempber 2002, May 2003 and August 2004) obtained during the measuring campaign between September 2002 and July 2006.

45

The maximum simulated daily discharge within this period reaches 2.74 m³/s, whereas the average low flow lies between 0.12 m³/s and 0.15 m³/s.

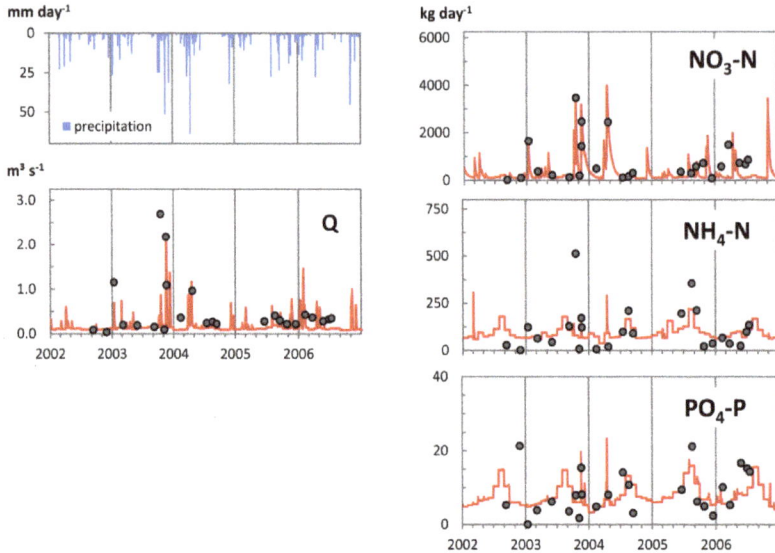

Figure 5. Comparison of simulated daily (red lines) and measured (dots) river discharge (Q), nitrate nitrogen (NO₃-N), ammonium nitrogen (NH₄-N) and phosphate phosphorus (PO₄-P) loads for the catchment of the Albujon wadi (including the diversion from Rambla de la Maraña) and the simulation period 2002–2006.

The model also simulates sufficiently well the annual dynamics of nitrate nitrogen (NO₃-N), ammonium nitrogen (NH₄-N) and phosphate phosphorus (PO₄-P) loads (Figure 5). The NO₃-N loads to the Mar Menor are dominated by agricultural activities. The peaks of both the simulated and the observed NO₃-N loads in the Albujon wadi occur during extreme precipitation events (storms), as a result of increased surface and subsurface runoff from the agricultural fields. The relationship between nitrogen enrichment and agriculture has been discussed in Lloret et al. [20] and Salas et al. [39], and also demonstrated by means of sampling in various studies on the Mar Menor and its catchment (e.g., [17,40]), and could also be nicely reproduced by our model.

In contrast to that, the NH₄-N and PO₄-P loads originate mainly from point source pollution (e.g., effluents from UWWTP) [19]. The simulated and observed NH₄-N and PO₄-P loads reach their maxima during the touristic peak in summer, when the population in the catchment increases by a factor of ten [17]. In most of the winter months, the observed NH₄-N and PO₄-P loads are close to zero, which does not comply with the chemical composition of the released effluents from the UWWTP (higher loads). The simulated winter loads correspond to the loads from the UWWTP, and are by that a bit higher than the observed ones. It is very likely that the actual released pollutants (ammonium and phosphate) are reduced through in-stream processes and plant uptake (e.g., by reeds) on their way from the UWWTP to the mouth of the Albujon wadi. In their study, for example, Álvarez-Rogel *et al.* [41] show that the coastal marshes of the Mar Menor play an important role in reducing phosphorus and nitrogen input to the lagoon. These processes are very complex and, at the current stage they are not included in the model, as a result of which the simulated nutrient loads at the mouth cannot be lower. On the other hand, skipping this step from the nutrient cycling would usually produce overestimated summer loads as well, which does not apply. The reasons for this discrepancy could be unknown small additional point sources in the catchment or incorrect information on the quality of the discharged effluents from the UWWTP.

The additional model validation using discharge data from Garcia-Pintado *et al.* [18] also shows quite satisfactory results (Figure 6). In the dry period, between June 2003 and September 2003 the model misses one small observed peak, which is most likely due to incorrect data from the UWWTP or the lack of information on other point sources in the catchment. Furthermore, the simulated peaks in winter 2003/2004 are slightly higher than the observed ones. At this point, it should be recalled that the curve adopted from Garcia-Pintado *et al.* [18] is based on 36 measurements only, while the simulated biweekly flow dynamics from SWIM are based on about 580 values. It is not unlikely that some of the actual peaks or low flow values were not recorded during the measuring campaign. This may, to some extent explain the discrepancies between both curves, which are in general small.

Table 3 summarizes the average annual water inflow and nutrient (NO_3-N, NH_4-N and PO_4-P) loads to the lagoon as simulated by SWIM and estimated by Garcia-Pintado *et al.* [18]. The projected total discharge of the Albujon wadi (5.51 hm³) for this period is nearly equal to the one estimated by Garcia-Pintado *et al.* [18] (5.46 hm³). At the same time it is considerably lower than the one estimated by Velasco *et al.* [17] (20.14 hm³) for the same period, which based their estimates on seven sporadic measurements only. This demonstrates nicely how uncertain estimates could be due to rare measurements, and how important it is to use sufficiently long and continuous time series for extrapolations of this type. The simulated total NO_3-N input to the lagoon is slightly higher than the one estimated by Garcia-Pintado *et al.* [18], whereas the loads of the two point-sources dominated nutrients (NH_4-N and PO_4-P) are very close to the observed ones.

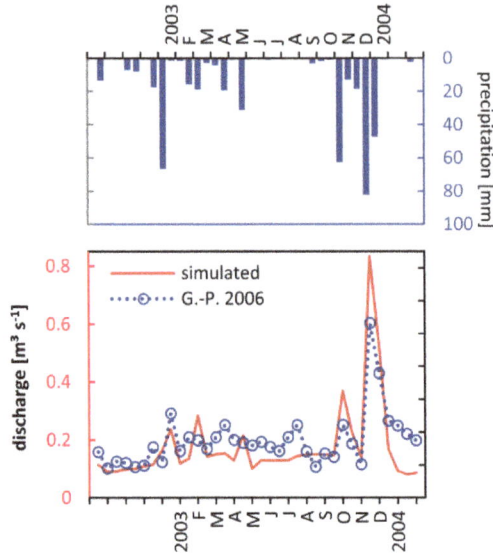

Figure 6. Comparison of simulated and estimated (Garcia-Pintado *et al.* [18]) biweekly discharge for the Albujon wadi catchment (including the diversion from Rambla de la Maraña).

Table 3. Comparison of simulated and estimated (Garcia-Pintado *et al.* [18]) average annual total inflow (Q) and nutrients (NO_3-N, NH_4-N and PO_4-P) input from the Albujon catchment (including the diversion from Rambla de la Maraña) to the Mar Menor.

Variable	G.-P. (2006) (October 2002–February 2004)	SWIM (October 2002–February 2004)
Q	5.46	5.51
NO_3-N	112.84 *	153.84 *
NH_4-N	29.4	31.55
PO_4-P	2.57	2.54

Note: * (February 2003–February 2004).

4.2. Water Inflow and Nutrients Input to the Lagoon

The simulated total annual inflow to the Mar Menor for the period 2002–2011 is about 8.7 hm³ (Table 4). The share of the effluent is about 35% and that of infiltrated irrigation water about 8%. According to our calculations, less than 1% of the applied water for irrigation reaches the Mar Menor, which implies that almost all of it is used in plant transpiration, soil evaporation processes and groundwater recharge.

Table 4. Values of long-term average annual (2002–2011) simulated water inflow from the total drainage area of the Mar Menor and from the catchment of the Albujon wadi to the lagoon, as well as of the amount of additional water added to the whole system (effluent and irrigation water).

Unit (Hm$^3 \cdot$a^{-1})	Mar Menor Drainage Area	Albujon Wadi Catchment *
Q	8.7	5.2
Q without irrigation	8.0	5.0
Effluent	3.0	3.0
Irrigation water	151	63

Note: * including Rambla de la Maraña.

In the catchment of the Albujon wadi (including diverted water from Rambla de la Marana) the share of natural flow is about 39% only. Almost 2/3 of the total discharge at the mouth of the wadi consists of infiltrated irrigation water (4%) and discharged effluents (60%).

The seasonal dynamics of the total water inflow to the Mar Menor (Figure 7) follows the precipitation dynamics in the catchment. We can observe higher flows during the wet period from September to April, reaching 0.4 m$^3 \cdot$s^{-1} on average. Since in our model setup irrigation is applied between March and September we can also observe some peaks, but less pronounced in the flow dynamics of the infiltrated irrigation water. In comparison, the effluent induced part to the total inflow is nearly constant over the year. The small rise in April, followed by a continuous increase until August represents the touristic activities and variations in water consumption in the catchment.

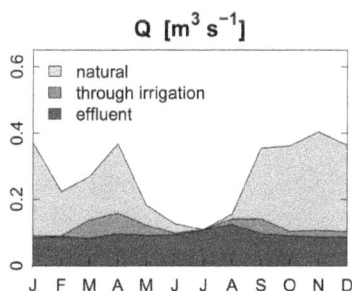

Figure 7. Long-term average (2002–2011) seasonal dynamics of total water inflow to the Mar Menor.

The total average annual nutrient inputs to the Mar Menor are about 192 t NO$_3$-N, 25 t NH$_4$-N and 2 t PO$_4$-P. Almost all of the NH$_4$-N and PO$_4$-P loads reaching the lagoon every year are coming from the released effluents. The point source contribution to the total NO$_3$-N input is about 14% only, whereas the diffuse sources from arable land account for about 72% of the total average annual NO_3-N input to the lagoon.

The simulated average annual groundwater recharge in the catchment is about 97 mm·y^{-1}. The mean evapotranspiration rate sums up to 412 mm·y^{-1} (potential evapotranspiration accounts for 882 mm·y^{-1}). The surface and subsurface flows (excluding UWWTP effluents) reaching the Mar Menor every year are about 4 mm·y^{-1} only. The mean recorded precipitation over the catchment for this period (2002–2011) is about 336 mm·y^{-1}.

4.3. Average Annual Impacts on Water Resources

The results of the impact assessment on the major water cycle components in the catchment and the major nutrient inputs to the lagoon are presented firstly for each parameter set for the climate change scenarios only, and for the combined scenarios subsequently.

4.3.1. Changes in Major Water Cycle Components

(a) Climate change impacts

The simulated impacts of climate change on average daily discharge (Q), average annual groundwater recharge (GWR) and average annual evapotranspiration (ETa) in the catchment correspond well to the observed climate change signals for precipitation.

The projections, driven by 15 climate scenarios show a moderate decrease of long-term average daily discharge to the lagoon for all three scenario periods compared to the reference period p0 (Figure 8). The disagreement among scenarios is the biggest for the first scenario period p1 and decreases visibly towards the end of the century. Moreover, the negative trend in daily discharge becomes more obvious for the last scenario period p3. The simulated median annual changes in Q are −1.0% for p1, −3.5% for p2 and −10.6% for p3.

Other studies on climate and land use change impacts on water resources in Mediterranean catchments have come to similar results, although, in general, the analyzed catchments were more natural and the effect of climate change was more evident. For example, Morán-Tejeda et al. [41] constructed future climate scenario data based on the information from three climate scenarios from the ENSEMBLES project (corresponding to s1, s5 and s13 scenarios in this study) and applied these to drive two different hydrological models, one of them being the SWAT model. Their results showed on average a decrease in water yield of 9%–15% for the period 2021–2050. Another study, carried out by Molina-Navarro et al. [42] estimated an average decrease in runoff of 22% for 2045–2064 using climate data based on the A1B emission scenario and a set of regional climate projections provided by the

Spanish meteorological service. D'Agostino *et al.* [43] used climate data based on the temperature and precipitation changes projected by a single GCM for their case study area and estimated as well an average decrease in streamflow of 16%–25% by the year 2050.

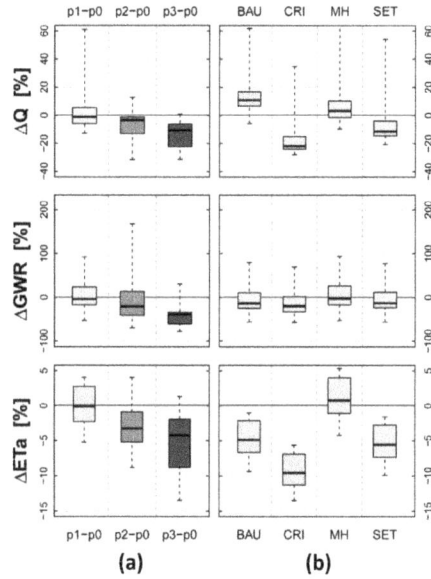

Figure 8. Long-term average annual changes in total inflow (Q) to the lagoon as well as in groundwater recharge (GWR) and actual evapotranspiration (ETa) in the Mar Menor catchment shown as boxplots. (**a**) Climate change impacts, showing results for the three future scenario periods (p1, p2 and p3) compared to the reference period (p0); (**b**) Combined, climate and socio-economic impacts showing results for each of the four socio-economic scenarios (BAU, CRI, MH and SET) compared to the reference conditions.

Similar as for Q, the average annual groundwater recharge in the Mar Menor catchment is also projected to decrease in periods p2 and p3. However, although GWR is strongly related to irrigation [28,44,45], it is still less influenced by water management (irrigation + effluents), and, in comparison to Q, more sensitive to climate change (variations in precipitation). The s1 scenario, for instance causes an average annual increase of 167% in p2, while the s4 scenario leads to a decrease of −78% in p3. Nevertheless, on average we can observe a decreasing trend in groundwater recharge for the three scenario periods (−4.1% for p1, −21.2% for p2 and −38.8% for p3).

These results are again similar to the values on changes in groundwater recharge reported in literature. Pulido-Velazquez *et al.* [46], for example, estimated an average decrease of 7% for 2010–2040, 16% for 2040–2070 and 30% for 2070–2100 compared to 1961–1990 for a mesoscale catchment located just north (in the Jucar basin) of our study area. D'Agostino *et al.* [43] projected a decrease of 21%–31% by the year 2050 for their case study area in Italy.

The average annual actual evapotranspiration in the catchment also follows the precipitation trend of climate scenarios, although the relative changes for ETa are clearly lower compared to that of the other two components Q and GWR. This is mainly because actual evapotranspiration, on the one hand decreases with decreasing water availability (less rainfall), but on the other hand increases with rising air temperatures, as projected by all 15 climate scenarios, and for all three scenario periods. According to our simulations ETa stays almost the same on average in period p1 (−0.06%) and decreases slightly in the last two scenario periods (−3.2% for p2 and −4.2% for p3). It should be also mentioned that the lower percentage changes for ETa are partly explained by its high absolute average values compared to Q and GWR.

(b) Combined scenario impacts

Depending on the assumed changes, the applied socio-economic scenarios reduce or intensify the projected climate change impacts on Q, GWR and ETa for the first scenario period (Figure 8b).

The average daily total discharge to the lagoon increases for the BAU (13.7%) and MH (7.6%) scenarios and decreases for the CRI (−16.5%) and SET (−5.8%) scenarios. The simulated changes are the result of changes in water management mainly, and only to some extent of changes in the land use patterns. An increase in population and tourism, as assumed for the two scenarios BAU and MH leads to an increase in external water transfer for drinking water supply and to higher effluent volumes released from the UWWTP to the Albujon wadi. The decreasing population and touristic activities in the CRI and SET scenarios on the contrary lead to a decrease of the released effluents and consequently to a decrease of the total inflow to the lagoon. The reduction (BAU, CRI and SET) and increase (MH) of the irrigation zone and the agricultural land have only a minor impact on Q.

In contrast to discharge, the combined impacts on groundwater recharge are clearly influenced by changes in the land use patterns. In the CRI scenario, the reduction of the irrigated area (−45%) as well as of the agricultural land (−30%) lead, in combination with climate change, to an average decrease in GWR by about 10%. The projected average reductions of GWR in the BAU and SET scenarios are, compared to that, relatively low (−1.6% and −0.7%). In the MH scenario, an increase of irrigated and agricultural area by 5% each is assumed. This leads to an increase of infiltrated irrigation water, and by that, to a slight intensification of the climate induced average trend for GWR to 10.4% on average.

The projected vague trend in actual evapotranspiration for p1 (-0.1% on average) is intensified in the BAU (-4.7%), CRI (-9.2%) and SET (-5.3%) scenarios and reversed in the MH scenario (1.1%). The reasons are similar, as for the observed changes in groundwater recharge. The assumed decrease in agricultural land reduces on average the transpiration rate of the vegetation in the catchment. In addition, the decrease of the irrigated area reduces the amount of water available for evapotranspiration. In the MH scenario, the assumed agricultural expansion (increase of agricultural land and of irrigated land) has exactly the opposite effect.

The simulated average annual spatial changes in runoff (RUN), groundwater recharge (GWR) and actual evapotranspiration (ETa) in the Mar Menor catchment (Figure 9) can be easily related to the implemented scenario specific land use changes (compare with Figure 2), especially to those concerning the Campo de Cartagena irrigation zone.

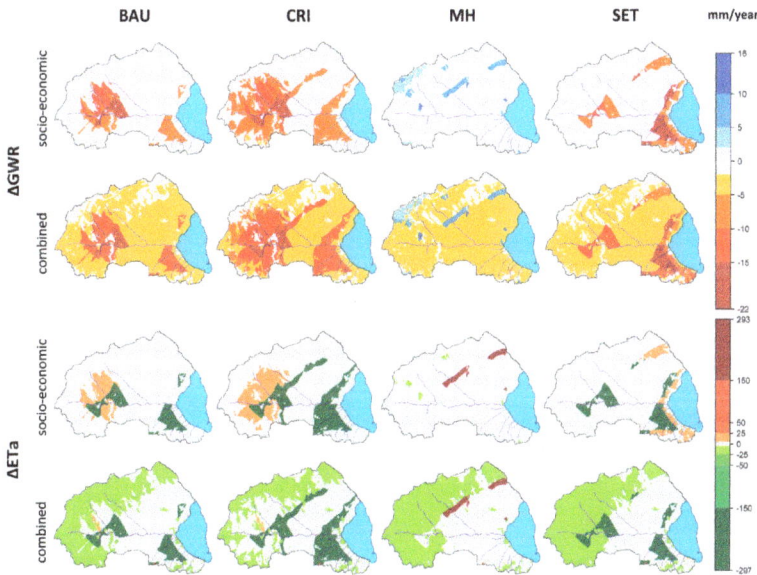

Figure 9. Maps of long-term average annual absolute spatial changes in groundwater recharge (GWR) and actual evapotranspiration (ETa) in the Mar Menor catchment showing the socio-economic impacts only as well as the combined climate and socio-economic impacts for each of the four socio-economic scenarios (BAU, CRI, MH and SET) compared to the reference conditions.

The obtained changes in average annual runoff are negligible, and range between -2 and 1 mm. Therefore, maps showing these results were excluded from further analysis and are not shown in this paper.

In general, groundwater recharge and actual evapotranspiration decrease on agricultural land that was excluded from the irrigation zone and increase on areas that were included into the new irrigation zone. In addition, the conversion of agricultural land into fallow leads to further reduction in GWR and at the same time to an increase in ETa.

We can observe that under the BAU, CRI and SET scenarios groundwater recharge decreases between -5 and -22 mm and actual evapotranspiration between -150 and -297 mm on areas affected by land use changes. On areas that are located outside the reference irrigation zone and which were converted to fallow ETa increased between 5 and 25 mm. This is mainly because the vegetation cover of fallow is permanent and the plant transpiration is on average higher than on cultivated land. This also means that less water remains available for groundwater recharge, which results in a reduced GWR compared to the reference conditions. The decrease is further intensified through the absence of additional irrigation water on areas that were excluded from the irrigation zone under the three scenarios. The actual evapotranspiration on those areas is also reduced compared to the reference conditions.

In the MH scenario the irrigation zone is extended, which leads to an increase in both groundwater recharge (5–10 mm) and actual evapotranspiration (150–293 mm) on the newly formed irrigated areas. The conversion of fallow and grassland into agricultural land outside the reference irrigation zone leads to an increase in GWR (2.5–16 mm) and a decrease in ETa (-5–25 mm).

In combination with climate, many of the observed negative changes in groundwater recharge are intensified, as precipitation is projected to decrease slightly (-1.6%) for the first scenario period. Moreover, an average decrease in GWR of -0.25–5 mm can be observed over the catchment. The effect of climate change on actual evapotranspiration is visible only on areas outside the irrigation zone. In general, ETa decreases as precipitation decreases. This leads to an average reduction in ETa of -5–25 mm on areas that were not affected by land use changes, and an intensification of the trend on areas with a negative change under the socio-economic scenarios. On the other hand, most of the land with a slight positive trend under the socio-economic scenarios shows a negligible change of ±5 mm under the combined scenarios.

Unlike the climate change impacts, this part of our results practically cannot be compared with other studies, as the applied socio-economic scenarios are unique and were developed exclusively for the drainage area of the Mar Menor. Moreover, to our knowledge, there are only few studies considering land use changes in combination with climate change in the Mediterranean region (e.g., [41–43,46]), however in none of these catchments, except for one, water management aspects were considered. Still, some similarities to these studies can be found. In the Mancha Oriental Aquifer [46] for example, the assumed increase of irrigation area leads, similar as in the Mar Menor catchment (MH scenario) to an increase in groundwater recharge. Nevertheless, the authors conclude that climate change will put additional stress on the system in future, although, currently (and similar as in the Mar Menor catchment) there is a stabilization of the groundwater levels.

4.3.2. Changes in Major Nutrients Loads

(a) Climate change impacts

The nutrient input to the Mar Menor can be subdivided into two groups: the input dominated by diffuse pollution (NO_3-N), and the input dominated by point source pollution (NH_4-N and PO_4-P). While diffuse pollution is highly sensitive to changes in precipitation and runoff, point source pollution is influenced by water management, and has no direct response to climate change. The results of the climate change impact assessment on nitrate nitrogen, ammonium nitrogen and phosphate phosphorus loads clearly reflect this behavior (Figure 10).

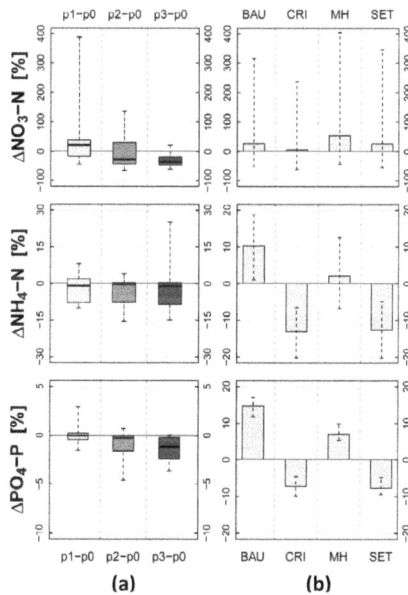

Figure 10. Long-term average annual changes in nitrate nitrogen (NO_3-N), ammonium nitrogen (NH_4-N) and phosphate phosphorus (PO_4-P) loads to the lagoon shown as boxplots. (a) Climate change impacts, showing results for the three future scenario periods (p1, p2 and p3) compared to the reference period (p0); (b) Combined, climate and socio-economic impacts showing results for each of the four socio-economic scenarios (BAU, CRI, MH and SET) compared to the reference conditions.

The average daily NO_3-N loads in the Mar Menor catchment are projected to increase in the first period (22% for p1) and then to decrease in the two following scenario periods (−27% for p2 and −36% for p3). For the period 2011–2040, simulations driven by 10 out of 15 climate scenarios project a positive change, although only eight climate scenarios have a positive precipitation signal for this period. Moreover, the simulations driven by the s4 scenario produce an increase of 389%, which is much higher than the increase in simulated average daily discharge (surface-, subsurface runoff and groundwater contribution) for this scenario (61%). Peaks in nitrate load like this can occur, when some of the projected extreme precipitation events take place exactly after fertilization, which causes higher nutrient concentrations in the generated runoff. For the second and third scenario periods, the agreement on future projections between climate scenarios is much higher and the direction of the trend is much clearer, the simulated loads decrease.

In contrast to NO_3-N, the range of projections for ammonium nitrogen and phosphate phosphorus loads is rather small. The changes of average daily NH_4-N input to the lagoon are negligible (−0.9% for p1, −0.2% for p2 and −1.0% for p3) and so are the simulated changes for daily PO_4-P loads (−0.03% for p1, −0.3% for p2 and −1.2% for p3). Besides the fact that both components originate mainly from the effluents of the UWWTP, which is not directly influenced by climate change, the positive ammonium and phosphate ions are absorbed by the negatively charged soil particles, which protect them from being washed out during heavy precipitation events.

(b) Combined scenario impacts

Similar, as for the major water flow components, the applied socio-economic scenarios reduce or intensify the projected climate change impacts on NO_3-N, NH_4-N and PO_4-P loads to the lagoon and show similar uncertainty ranges as for the climate change impacts for the first scenario period (Figure 10b). However, unlike climate, the assumed socio-economic changes have a significant impact on both, the nutrients input from diffuse pollution (e.g., through changes in fertilization) and the input dominated by point source pollution (e.g., through changes of the effluents). The nitrate nitrogen loads for instance are strongly influenced by changes related to fertilization (amount of applied fertilizer and size of fertilized area), whereas the ammonium nitrogen and phosphate phosphorus loads are affected mostly by variations in the amount and chemical composition of the urban effluents.

In the case of nitrate nitrogen the BAU, CRI and SET scenarios reduce the projected climate change impact on the total NO_3-N input to the lagoon, while the MH scenario intensifies the climate-induced change (50.1% on average). The simulated average daily NO_3-N load increases the least for the CRI scenario (5.3%), in which the strongest decrease in agricultural land (−30%) and irrigated area (−45%)

as well as the highest reduction of applied mineral and organic fertilizers (by -20% each) were assumed. The loads increase the most for the MH scenario (55.3%), as this scenario assumes an increase in both, agricultural land (5%) and irrigated area (5%). In the BAU and SET scenarios the NO_3-N loads increase by 27.1% and 26.9%. The agricultural land and irrigated area are reduced by a similar factor in both scenarios (BAU: -14% and -22%; SET: -15% and -25%) and lead therefore to a similar change in the total NO_3-N input.

According to these results the conversion of agricultural land to fallow in close proximity to the lagoon (SET scenario) does not have the desired effect of acting like a buffer strip and reducing notably the amount of generated nutrient load from the agricultural fields.

The simulated average daily ammonium nitrogen and phosphate phosphorus inputs to the Mar Menor increase in the BAU and MH scenarios, and decrease in the CRI and SET scenarios. These changes can be related to both, the assumed changes in population and tourism and changes in the agricultural practices. As population and tourism increase/decrease, the nutrients released with urban effluents into the Albujon wadi also increase/decrease. In addition, the reduction of mineral and organic fertilizers, as assumed for the CRI (only mineral), MH and SET scenarios reduce to some extent the total NH_4-N and PO_4-P loads in the catchment.

We can observe a stronger increase of average daily NH_4-N and PO_4-P loads in the BAU scenario (10.3% and 14.8%) compared to the MH (2.7% and 7.1%) scenario, although the increase of point source pollution is higher in the MH scenario. However, this increase is partly compensated by a decrease in mineral fertilization, which has not been assumed in the BAU scenario. There is not much difference between the simulated changes in NH_4-N and PO_4-P loads in the CRI (-13.1% and -7.2%) and SET scenarios (-12.7% and -7.7%), which was to be expected, as these two scenarios are very similar regarding the assumed changes.

4.4. Impacts on Seasonal Dynamics

The combined impacts on seasonal dynamics of total inflow (Q), groundwater recharge (GWR), actual evapotranspiration (ETa) and nutrient loads (NO_3-N, NH_4-N and PO_4-P) are presented in Figure 11. The graphs show the long-term average (mean of SWIM outputs driven by 15 climate scenarios) monthly combined (black lines) and socio-economic (green lines) differences between each of the four socio-economic scenarios and the reference scenario, as well as an outer uncertainty band (light grey), defined by the maximum and minimum values of all model outputs and an inner range (dark grey), representing the 25th and 75th percentiles of all results driven by the 15 climate scenarios.

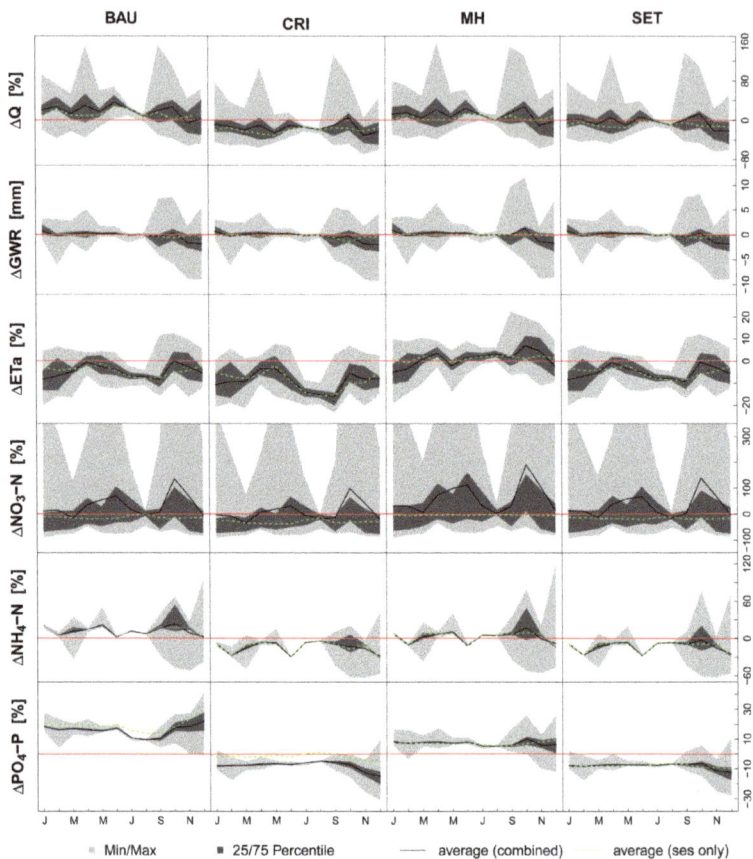

Figure 11. Long-term average monthly impacts of combined climate and socio-economic changes (black line) and of socio-economic changes only (green dashed line) for each of the four socio-economic scenarios (BAU, CRI, MH and SET) on water inflow (Q), groundwater recharge (GWR), actual evapotranspiration (ETa), nitrate nitrogen (NO_3-N), ammonium nitrogen (NH_4-N) and phosphate phosphorus (PO_4-P) shown with uncertainty bands representing the minimum and maximum values (light grey) as well as the 25th and 75th percentiles (dark grey) of the results obtained from 15 climate scenarios.

On average, the combined impacts on total inflow to the lagoon show a moderate variation throughout the year for all four scenarios. The uncertainty between projections is very high during the wet periods in spring and autumn (e.g., between −40% and 161% for the MH scenario in April), which reflects the disagreement between the 15 climate scenarios on future precipitation trends. The inner uncertainty band of model outputs (25th/75th percentiles) is considerably smaller and shows nearly the same dynamics as the socio-economic impacts only. In all four scenarios, the agreement among projections is the strongest for the period between June and August, when water flow in the catchment is mainly influenced by the inputs from the UWWTP and the infiltrated irrigation water. The disagreement among scenarios is higher during the wet periods, when the natural flow, which is influenced exclusively by climate, has a large share in the total water inflow to the lagoon.

The impacts on groundwater recharge are shown in absolute values, as their relative changes are too high in some months, due to small absolute values. Similar as for Q, the uncertainty among projections is stronger for the wet periods and much lower for the dry period. The seasonal dynamics of combined changes in GWR are practically the same for all four scenarios, and the impacts of socio-economic changes only are negligible.

The changes in seasonal dynamics of actual evapotranspiration are shown in the third row (Figure 11). In the BAU, CRI and SET scenarios, the simulated changes show on average a decrease between -16% (September, CRI) and -0.4% (April, BAU). The decrease is the smallest in April and October, when some of the ENSEMBLES scenarios show a strong increase in precipitation that in turn leads to an increase in actual evapotranspiration. In the MH scenario, the applied changes in land use and irrigation act in most months against the projected climate change trends for ETa and the actual evapotranspiration increases between 1% and 7% in all months except for January, February, May and December, when climate change has still a higher impact. The uncertainty ranges do not show similar distinct peaks in wet periods as observed for Q or GWR, due to a strong dependency of ETa on irrigation in addition to climate.

The uncertainty of projected seasonal changes in total nitrate nitrogen (NO_3-N) inputs to the lagoon is among the highest from the analyzed components. The maximum simulated monthly changes reach 987% in the CRI scenario and even 1550% in the MH scenario. This is the result of extreme precipitation events projected especially by one of the climate scenarios (s5). During such events high amounts of NO_3-N are washed from the soils and transported via surface, subsurface and groundwater flow to the streams flowing into the Mar Menor. In the CRI scenario, in which the rate of applied mineral and organic fertilizers was reduced by 20%, the simulated NO_3-N peaks are is the smallest, whereas they are the highest in the BAU and especially MH scenarios. The socio-economic impacts only correspond well to the assumed changes in land use and agricultural practices, and are nearly constant throughout the year.

The relative monthly changes in ammonium nitrogen (NH_4-N) and phosphate phosphorus (PO_4-P) loads follow mainly the estimated relative changes in NH_4-N and PO_4-P inputs from the urban effluents. Therefore, the uncertainty related to climate change is relatively low compared to the other components and mainly visible during the wet periods in autumn and spring, when, similar as for NO_3-N, fertilizers surplus from the agricultural fields are washed to the lagoon. The seasonal dynamics of the combined impacts on both components however are still quite uncertain, as the annual distribution of the estimated effluent changes is based on very simple assumptions.

5. Summary and Conclusions

We assessed the combined impacts of medium term socio-economic changes and climate change on water resources in the Mar Menor catchment by applying the eco-hydrological model SWIM driven by a set of 15 regional climate scenarios, and in combination with a set of four different management settings and land use maps. Our results show that potential socio-economic changes can further intensify or reduce the climate induced impacts on total water inflow and nutrient input to the lagoon, as well as on groundwater recharge and actual evapotranspiration in the catchment in the near future (around the year 2030). This is due to the fact that the Mar Menor catchment is highly human influenced through intensive irrigation and mass tourism (significant point source pollution).

The climate change signals of the 15 applied climate scenarios suggest a warmer and dryer future for the Mar Menor catchment, which in turn causes a decreasing trend in all six analyzed components (Q, GWR, ETa, NO_3-N, NH_4-N and PO_4-P) by the end of the century. The projected changes in precipitation are quite mixed for the first scenario period (2011–2040), and so are the simulated impacts on all investigated components. Looking at outputs averaged over 15 scenarios, we can see that NO_3-N loads show some increase, and all other variables remain practically unchanged. The projected negative trends of potential climate change impacts on water resources become pronounced in the second (2041–2070) and third (2071–2098) scenario periods, and also the uncertainty of the results increases with time. It is worth mentioning that NH_4-N and PO_4-P loads in the Mar Menor catchment are less vulnerable to changes in precipitation and show considerably lower climate change impacts compared with the other components. The reason is that they are strongly human influenced and depend mainly on changes in the released urban effluents. As intense and strong precipitation events are likely to increase in future [6] the outputs of model simulations driven by some of the 15 climate scenarios reach in these cases (Q and NO_3-N in p1, GWR in p2) extremely high maximum values. The combination of climate change and socio-economic scenarios revealed that the two least desirable scenarios for the near future from the economic point of view, the crisis and set-aside scenarios, could be beneficial for the lagoon and its catchment from the ecological point of view, whereas the opposite is the case for the business as usual and the managed horizons scenarios.

The CRI and SET scenarios assume a strong reduction of the agricultural land, along with a reduction of the applied fertilizers. These measures reduce drastically the nutrient load from diffuse sources, and by that the nitrate nitrogen input to the Mar Menor. In addition, the assumed decrease in population and tourism reduces the nutrient contribution from point sources (UWWTP) and decreases further the nutrient loads to the lagoon.

Being concerned about the diffuse pollution only (NO_3-N), the business as usual scenario can be also awarded an environmental friendly scenario with a positive effect on NO_3-N input to the lagoon. However, the assumption of agricultural reduction in the BAU scenario is accompanied by an assumed increase in tourism, which leads to a significant increase in ammonium nitrogen and phosphate phosphorus loads.

The population and tourism changes in the managed-horizons scenario are less dramatic regarding point source pollution (lower increase compared to BAU), but instead rather problematic with regard to diffuse pollution. The MH scenario is the only scenario assuming an intensification of the current agricultural practices, which contributes significantly to a NO_3-N enrichment in the catchment.

Apart from their impacts on nutrient loads, the socio-economic scenarios also influence water inflow to the lagoon as well as groundwater recharge and actual evapotranspiration in the catchment. It must be noted that the reduction of irrigation and urban effluents in the CRI and SET scenarios, leads not only to a reduction of diffuse and point source pollution but also to a decrease in average daily discharge to the Mar Menor, which is not necessarily beneficial for the lagoon. A decrease in water inflow can cause changes in the salinity level of the lagoon, which in turn may lead to shifts in the lagoon's biological community. Furthermore, the conversion of agricultural land to a vegetation type with a higher annual transpiration rate (BAU, CRI and SET scenarios) leads, on average, to lower groundwater recharge rates in the catchment. This effect is enhanced when land is excluded from the irrigation zone and might, on the long-term, lead to a problematic drop of the phreatic levels in the catchment.

The results of this study have shown that certain measures can reduce the negative impacts caused by climate change in the near future, while others are less recommendable as they would intensify the existing problems in the catchment. It should be kept in mind that the assumed future changes in water supply from the Tagus-Segura IBT, which is one of the key factors regarding agricultural productivity and touristic development in the region, are extremely uncertain. The amount of water transferred to the catchment and the allocation of this water to the different sectors (agriculture, domestic use, *etc.*) depends, of course, on future climatic conditions but also, and very strongly on political decisions that are practically impossible to predict. Therefore, and keeping in mind other sources of uncertainty (e.g., one emission scenario only, static socio-economic scenarios, fixed management and land use changes, *etc.*) our results should always be seen in the context of the applied methodology in this study.

The described model outputs of SWIM were used as input in a carry on study [47], which first analyzed the physiochemical and biological changes in the Mar Menor and then assessed their potential implications for the ecological status of the lagoon and the ecosystem services it supports. Furthermore, they are used, in combination with the results of other scientific disciplines (biology, sociology, legal science and others), to develop a framework for an integrated management of the lagoon under the context of climate change [48].

Although the results of this study have already found reasonable applications, they can be improved and extended. For example, coastal marshes could be implemented into the model, in order to investigate their effect on reducing the nutrient input to the lagoon. Moreover, assumptions on the potential changes in water supply from the Tagus-Segura IBT could be based on a hydrological impact study on the capacities and future operations of relevant reservoirs in the Tagus River Basin. This would allow implementing water management changes in a dynamic way that is scientifically underpinned and linked directly to climate change. Besides, a land use scenario assuming dry-crop farming instead of the intensively irrigated horticulture, and an assessment of its implications for the catchment and the lagoon, is recommended, as such scenario could be become unavoidable in case of a strong decrease in the supplied water from the IBT.

Acknowledgments: This research was performed in the framework of the LAGOONS project, which was funded by the European Commission 7th Framework Programme (EU FP7) under grant agreement No. 283157. The climate scenario data used in this study were produced in the EU FP6 Integrated Project ENSEMBLES (No. 505539). The applied land use maps were created by Stein Turtumøygard from the Norwegian Institute for Agricultural and Environmental Research. The authors would also like to acknowledge the LAGOONS partners from the University of Dundee for their work in developing the framework and methodologies that were used to create the qualitative socio-economic scenarios as well as the LAGOONS case study partners from the University of Murcia for providing data for the model setup as well as for organizing and running the stakeholder workshops in the Mar Menor catchment. The authors also thank Vanessa Wörner for her support in data management.

Author Contributions: The methodology described in this paper was planned and designed by all authors. The model setup, calibration and validation, as well as the scenario simulations were performed by Anastassi Stefanova. The analysis and interpretation of climate scenario evaluation and model outputs was done by all authors. The manuscript, including all text, tables and figures was prepared by Anastassi Stefanova and supervised by Valentina Krysanova.

Conflicts of Interest: The authors declare no conflict of interest.

References

1. Lionello, P.; Malanotte-Rizzoli, P.; Boscolo, R. Mediterranean climate variability over the last centuries: A review. In *The Mediterranean Climate: An Overview of the Main Characteristics and Issues*; Elsevier: Amsterdam, The Netherlands, 2006.
2. Giorgi, F. Climate change hot-spots. *Geophys. Res. Lett.* **2006**, *33*.

3. Diffenbaugh, N.S.; Giorgi, F. Climate change hotspots in the CMIP5 global model ensemble. *Clim. Chang.* **2012**, *114*, 813–822.

4. Van Oldenborgh, G.J.; Collins, M.; Arblaster, J.; Christensen, J.H.; Marotzke, J.; Power, S.B.; Rummukainen, M.; Zhou, T. Annex I: Atlas of global and regional climate projections. In *Climate Change 2013: The Physical Science Basis. Contribution of Working Group I to the Fifth Assessment Report of the Intergovernmental Panel on Climate Change*; Stocker, T.F., Qin, D., Plattner, G.K., Tignor, M., Allen, S.K., Boschung, J., Nauels, A., Xia, Y., Bex, V., Midgley, P.M., Eds.; Cambridge University Press: Cambridge, UK, 2013.

5. Giannakopoulos, C.; Bindi, M.; Moriondo, M.; LeSager, P.; Tin, T. Climate Change impacts in the Mediterranean resulting from a 2 °C global temperature rise. In *WWF Report for a Living Planet*; The Global Conservation Organization: Gland, Switzerland, 2005.

6. Giorgi, F.; Lionello, P. Climate change projections for the Mediterranean region. *Glob. Planet. Chang.* **2008**, *63*, 90–104.

7. Planton, S.; Lionello, P.; Artale, V.; Aznar, R.; Carrillo, A.; Colin, J.; Congedi, L.; Dubois, C.; Elizalde, A.; Gualdi, S.; *et al.* The climate of the Mediterranean region in future climate projections. In *The Climate of the Mediterranean Region*; Lionello, P., Ed.; Elsevier: Amsterdam, The Netherlands, 2012.

8. Gualdi, S.; Somot, S.; Li, L.; Artale, V.; Adani, M.; Bellucci, A.; Braun, A.; Calmanti, S.; Carillo, A.; Dell'Aquila, A.; *et al.* The CIRCE simulations: Regional climate change projections with realistic representation of the Mediterranean Sea. *Bull. Am. Meteor. Soc.* **2013**, *94*, 65–81.

9. Navara, A.; Tubiana, L. Regional assessment of climate change in the Mediterranean. In *Advances in Global Change Research*; Springer: Heidelberg, Germany, 2013; Volume 3.

10. Kovats, R.S.; Valentini, R.; Bouwer, L.M.; Georgopoulou, E.; Jacob, D.; Martin, E.; Rounsevell, M.; Soussana, J.F. Europe. In *Climate Change 2014: Impacts, Adaptation, and Vulnerability. Part B: Regional Aspects. Contribution of Working Group II to the Fifth Assessment Report of the Intergovernmental Panel on Climate Change*; Barros, V.R., Field, C.B., Dokken, D.J., Mastrandrea, M.D., Mach, K.J., Bilir, T.E., Chatterjee, M., Ebi, K.L., Estrada, Y.O., Genova, R.C., Eds.; Cambridge University Press: Cambridge, UK, 2013; pp. 1267–1326.

11. Ulbrich, U.; Lionello, P.; Belušić, D.; Jacobeit, J.; Knippertz, P.; Kuglitsch, F.G.; Leckebusch, G.C.; Luterbacher, J.; Maugeri, M.; Maheras, P.; *et al.* 5–Climate of the Mediterranean: Synoptic Patterns, Temperature, Precipitation, Winds, and Their Extremes. In *The Climate of the Mediterranean Region*; Elsevier: Oxford, UK, 2012; pp. 301–346.

12. *Atlas Climático Ibérico—Iberian Climate Atlas*; Agencia Estatal de Meteorología (AEMET): Madrid, Spain, 2011; (In Spanish, Portuguese and English).

13. *Water Resources across Europe—Confronting Water Scarcity and Drought*; Report No 2/2009; European Environment Agency: Copenhagen, Denmark, 2009.

14. Ruiz, M.; Velasco, J. Nutrient bioaccumulation in *Phragmites australis*: Management tool for reduction of pollution in the Mar Menor. *Water Air Soil Pollut.* **2009**, *205*, 173–185.

15. LAGOONS 2012: Factsheet Mar Menor, Socio-Economic and Policies Issues. Available online: http://lagoons.web.ua.pt (accessed on 6 January 2015).

16. Comunidad de Regantes Campo de Cartagena (CRCC). Available online: http://www.crcc.es/informacion-general/informacion-c-r-c-c/ (accessed on 7 January 2015).

17. Velasco, J.; Lloret, J.; Millán, A.; Marín, A.; Barahona, J.; Abellán, P.; Sánchez-Fernández, D. Nutrient and particulate inputs into the Mar Menor lagoon (SE Spain) from an intensive agricultural watershed. *Water Air Soil Pollut.* **2006**, *176*, 37–56.

18. García-Pintado, J.; Martínez-Mena, M.; Barberá, G.G.; Albaladejo, J.; Castillo, V.M. Anthropogenic nutrient sources and loads from a Mediterranean catchment into a coastal lagoon: Mar Menor, Spain. *Sci. Total Environ.* **2007**, *373*, 220–239.

19. Gilaber, J. Seasonal plankton dynamics in a Mediterranean hypersaline coastal lagoon: The Mar Menor. *J. Plankton Res.* **2001**, *23*, 207–217.

20. Lloret, J.; Marin, A.; Marin-Guirao, L.; Velasco, J. Changes in macrophytes distribution in a hypersaline coastal lagoon associated with the development of intensively irrigated agriculture. *Ocean Coast. Manag.* **2005**, *48*, 828–842.

21. Lloret, J.; Marín, A. The contribution of benthic macrofauna to the nutrient filter in coastal lagoons. *Mar. Pollut. Bull.* **2011**, *62*, 2732–2740.

22. Lloret, J.; Marin, A.; Marin-Guirao, L. Is coastal lagoon eutrophication likely to be aggravated by global climate change? *Estuar. Coast. Shelf Sci.* **2008**, *78*, 403–412.

23. Kilsby, C.G.; Tellier, S.S.; Fowler, H.J.; Howels, T.R. Hydrological impacts of climate change on the Tejo and Guadian Rivers. *Hydrol. Earth Syst. Sci.* **2007**, *11*, 1175–1189.

24. Mariotti, A.; Zeng, N.; Yoon, J.H.; Artale, V.; Navarra, A.; Alpert, P.; Li, L.Z.X. Mediterranean water cycle changes: Transition to drier 21st century conditions in observations and CMIP3 simulations. *Environ. Res. Lett.* **2008**, *3*.

25. Estrela, T.; Pérez-Martin, M.A.; Vargas, E. Impacts of climate change on water resources in Spain. *Hydrol. Sci. J.* **2012**, *57*, 1154–1167.

26. Mariotti, A.; Pan, Y.; Zeng, N.; Alessandri, A. Long-term climate change in the Mediterranean region in the midst of decadal variability. *Clim. Dyn.* **2015**, *44*, 1437–1456.

27. Baker, N.C.; Huang, H.P. A Comparative Study of precipitation and Evaporation between CMIP3 and CMIP5 Climate Model Ensembles in Semiarid Regions. *J. Clim.* **2014**, *27*, 3731–3749.

28. Krysanova, V.; Wechsung, F.; Arnold, J.; Srinivasan, R.; Williams, J. *SWIM (Soil and Water Integrated Model): User Manual*; PIK Report No. 69; Potsdam Institute for Climate Impact Research: Potsdam, Germany, 2000.

29. Van der Linden, P.; Mitchell, J.F.B. *ENSEMBLES: Climate Change and Its Impacts: Summary of Research and Results from the ENSEMBLES Project*; Met Office Hadley Centre: Exeter, UK, 2009; p. 160.

30. Poch-Massegú, R.; Jiménez-Martínez, J.; Wallis, K.J.; Ramírez de Cartagena, F.; Candela, L. Irrigation return flow and nitrate leaching under different crops and irrigation methods in Western Mediterranean weather conditions. *Agric. Water Manag.* **2014**, *134*, 1–13.

31. Ad-hoc-Arbeitsgruppe Boden. *Bodenkundliche Kartieranleitung*, 5th ed.; Bundesanstalt für Geowissenschaften und Rohstoffe in Zusammenarbeit mit den Staatlichen Geologischen Diensten: Hannover, Germany, 2005; p. 438. (In German)

32. Stefanova, A.; Krysanova, V.; Hesse, C.; Lillebø, A. Climate change impact assessment on water inflow to a coastal lagoon: Ria de Aveiro watershed, Portugal. *Hydrol. Sci. J.* **2014**.

33. Hesse, C.; Krysanova, V.; Stefanova, A.; Bielecka, M.; Domnin, D. Assessment of climate change impacts on water quantity and quality of the multi-river Vistula Lagoon catchment. *Hydrol. Sci. J.* **2014**.

34. Teutschbein, C.; Seibert, J. Regional Climate Models for Hydrological Impact Studies at the Catchment Scale: A Review of Recent Modeling Strategies. *Geogr. Compass* **2010**, *4*, 834–860.

35. Christensen, J.H.; Hewitson, B.; Busuioc, A.; Chen, A.; Gao, X.; Held, I.; Jones, R.; Kolli, R.K.; Kwon, W.T.; Laprise, R.; *et al.* Regional Climate Projections. In *Climate Change 2007: The Physical Science Basis. Contribution of Working Group I to the Fourth Assessment Report of the Intergovernmental Panel on Climate Change*; Solomon, S., Qin, D., Manning, M., Chen, Z., Marquis, M., Averyt, K.B., Tignor, M., Miller, H.L, Eds.; Cambridge University Press: Cambridge, UK, 2007.

36. Giorgi, F.; Hewitson, B.; Christensen, J.H.; Hulme, M.; von Storch, H.; Whetton, P.; Jones, R.; Mearns, L.; Fu, C. Regional Climate Information—Evaluation and Projections, Chapter 10, Climate Change 2001: The Scientific Basis. In *Contribution of Working group I to the Third Assessment Report of the Intergovernmental Panel on Climate Change*; Cambridge University Press: Cambridge, UK, 2001; pp. 583–638.

37. Baggett, S.; Gooch, G.D. *Activities Report: Report on Raising Public Participation and Awareness Including Design of Uptake and Capacity Building Activities*; LAGOONS Report D4.1; European Commission: Brussels, Belgium, 2014; p. 61.

38. Baggett, S.; Gooch, G.D. *Final Scenarios*; LAGOONS Report D4.2; European Commission: Brussels, Belgium, 2014; p. 45.

39. Salas, F.; Teixeira, H.; Marcos, C.; Marques, J.C.; Pérez-Ruzafa, A. Applicability of the trophic index TRIX in two transitional ecosystems: The Mar Menor lagoon (Spain) and the Mondego estuary (Portugal). *ICES J. Mar. Sci.* **2008**, *65*, 1442–1448.

40. Marin-Perez, R.; García-Pintado, J.; Gómez, A.S. A Real-Time Measurement System for Long-Life Flood, Monitoring and Warning Applications. *Sensors* **2012**, *12*, 4213–4236.

41. Morán-Tejeda, E.; Zabalza, J.; Rahman, K.; Gago-Silva, A.; López-Moreno, J.I.; Vincente-Serrano, S.; Lehmann, A.; Tague, C.L.; Benistn, M. Hydrological impacts of climate and land-use changes in a mountain watershed: Uncertainty estimation based on model comparison. *Ecohydrology* **2014**.

42. Molina-Navarro, E.; Trolle, D.; Martínez-Pérez, S.; Sastre-Merlín, A.; Jeppesen, E. Hydrological and water quality impact assessment of a Mediterranean limno-reservoir under climate change and land use management scenarios. *J. Hydrol.* **2014**, *509*, 354–366.

43. D'Ágostino, D.R.; Trisorio, L.G.; Lamaddalena, N.; Ragab, R. Assessing the results of scenarios of climate and land use changes on the hydrology of an Italian catchment: Modelling study. *Hydrol. Process.* **2010**, *24*, 2693–2704.

44. Álvarez-Rogel, J.; Jiménez-Cárceles, F.; Egea Nicolás, C. Phosphorus and nitrogen content in the water of a coastal wetland in the Mar Menor Lagoon (SE Spain): Relationships with effluents from urban and agricultural areas. *Water Air Soil Pollut.* **2006**, *173*, 21–38.

45. Baudron, P.; Barbecot, F.; García Aróstegui, J.L.; Leduc, C.; Travi, Y.; Martinez-Vicente, D. Impacts of human activities on recharge in a multi-layered semiarid aquifer (Campo de Cartagena, SE Spain). *Hydrol. Process.* **2013**, *28*, 2223–2236.

46. Pulido-Velazquez, M.; Peña-Haro, S.; Garcia-Prats, A.; Mocholi-Almudever, A.F.; Henriquez-Dole, L.; Macian-Sorribes, H.; Lopez-Nicolas, A. Integrated assessment of the impact of climate and land use changes on groundwater quantity and quality in Mancha Oriental (Spain). *Hydrol. Earth Syst. Sci.* **2014**, *11*, 10319–10364.

47. *Results of Combined Climate and Ecosystem Processes: Report Describing Results of Combined Climate and Ecosystem Processes*; LAGOONS Report D6.3; European Commission: Brussels, Belgium, 2014; p. 139.

48. Lillebø, A.; Stålnacke, P.; Gooch, G.D. *Coastal Lagoons in Europe: Integrated Water Resource Strategies*; International Water Association: London, UK, 2015; p. 250.

Short-Term Forecasting of Water Yield from Forested Catchments after Bushfire: A Case Study from Southeast Australia

Mana Gharun, Mohammad Azmi and Mark A. Adams

Abstract: Forested catchments in southeast Australia play an important role in supplying water to major cities. Over the past decades, vegetation cover in this area has been affected by major bushfires that in return influence water yield. This study tests methods for forecasting water yield after bushfire, in a forested catchment in southeast Australia. Precipitation and remotely sensed Normalized Difference Vegetation Index (NDVI) were selected as the main predictor variables. Cross-correlation results show that water yield with time lag equal to 1 can be used as an additional predictor variable. Input variables and water yield observations were set based on 16-day time series, from 20 January 2003 to 20 January 2012. Four data-driven models namely Non-Linear Multivariate Regression (NLMR), K-Nearest Neighbor (KNN), non-linear Autoregressive with External Input based Artificial Neural Networks (NARX-ANN), and Symbolic Regression (SR) were employed for this study. Results showed that NARX-ANN outperforms other models across all goodness-of-fit criteria. The Nash-Sutcliffe efficiency (NSE) of 0.90 and correlation coefficient of 0.96 at the training-validation stage, as well as NSE of 0.89 and correlation coefficient of 0.95 at the testing stage, are indicative of potentials of this model for capturing ecological dynamics in predicting catchment hydrology, at an operational level.

Reprinted from *Water*. Cite as: Gharun, M.; Azmi, M.; Adams, M.A. Short-Term Forecasting of Water Yield from Forested Catchments after Bushfire: A Case Study from Southeast Australia. *Water* **2015**, *7*, 599–614.

1. Introduction

Forested catchments of southeast Australia supply most of the water for at least 25% of Australia's population, as well as for nationally significant industries including agriculture [1,2]. Underlying interactions between soil-plant-atmosphere make the relationship between rainfall and runoff non-linear and complex to model in forested catchments. In addition, catchment vegetation cover is continuously affected by activities such as logging, long-term land use and changes, and major disturbances (e.g., bushfires) that in return influence the catchment water yield [3,4]. Bushfires affect catchment hydrology through changes in the structure and density of the vegetation cover, leading to changes in evapotranspiration and water yield

67

over the regeneration period that last sometimes for several decades. For example a 5% change in forest evapotranspiration due to changes in the vegetation cover affects mean water yield by 20% in southeast Australia [5]. In the state of Victoria, the impact of the 1939 bushfires on water yield from mountain ash forests peaked around 20 years after the fire following the relationship that exists between changes in the leaf area index [6] and catchment water yield [3,6].

Climate and vegetation cover characteristics have commonly been used as key predictors of post-fire runoff from forested catchments in Australia [3,7–10]. Previously, short-term estimation and forecast of catchment water yield were based on a range of methods, from purely empirical simple models to highly sophisticated distributed process-based models defined by partial differential equations (e.g., the Systeme Hydrologique Europeen model [11] or the Macaque model [9]). Over the past decades, data-driven models have become increasingly useful in hydrological forecasting on the basis that they avoid having to address the problems of the spatial and temporal variability, and the uncertainty of the inputs and the parameters, as opposed to the physically-based models that require a wide range of catchment and climate information [12–14].

The most commonly applied data-driven methods in short-term estimating and forecasting of catchment water yield are the parametric regressions such as the multivariate regressions [15–17], nonparametric regressions such as the K-nearest neighbor method [18–20], symbolic regressions such as genetic programming [21–23], and artificial intelligence based methods such as neural networks [24–26]. Hydrological processes contain non-linearities that are commonly modeled with data-driven techniques as an alternative to linear regression methods. Numerous studies have compared data-driven techniques with regression models (linear or non-linear) and have underlined the interest in using data-driven methods (see for example [27–29]). In this study, we employ four data-driven techniques, namely Nonlinear Multivariate Regression (NLMR), K-Nearest Neighbor (KNN), Nonlinear Autoregressive with External Input based Artificial Neural Networks (NARX-ANN), and Symbolic Regression (SR) to explore their potential for capturing ecological dynamics in predicting catchment hydrology at an operational level.

2. Materials and Methods

2.1. Case Study and Data Sources

The Corin catchment is located in the Namadgi National Park and is part of the Cotter river catchment in the Australian Capital Territory (ACT). The catchment lies about 50 km west of Canberra, at the northern end of the Australian Alps (35°39'25" S, 148°49'53" E), and encompasses an area of 148 km^2 (Figure 1). The catchment is covered by native eucalypt forests and soils of the area are derived from highly weathered Ordovician sediments and are acidic and duplex in structure [30]. The underlying rock types are granite, limestone and shale, and the topography is mountainous (steep with rocky outcrops). Summers are characterized as warm and often hot, with dry periods of between six and eight weeks. In winter (July), mean daily maximum and minimum temperatures in sheltered locations (mid-slope) are 14 and −1 °C respectively, while in summer (January) the respective temperatures are 24 °C and 10 °C. Mean annual rainfall is approximately 1150 mm. Annual evaporation and seepage losses from the catchment are estimated to be 630 mm and stream discharge typically peaks between August and September and reaches a minimum between March and May [31] (Figure 2).

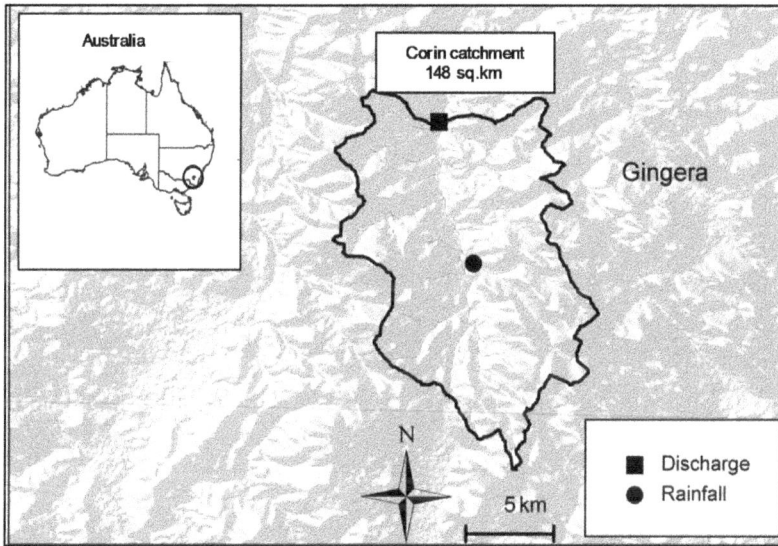

Figure 1. Study area and location of hydrometry station and the precipitation gauge.

Identification and documentation of fires within the Australian Capital Territory (ACT) date back to 1730. In the past 100 years, there has been major bushfires in the catchment and surrounding areas in the summers of 1920, 1926, 1939, 1983 and 2003. With the exception of the 1920 fire, all have followed severe droughts where rainfall in the months preceding the fire was well below average [32]. Most recently, bushfire in January 2003 affected nearly 100% of the catchment.

We used Normalized Difference Vegetation Index (NDVI) as a proxy for vegetation cover [33]. Our time series data encompass a period between the latest bushfire in the catchment of study (January 2003) and January 2012. Daily time series of water yield and rainfall (verified for gaps and aggregated from hourly measurements) were provided by Actew-AGL (the utility that supplies water to

Canberra). Water yield was measured at a gauging station upstream of the Corin reservoir, and rainfall was measured at a weather station in the middle of the catchment (Figure 1). An annual hydrograph of the inflow to the Corin Dam and a hyetograph at the Mount Ginini weather station (closest synoptic station to Corin dam) are presented in Figure 2.

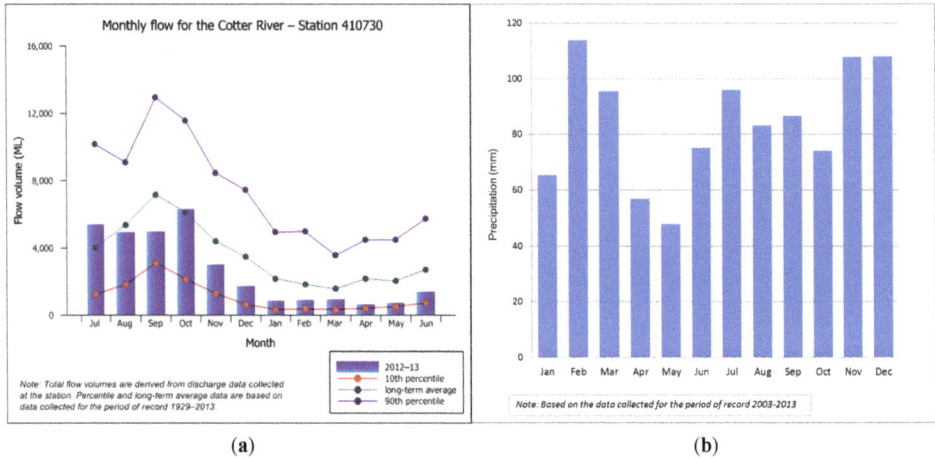

(a)   (b)

Figure 2. Annual hydrograph at the inflow of the Corin dam provided by the Bureau of Meteorology of Australia (**a**); and hyetograph at of the Mount Ginini automatic weather station (**b**).

Time series of MODIS NDVI version 5 derived from red (0.62–0.67 μm) and near-infrared (0.841–0.876 μm) reflectance data were extracted and used directly. MODIS NDVI was selected because: (1) it provides an estimate of vegetation ecological changes over time [34] and fire damage [35], and is directly related to leaf area index (LAI) [36,37]; (2) remotely sensed NDVI is available at a higher spatial resolution than LAI (250 m instead of 1 km); (3) there is less uncertainty associated with estimation of NDVI by satellite data, compared to LAI of eucalypt forests in Australia [38]. Data obtained from the MODIS-TERRA sensor (MOD13Q1 product version 5) has a 250 m spatial resolution and is a composited output over 16 days. Version-5 MODIS/Terra NDVI is validated over a widely distributed set of locations and time periods [37]. We downloaded scene h30v12 of the product, from the NASA Land Processes Distribution Active Archive Centre Data Pool. Using the MODIS reprojection tool (Version 4.1; USGS Earth Resources Observation and Science Center, Sioux Falls, SD, USA) NDVI scenes were cut to the study area and average values for the catchment were used. Average NDVI for the catchment was used since the catchment of study has a relatively homogenous forest cover in the sense that the native eucalypt forests that cover the catchment did not differ widely in their canopy characteristics.

A short-term forecast (intra-month) of water yield is worthwhile information for water allocation as well as water stress assessments, especially since immediate bushfire impacts on water yield are relatively unknown. NDVI is only available every 16 days, therefore water yield and precipitation variables were reformed to 16-day intervals by accumulating their daily values.

2.2. Standardization and Goodness of Fit Criteria

Data standardization adjusts all data so that they fall within a prescribed range and have common basic statistical characteristics. The result of standardization is a data space without the bias, which appears usually as a result of scale, and consequently all input variables are treated equally. Given the natural range of the dependent and independent variables being equal to or greater than 0, the most simple and efficient method for standardizing variables was:

$$y_{stan} = \frac{y}{Max(y)} \tag{1}$$

where y_{stan} is the amount of variable y after standardization; and $Max(y)$ is the maximum value of y within the time series. Following goodness of fit criteria were used for comparing the results of different data driven methods:

- Root Mean Square Error (RMSE)

$$RMSE = \frac{\sqrt{\sum_{i}^{n} (obs_i - for_i)^2}}{n} \tag{2}$$

where obs_i and for_i are observed and forecasted value of the dependent variable at time step i, respectively; and n is the total number of time steps.

- Volume Error (VE)

$$VE = \frac{\sum_{i}^{n} \left| \frac{obs_i - for_i}{obs_i} \right|}{n} \tag{3}$$

- Correlation (Corr)

$$Corr\% = \frac{Cov(obs, for)}{\sigma_{obs} \times \sigma_{est}} \tag{4}$$

where $Cov(obs, for)$ is the covariance between observed and forecasted values; and σ_{obs} and σ_{est} are standard deviation of observed and estimated values, respectively.

- *Nash-Sutcliffe Efficiency (NSE)*

NSE is used to evaluate the estimative power of hydrological models [39]:

$$NSE = 1 - \frac{\sum_{i=1}^{n}(obs_i - for_i)^2}{\sum_{i=1}^{n}(obs_i - \overline{obs})^2} \tag{5}$$

where $\overline{obs}$ is the average of observed values from $i = 1{:}n$. NSE can range from $-\infty$ to 1. NSE = 1 corresponds to a perfect match of modeled discharge to the observed data. NSE = 0 indicates that the model forecasts are as accurate as the mean of the observed data, and NSE < 0 is an indication that observed mean is a better predictor than the model or in other words, when the residual variance is larger than the data variance.

2.3. Data-Driven Methods

Data-driven methods that were applied in this study were:

- *Non-Linear Multivariate Regression (NLMR)*

NLMR estimates unknown coefficients of predictors based on a non-linear optimization approach in the following form:

$$Min\ Z = \frac{Z_{calibration}}{Z_{validation}} + \frac{Z_{validation}}{Z_{calibration}}$$

$$Z_{tcalibration} = \sum_{i=1}^{m}\left|Q_{obs}(i) - Q_{for}(i)\right|$$

$$Z_{validation} = \sum_{i=m+1}^{k}\left|Q_{obs}(i) - Q_{for}(i)\right| \tag{6}$$

$$Q_{for}(i) = \left[\left(\sum_{j=1}^{n}c_j x_j^{b_j}\right) + c_{n+1}\right]_i$$

$$Q_{for} \geqslant 0$$

where m is the number of samples used for calibration; k is the number of samples used for both calibration and validation; $Q_{obs}(i)$ and $Q_{for}(i)$ are observed and forecasted water yields at time i; x_1 to x_n are n predictor variables; c_j and b_j are coefficients of predictors and c_{n+1} is the constant of the model. Here we used "Lingo optimization package" [40] to optimize the model for coefficients and constant values.

- *K-Nearest Neighbor (K-NN)*

The K-NN method develops a distribution function of estimated values using a nonparametric kernel distribution function. The concept is based on observing estimator variable values at a given time and searching for similar conditions in the past, within the time series. These similar conditions can be considered as possible solutions depending on the degree of similarity between estimator variables at current and past time points [18,20]. General algorithm for K-NN would be:

(1) setting a matrix with m columns (number of predictors) and $n + 1$ rows (length of time series).

(2) last row of the above-mentioned matrix is assumed as a vector of predictors at current time ($x_{j,t}$ $j = 1{:}m$).

(3) remaining rows are assumed as a matrix of predictors at historical time series ($x_{j,t-i}$ $j = 1{:}m$ $i = 1{:}n$).

(4) vector Q is defined with n rows of independent variable values from $t - n$ to $t - 1$.

(5) using a distance function, distances between $x_{j,t}$ and $x_{j,(t-i)}$ are calculated.

$$Dist(t - i) = f(w_j, x_{j,(t-i)}, x_{jt}) \tag{7}$$

where w_j are weights of predictor variables at the distance function. We chose the Euclidean function as the distance function with equal weights to predictor variables.

(6) distance vector ($Dist$) is sorted from minimum to maximum ($SDist$) and vector Q is assorted based on $SDist$.

(7) best number of neighbors (k) are specified based on a variety of methods. Here we have used the empirical equation $K = \sqrt{n}$ in which n is the length of the time series which is used as historical data for calibration and validation stages [41].

(8) a discrete Kernel function is used to give weights to k neighbors [42].

$$S(e) = \frac{1/SDist(t - e)}{\sum\limits_{e=1}^{k} 1/SDist(t - e)} \quad e = 1...k \tag{8}$$

(9) forecast value at current time is calculated as:

$$Forecast = S \times Q^T \tag{9}$$

where T is the transpose operation.

- *Nonlinear Autoregressive with External Input Based Artificial Neural Networks (NARX-ANN)*

NARX-ANN is able to model nonlinear autoregressive time series and it is quite appropriate to identify nonlinear dependencies among dependent and estimator variables [43,44]. This model is a recurrent dynamic network, with feedback connections compassing several layers of the network. The defining equation for the NARX-ANN model is [45]:

$$y(t) = f[y(t-1), ..., y(t-d), x(t-1), ..., x(t-d)] + e(t) \tag{10}$$

where $y(t)$ is dependent time series at time t; f is a set of nonlinear functions; x is matrix of independent variables; $e(t)$ is the white noise residuals; and d is number of delays. NARX-ANN was coded and run in "MATLAB R2013a" programming package.

- *Symbolic Regression (SR)*

SR searches the space of mathematical equations to find an equation, which appropriately fits the data, by changing both the type of mathematical functions as well as the value of the parameters. This process starts with choosing initial expressions that randomly couple mathematical building blocks. Then, latter equations are shaped by reincorporating previous equations and altering their sub-expressions via an evolutionary algorithm similar to genetic programming; and ultimately final mathematical equations are ranked using a ratio of accuracy and equation complexity [46]. In this study we used "Eureqa Formulize" software [47] to form SRs. Recently this software has been increasingly applied by researchers in different environmental studies as a reliable tool for analyzing SR-based issues [48–51].

3. Results and Discussion

Cross-correlation (based on Pearson Correlation) with 95% confidence interval was undertaken on standardized time series to find the most appropriate time lags (TL) between all independent and dependent variables. Results shows that TL = 1 has the highest correlations in all cases with the exception of zero. This means for forecasting standardized water yield at time $t + 1$ by three methods of NLMR, K-NN and SR, the values of standardized precipitation, standardized NDVI and standardized water yield at time t should be used; however in NARX-ANN best TL would be derived based on a trial-error process which will be explained.

At the next step, data were split into three blocks of (1) calibration including 70% of data; (2) validation including 15% of data; and (3) verification including 15% of data. While data were randomly distributed into three blocks, we ensured that each block included extreme events, in addition to more normal values. During the calibration stage (also known as training) 70% of our total data is calibrated for the model parameters. The model is then tested during the validation stage with an unused 15% of the total data. This process is repeated until total error for the calibration/validation stages is minimum. Once the optimum model is selected, an independent 15% of the data set that has not been used in the modeling process is used to examine model's accuracy. This is called the verification stage.

As for the K-NN method, best number of neighbors was 13 based on the number of data, which were located in the calibration and validation blocks. The NLMR model for short-term forecasting of water yield was derived as following (all used variables in Equation (11) are standardized by Equation (1), and therefore are dimensionless):

$$WY_i = (251.77 \times Prec_{i-1}^{-4.7 \times E - 0.005}) + (-59.125 \times NDVI_{i-1}^{-0.001}) + (-485 \times WY_{i-1}^{0.00001}) + 292.65 \quad (11)$$

Based on the recommendations of previous studies [44,52], here in the three-layer NARX-ANNs model, "Levenberg–Marquardt" algorithm function was used as back-propagation calibration-validation algorithm, and "Tangent sigmoid" function was applied for the hidden layer neurons, and finally "Linear Transfer" function was employed in the output layer neuron. Further, a range of 1 to 5 for number of delays as well as 1 to 20 for number of neurons in the hidden layer were examined to reach the best neural network's architecture. Considering the results of the objective function of mean square error, the best number of neurons was found to be 10; moreover, the best number of delays for precipitation and NDVI ("inputDelays") equals 1 and ("feedbackDelays") equals 2 for the water yield.

As for SR, five main mathematical operators, exponential and trigonometry functions were considered for building the mathematical blocks. Figure 3 shows the mathematical solution's accuracy *vs.* its complexity. Here we considered the solution with a mean absolute error 0.08 and complexity of 33 as the optimum point on the frontier. After this point by increasing the complexity the amount of errors had ignorable discrepancies. The stability and maturity of final solutions after 4.7×10^7 generations were 0.77% and 98.6%, respectively. Ultimately based on the optimum point, the proposed mathematical solution can be presented as (all used variables in Equation (12) are standardized by Equation (1), and therefore are dimensionless):

$$WY_i = 0.64 + (22.72 \times WY_{i-1} \times NDVI_{i-1}^{28.78}) - (0.70 \times NDVI_{i-1}^{(NDVI_{i-1} + 9.44 \times Prec_{i-1} \times WY_{i-1}^2 - WY_{i-1})}) \quad (12)$$

where i is time step; WY is water yield; and *Prec* is precipitation.

Table 1 presents a sensitivity analysis over Equation (12). Here, sensitivity means the relative impact that a predictor has on the target variable (streamflow). In Table 1 "%Positive" is the likelihood that increasing this variable will increase the target variable; "Positive Magnitude" is when increase in this variable lead to increase in the target variable, this is generally how big the positive impact is. "%Negative" is the likelihood that increasing this variable will decrease the target variable, and finally "Negative Magnitude" is when increase in this variable leads to decrease in the target variable and this is generally how big the negative impact is. According to Table 1, the model is most sensitive to NDVI (sensitivity = 1.31) even more than to precipitation (sensitivity = 0.08). Results show that at 31% of times, NDVI will have positive impact with magnitude of 3.81, and at 55% of times, streamflow with time lag 1 will have positive impact on forecasting streamflow with a magnitude of 0.7.

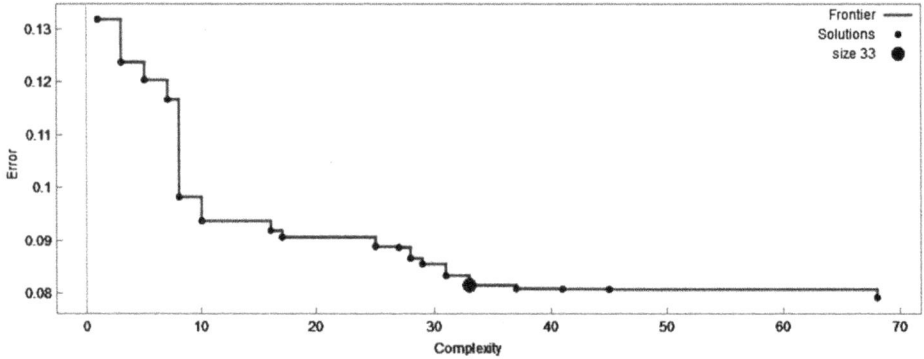

Figure 3. Mathematical solution accuracy *vs.* model complexity in the SR method. Y-axis shows mean absolute error of the forecast *vs.* observations for the entire data.

Table 1. Sensitivity analysis of the optimum solution derived from SR.

Variable	Sensitivity	% Positive	Positive Magnitude	% Negative	Negative Magnitude
NDVI$_{i-1}$	1.31	31%	3.81	69%	0.21
WY$_{i-1}$	0.70	55%	1.16	45%	0.14
Preci$_{i-1}$	0.08	99%	0.08	0%	0

Table 2 summarizes final results of the goodness of fit criteria for all models, at three stages of Calibration-Validation, Verification, and the Entire Data. NARX-ANN produced the best correlation coefficient (at the Calibration-Validation, 0.91) and K-NN the worst (at the entire data, 0.63). Using the RMSE criterion, NARX-ANN had the best performance at the Calibration-Validation stage (0.07) and K-NN had the worst performance at the Verification stage (0.21). As for VE, SR performed the best (Calibration-Validation stage, 1.09) and K-NN the worst (Verification stage, 3.57), and finally considering the NSE criterion, NARX-ANN showed the best performance (Verification stage, 0.80) and K-NN the worst performance (Verification stage, −3.54). In a pair-wise comparison between different methods, the ranking of performances is: (1) NARX-ANN; (2) SR; (3) NLMR; and (4) K-NN.

Table 2. Performance statistics for the tested data-driven methods.

Methods	Calibration-Validation (85% of Data)				Verification (15% of Data)				Entire Data			
	Corr	RMSE	VE %	NSE	Corr	RMSE	VE %	NSE	Corr	RMSE	VE %	NSE
K-NN	0.64	0.16	3.42	−0.10	0.74	0.21	3.57	−3.54	0.63	0.17	3.44	−0.33
NLMR	0.79	0.10	1.51	0.60	0.78	0.20	1.50	0.40	0.76	0.12	1.51	0.55
NARX-ANN	0.91	0.07	1.20	0.80	0.90	0.11	1.50	0.80	0.90	0.08	1.24	0.80
SR	0.82	0.09	1.09	0.67	0.80	0.16	1.20	0.63	0.82	0.10	1.16	0.67

According to Figure 4, K-NN has overestimated the low streamflows, and underestimated the high streamflow values while with NLMR (Figure 5) only underestimation of high streamflow values is remarkable. NARX-ANN by far has forecasted a wide range of streamflows much more accurately in comparison with other three models (Figure 6). Finally, considering Figure 7 the forecast errors of low, moderate and high streamflows at SR are rationally similar, and it is hard to distinguish the advantages of this model in forecasting a specific range of streamflow values. For comparison purposes we also parameterized a conventional linear regression model of form $y = a + bx_1 + cx_2 + dx_3$ and found its performance very poor (Corr = 0.45) and the technique inappropriate compared to the data-driven models.

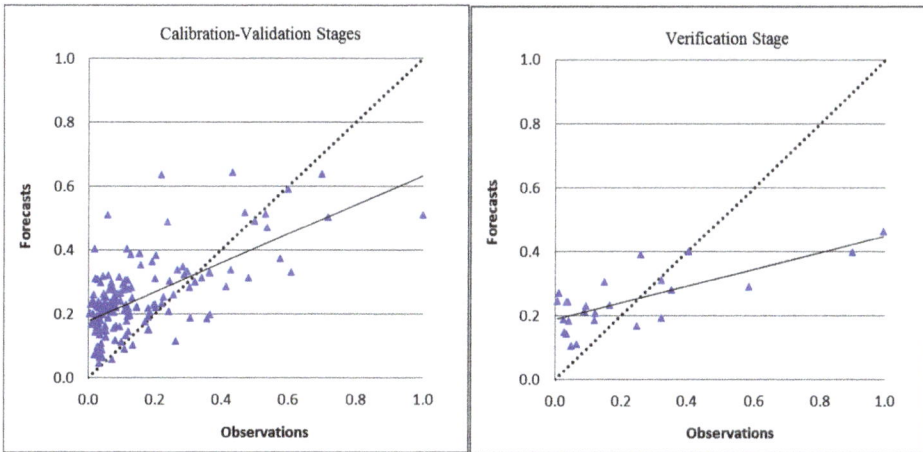

Figure 4. Final results of standardized forecasts *vs.* standardized observations using the K-NN method.

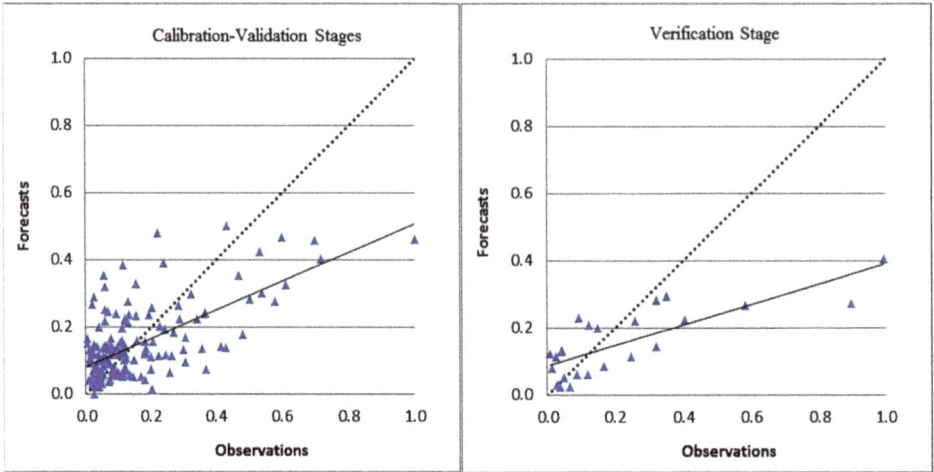

Figure 5. Final results of standardized forecasts *vs.* standardized observations using the NLMR method.

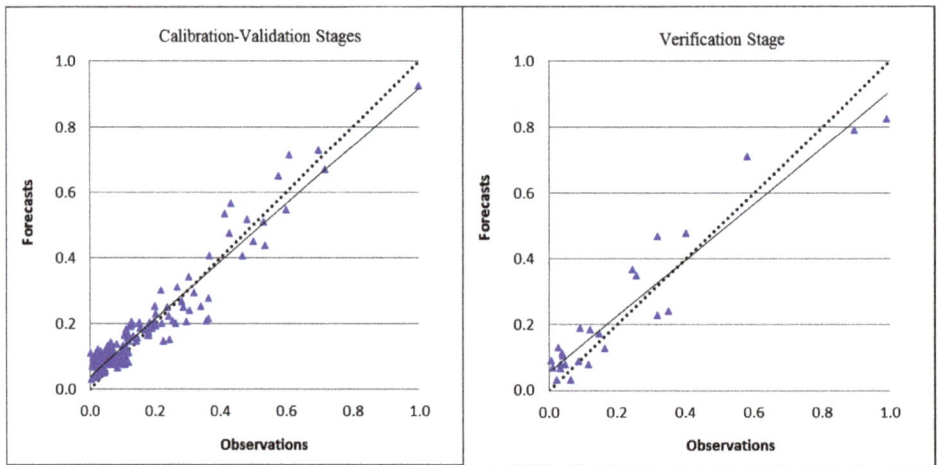

Figure 6. Final results of standardized forecasts *vs.* standardized observations using the NARX-ANN method.

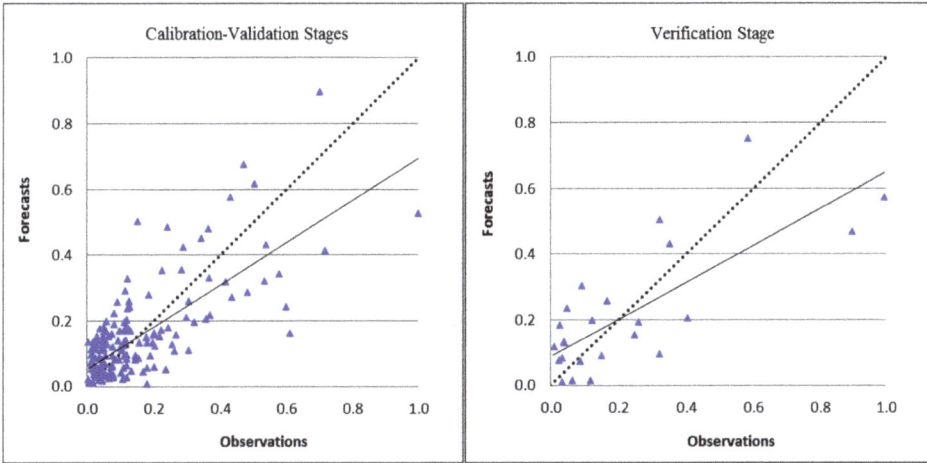

Figure 7. Final results of standardized forecasts *vs.* standardized observations using the SR method.

An encouraging aspect of the NLMR and SR models was that they presented mathematical equations to be used in short-term forecasting of water yield. Of these two models, SR outperformed NLMR especially in extreme events. Obviously, models presented here could further be improved by using longer periods of continuous data in the analysis.

In case top priority is given to forecasting extreme events, the NARX-ANN model can be improved by including new performance function networks [53], and the SR model may be improved by dividing the data into extreme and normal groups first, and then modeling each group separately, as proposed by Charhate *et al.* [54]. In this study, a 16-day interval was considered due to limited NDVI data availability. In case NDVI values are available with a higher temporal resolution, a shorter time interval (e.g., daily) might provide a more realistic short term forecast of the stream flow, because flood events that occur during this time interval will not be smoothed anymore, and because precipitation and discharge data generally have more heterogeneity than NDVI values.

Concluding, this study shows that in this catchment in southeast Australia, different data-driven models perform differently. The NARX-ANN model is superior to the rest of the techniques and would be a suitable tool for catchment managers and water utilities in the absence of extensive climate, soil and vegetation data.

4. Conclusions

Hydrological processes contain non-linearities that are commonly modeled with data-driven techniques as an alternative to conventional regression and process-based methods. Nevertheless, the lumped data-driven models have limitations in delivering hydrological insights. An underlying justification for the variability in the fit metric values in this study could be the physical processes that are under-presented with the current inputs. For example, hydrological processes such as snowmelt at higher elevations within the catchment impact the timing of the discharge and are inadequately presented via precipitation inputs. When limited data is available, uncertainties associated with hydrological data exert even larger limitations on the model (e.g., rainfall uncertainties driven by spatial scale or discharge uncertainties dominated by flow condition and gauging method [55]). While a varied mix of data-driven techniques have emerged for modeling hydrological time series, limitations to the mechanistic and physical rationale that can be afforded to the internal structure and behaviors of such models still need to be considered.

Acknowledgments: We thank ActewAGL for providing us with the rainfall and streamflow data. This study is part of an Australian Research Council (ARC) Linkage Project LP0989881.

Author Contributions: Mark Adams designed and led the ARC research project. All authors contributed to the design of the research described here. Mana Gharun and Mohammad Azmi conducted the analyses and wrote the manuscript.

Conflicts of Interest: The authors declare no conflict of interest.

References

1. Langford, K.J. Changes in yield of water following a bushfire in a forest of *Eucalyptus regnans*. *J. Hydrol.* **1976**, *29*, 87–114.
2. Viggers, J.I.; Weaver, H.J.; Lindenmayer, D.B. *Melbourne's Water Catchments: Perspectives on a World-Class Water Supply*; CSIRO Publishing: Collingwood, Australia, 2013.
3. Kuczera, G. Prediction of water yield reductions following a bushfire in ash-mixed species eucalypt forest. *J. Hydrol.* **1987**, *94*, 215–236.
4. McMichael, C.E.; Hope, A.S. Predicting streamflow response to fire-induced landcover change: Implications of parameter uncertainty in the MIKE SHE model. *J. Environ. Manag.* **2007**, *84*, 245–256.
5. Marcar, N.E.; Benyon, R.G.; Polglase, P.J.; Paul, K.I.; Theiveyanathan, S.; Zhang, L. *Predicting Hydrological Impacts of Bushfire and Climate Change in Forested Catchments of the River Murray Uplands: A Review*; CSIRO, Water for a Healthy Country National Research Flagship: Canberra, Australia, 2006.
6. Watson, F.G.R.; Vertessy, R.A.; Grayson, R.B. Large-scale modelling of forest hydrological processes and their long-term effect on water yield. *Hydrol. Process.* **1999**, *13*, 689–700.
7. Green, E.P.; Mumby, P.J.; Edwards, A.J.; Clark, C.D.; Ellis, A.C. Estimating leaf area index of mangroves from satellite data. *Aquat. Bot.* **1997**, *58*, 11–19.

8. Vertessy, R.A.; Watson, F.G.R.; O'Sullivan, S.K.; Davis, S.; Campbell, R.; Benyon, R.; Haydon, S.R. *Predicting Water Yield from Mountain Ash Forest Catchments*; Report 98/4; Cooperative Research Centre for Catchment Hydrology: Clayton, Australia, 1998.

9. Turner, D.P.; Cohen, W.B.; Kennedy, R.E.; Fassnacht, K.S.; Briggs, J.M. Relationships between leaf area index and Landsat TM spectral vegetation indices across three temperate zone sites. *Remote Sens. Environ.* **1999**, *70*, 52–68.

10. Vertessy, R.A.; Watson, F.G.R.; O'Sullivan, S.K. Factors determining relations between stand age and catchment water balance in mountain ash forests. *For. Ecol. Manag.* **2001**, *143*, 13–26.

11. Abbott, M.B.; Bathurst, J.C.; Cunge, J.A.; O'Connell, E.; Rasmussen, J. An introduction to the European System: Systeme Hydrologique Europeen (SHE). *J. Hydrol.* **1986**, *87*, 61–77.

12. Muluye, G.Y.; Coulibaly, P. Seasonal reservoir inflow forecasting with low frequency climatic indices: A comparison of data-driven methods. *Hydrol. Sci. J.* **2007**, *52*, 508–522.

13. Londhe, S.; Charhate, S. Comparison of data-driven modelling techniques for river flow forecasting. *Hydrol. Sci. J.* **2010**, *55*, 1163–1174.

14. Noh, S.J.; Tachikawa, Y.; Shiba, M.; Kim, S. Ensemble kalman filtering and particle filtering in a lag-time window for short-term streamflow forecasting with a distributed hydrologic model. *J. Hydrol. Eng.* **2013**, *18*, 1684–1696.

15. Lee, H.; McIntyre, N.R.; Wheater, H.S.; Young, A.R. Predicting runoff in ungauged UK catchments. *Proc. ICE Water Manag.* **2006**, *159*, 129–138.

16. Asquith, W.H.; Herrmann, G.E.; Cleveland, T.G. Generalized additive regression models of discharge and mean velocity associated with direct-runoff conditions in Texas: Utility of the U.S. geological survey discharge measurement database. *J. Hydrol. Eng.* **2013**, *18*, 1331–1348.

17. Asquith, W.H. Regression models of discharge and mean velocity associated with near-median streamflow conditions in Texas: Utility of the U.S. geological survey discharge measurement database. *J. Hydrol. Eng.* **2014**, *19*, 108–122.

18. Azmi, M.; Araghinejad, S.; Kholghi, M. Multi model data fusion for hydrological forecasting using K-nearest neighbour method. *Iran. J. Sci. Technol.* **2010**, *34*, 81–92.

19. Akbari, M.; Overloop, P.J.V.; Afshar, A. Clustered K nearest neighbor algorithm for daily inflow forecasting. *Water Resour. Manag.* **2011**, *25*, 1341–1357.

20. Saghafian, B.; Anvari, S.; Morid, S. Effect of Southern Oscillation Index and spatially distributed climate data on improving the accuracy of Artificial Neural Network, Adaptive Neuro-Fuzzy Inference System and K-Nearest Neighbour streamflow forecasting models. *Expert Syst.* **2013**, *30*.

21. Nourani, V.; Komasi, M.; Alami, M.T. Hybrid wavelet-genetic programming approach to optimize ANN modeling of rainfall-runoff process. *J. Hydrol. Eng.* **2012**, *17*, 724–741.

22. Danandeh Mehr, A.; Kahya, E.; Olyaie, E. Streamflow prediction using linear genetic programming in comparison with a neuro-wavelet technique. *J. Hydrol.* **2013**, *505*, 240–249.

23. Yilmaz, A.G.; Muttil, N. Runoff estimation by machine learning methods and application to the euphrates basin in Turkey. *J. Hydrol. Eng.* **2014**, *19*, 1015–1025.

24. Araghinejad, S.; Azmi, M.; Kholghi, M. Application of artificial neural network ensembles in probabilistic hydrological forecasting. *J. Hydrol.* **2011**, *407*, 94–104.

25. Singh, A.; Imtiyaz, M.; Issac, R.K.; Denis, D.M. Comparison of artificial neural network models for sediment yield prediction at single gauging station of watershed in eastern India. *J. Hydrol. Eng.* **2013**, *18*, 115–120.

26. Budu, K. Comparison of wavelet-based ANN and regression models for reservoir inflow forecasting. *J. Hydrol. Eng.* **2014**, *19*, 1385–1400.

27. Manel, S.; Dias, J.M.; Ormerod, S.J. Comparing discriminant analysis, neural networks and logistic regression for predicting species distributions: A case study with a Himalayan river bird. *Ecol. Model.* **1999**, *120*, 337–347.

28. Paruelo, J.; Tomasel, F. Prediction of functional characteristics of ecosystems: A copmarison of artificial neural networks and regression models. *Ecol. Model.* **1997**, *98*, 173–186.

29. Shamshiry, E.; Bin Mokhtar, M.; Abdulai, A. *Comparison of Artificial Neural Network (ANN) and Multiple Regression Analysis for Predicting the Amount of Solid Waste Generation in a Tourist and Tropical Area—Langkawi Island*; International Biological, Civil and Environmental Engineering: Dubai, United Arab Emirates, 2014.

30. Talsma, T. Soils of the Cotter catchment area, ACT: Distribution, chemical and physical properties. *Aust. J. Soil Res.* **1983**, *21*, 241–255.

31. White, I.; Wade, A.; Daniell, T.M.; Mueller, N.; Worthy, M.; Wasson, R. The vulnerability of water supply catchments to bushfires: Impacts of the January 2003 wildfires on the Australian Capital Territory. *Aust. J. Water Resour.* **2006**, *10*, 179–194.

32. Sparks, T.; Muller, N.; O'Shannassy, K.; Taylor, P. *Cotter Catchment Landscape Analysis*; ACTEW Corporation: Canberra, Australia, 2008.

33. Xavier, A.C.; Vettorazzi, C.A. Monitoring leaf area index at watershed level through NDVI from Landsat-7/ETM+ data. *Sci. Agric.* **2004**, *61*, 243–252.

34. Huete, A.; Justice, C.; van Leeuwen, W. *MODIS Vegetation Index (MOD 13) Algorithm Theoretical Basis Document ATBD13*; The University of Arizona: Tucson, AZ, USA, 1999.

35. Sever, L.; Leach, J.; Bren, L. Remote sensing of post-fire vegetation recovery; A study using Landsat 5 TM imagery and NDVI in North-East Victoria. *J. Spat. Sci.* **2012**, *57*, 175–191.

36. Coops, N.; Delahaye, A.; Pook, E. Estimation of eucalypt forest leaf area index on the south coast of New South Wales using Landsat MSS data. *Aust. J. Bot.* **1997**, *45*, 757–769.

37. Solano, R.; Didan, K.; Jacobson, A.; Huete, A. *MODIS Vegetation Index User's Guide*; MOD13 Series; Collection 5; Vegetation Index and Phenology Lab, The University of Arizona: Tucson, AZ, USA, 2010.

38. Hill, M.J.; Senarath, U.; Lee, A.; Zeppel, M.; Nightingale, J.M.; Williams, R.D.J.; McVicar, T.R. Assessment of the MODIS LAI product for Australian ecosystems. *Remote Sens. Environ.* **2006**, *101*, 495–518.

39. Nash, J.E.; Sutcliffe, J.V. River flow forecasting through conceptual models part I: A discussion of principles. *J. Hydrol.* **1970**, *10*, 282–290.

40. Schrage, L. *Optimization Modelling with LINGO*, 5th ed.; LINDO System Inc.: Chicago, IL, USA, 1999.

41. Tarboton, D.G.; Sharma, A.; Lall, U. The use of non-parametric probability distribution in streamflow modeling. In Proceedings of the Sixth South African National Hydrological Symposium, University of Natal, Pietermaritzburg, South Africa, 4–6 September 1993; pp. 315–327.

42. Yakowitz, S.J. Nonparametric density estimation, prediction, and regression for Markov sequences. *J. Am. Stat. Assoc.* **1985**, *80*, 215–221.

43. Yan, L.; Elgamal, A.; Cottrell, G.W. Substructure vibration NARX neural network approach for statistical damage inference. *J. Eng. Mech.* **2013**, *139*, 737–747.

44. AlHamaydeh, M.; Choudhary, I.; Assaleh, K. Virtual testing of buckling-restrained braces via nonlinear autoregressive exogenous neural networks. *J. Comput. Civil Eng.* **2013**, *27*, 755–768.

45. Ruslan, F.A.; Samad, A.M.; Zain, Z.M.; Adnan, R. Flood prediction using NARX neural network and EKF prediction technique: A comparative study. In Proceedings of the IEEE 3rd International Conference on System Engineering and Technology, Shah Alam, Malaysia, 19–20 August 2013.

46. Koza, J.R.; Keane, M.A.; Streeter, M.J. *Evolving Inventions*; Scientific American: New York, NY, USA, 2003; pp. 40–47.

47. Schmidt, M.; Lipson, H. Distilling free-form natural laws from experimental data. *Science* **2009**, *324*, 81–85.

48. Abrahart, R.J.; Beriro, D.J. How much complexity is warranted in a rainfall-runoff model? Findings obtained from symbolic regression, using Eureqa. In Proceedings of EGU General Assembly, Vienna, Austria, 22–27 April 2012.

49. Grelle, G.; Bonito, L.; Revellino, P.; Guerriero, L.; Guadagno, F.M. A hybrid model for mapping simplified seismic response via a GIS-metamodel approach. *Nat. Hazards Earth Syst. Sci.* **2014**, *2*, 963–997.

50. Larsen, P.E.; Cseke, L.J.; Miller, R.M.; Collart, F.R. Modeling forest ecosystem responses to elevated carbon dioxide and ozone using artificial neural networks. *J. Theor. Biol.* **2014**, *359*, 61–71.

51. Drobot, R.; Dinu, C.; Draghia, A.; Adler, M.J.; Corbus, C.; Matreata, M. Simplified approach for flood estimation and propagation. In Proceedings of the IEEE International Conference on Automation, Quality and Testing, Robotics, Cluj-Napoca, Romania, 22–24 May 2014.

52. Hamdan, M.A.; Bardan, A.A.; Abdelhafez, E.A.; Hamdan, A.M. Comparison of neural network models in the estimation of the performance of solar collectors. *J. Infrastruct. Syst.* **2014**.

53. Coulibaly, P.; Bobee, B.; Antcil, F. Improving extreme hydrologic eventsforecasting using a new criterion for artificial neural network selection. *Hydrol. Process.* **2001**, *15*, 1533–1536.

54. Charhate, S.B.; Dandawate, Y.H.; Londhe, S.N. Genetic programming to forecast stream flow. *Adv. Water Resour. Hydraul. Eng.* **2009**, *1–6*, 29–34.
55. McMillan, H.; Krueger, T.; Freer, J. Benchmarking observational uncertainties for hydrology: Rainfall, river discharge and water quality. *Hydrol. Process.* **2012**, *26*, 4078–4111.

Comparison of Water Flows in Four European Lagoon Catchments under a Set of Future Climate Scenarios

Cornelia Hesse, Anastassi Stefanova and Valentina Krysanova

Abstract: Climate change is supposed to remarkably affect the water resources of coastal lagoons as they are highly vulnerable to changes occurring at their catchment and/or ocean or sea boundaries. Probable impacts of projected climate changes on catchment hydrology and freshwater input were assessed using the eco-hydrological model SWIM (Soil and Water Integrated Model) for the drainage areas of four European lagoons: *Ria de Aveiro* (Portugal), *Mar Menor* (Spain), *Tyligulskyi Liman* (Ukraine) and *Vistula Lagoon* (Poland/Russia) under a set of 15 climate scenarios covering the time period until the year 2100. Climate change signals for all regions show continuously increasing trends in temperature, but various trends in precipitation. Precipitation is projected to decrease in two catchments on the Iberian Peninsula and increase in the Baltic region catchment, and does not show a clear trend in the catchment located near the Black Sea. The average projected changes in freshwater inputs reflect these changes in climate conditions, but often show variability between the scenarios, in future periods, and within the catchments. According to the individual degrees of water management influences in the four drainage basins, the climate sensitivity of river inflows is differently pronounced in each.

Reprinted from *Water*. Cite as: Hesse, C.; Stefanova, A.; Krysanova, V. Comparison of Water Flows in Four European Lagoon Catchments under a Set of Future Climate Scenarios. *Water* **2015**, *7*, 716–746.

1. Introduction

Positive trends in temperature and diverse changes in precipitation, affecting water balance components and regional water resources, have been observed worldwide over the past decades [1–5]. The observed annual average temperature increase across European land areas is higher than the global scale average [6] and amounts to 0.9 °C for 1901 to 2005, with some variability between regions and seasons [7]. The greatest warming over the past 30 years was detected over Scandinavia, especially in winter, whereas the Iberian Peninsula warmed mostly in summer [8]. It is very likely that the temperature increase since the mid of the 20th century is due to the increase in greenhouse gas concentrations in the atmosphere resulting from anthropogenic activities, and it is expected that the climate warming will continue [1,2]. Observed precipitation changes during the last decades show higher temporal and spatial variability compared to the changes of temperature. Annual precipitation has generally increased in Northern Europe and decreased in most parts of Southern Europe [6]. The effects of altered climate conditions on diverse ecosystemic and social functions can already be detected, such as a longer plant-growing season, changes in species distribution and biodiversity, retreating of glaciers, and humans suffering from heat waves (e.g., [9–12]).

According to Alcamo *et al.* [7], potential warming in Europe could reach values from +1 to +6 °C by the end of this century. The trends may noticeably vary in different European regions. Not only are changing temperatures expected in the future, but also shifting rainfall patterns and altered river runoff, less snow cover in extent and duration, rising sea levels, and continuously melting glaciers. Projections indicate a general increase in annual precipitation varying between +10% and +20% in Northern Europe and a decrease between −5% to −20% in Southern Europe and the Mediterranean region [6]. All these changes can have various effects on water resources and affect ecology and society as well as the ecosystem services of different regions.

Expected climate changes are also supposed to impact the coastal lagoon ecosystems in Europe. Changes at its ocean and/or catchment boundary conditions may cause changes in the lagoon's ecological status and in its ability to serve as a recreation area, living environment and source of livelihood. The potential impacts could differ considerably at regional and local scales, and have to be studied in order to help people prepare and improve the adaptive capacity of particular lagoon ecosystems [13,14]. Many authors emphasize the importance of developing adequate lagoon management plans and implementing proactive adaptation measures [15,16].

Climate impact studies for lagoons and coastal areas worldwide most often deal with the direct impacts of climate change on their water bodies and ecosystems (e.g., [17–19]). But the indirect impacts should not be forgotten as climate change can also cause variations in river runoff and freshwater inflow from the drainage areas to the lagoons (e.g., [20–22]). This certainly may affect the lagoon's condition by, for example, shifting its salinity, biodiversity or eutrophication status. Climate change impact assessment allows us to use models to study possible changes in river discharge under different climate conditions.

A common approach for hydrological impact studies at the catchment scale is to use climate parameters provided by climate models as input for calibrated and validated hydrological models [23]. As global climate models (GCM) have resolutions too coarse for regional eco-hydrological studies, downscaling is usually needed to get more reliable input data for the hydrological models. It is normally provided either by Regional Climate Models (RCM) driven by GCMs or by using statistical downscaling methods. Using climate scenarios from several driving RCMs (multi-model approach) is preferable in comparison to applying just one climate model as a driver to investigate the range of uncertainty for impact projections [24–27]. Such an approach is also applied in this study.

The main objective of the study was to perform a climate impact assessment for water flows in the catchment areas of four European lagoons—*Ria de Aveiro* (Portugal), *Mar Menor* (Spain), *Tyligulskyi Liman* (Ukraine) and *Vistula Lagoon* (Poland/Russia)—until the end of the 21st century. For that purpose, the eco-hydrological model SWIM (Soil and Water Integrated Model, [28]) was applied, and a set of 15 climate scenarios from the ENSEMBLES project [29] served as drivers.

The explicit research questions concerning model applications in the case study areas in this publication are as follows:

- Is the SWIM model able to sufficiently simulate the hydrology of the four chosen multi-river European lagoon catchments?
- What future climate changes can be expected in the four selected lagoon catchments?
- How are the river discharge and catchment runoff impacted by possible changed climate conditions?
- Is there a spatial heterogeneity of impacts between the catchments in Europe or within single catchments?
- Which climate parameter is most important in terms of influencing future river runoff?
- What suggestions can be made for the management of the four lagoon catchments, and what are the implications of this work for other lagoons and coastal systems?

The outputs of the SWIM catchment model driven by climate scenarios could also be used as inputs for lagoon models to simulate the responses of water bodies to altered freshwater inputs from the catchments resulting from changed climate conditions, as was done in a subsequent separate climate impact assessment.

Our catchment-to-coast modelling is the first step in an overall catchment-lagoon impact assessment study supporting the development of a pan-European strategy plan for coastal areas under a changing climate. The study is done within the European FP7 project LAGOONS. The results would enable a better understanding of potential future developments in lagoon catchments, and contribute to creating appropriate adaptation strategies for these regions in view of climate change.

2. Description of the Case Study Areas

Climate and land use change impact assessments were conducted for the catchments of four lagoons located in different regions of Europe and connected to four different seas:

(1) The **Ria de Aveiro** is located in Portugal and connected to the Atlantic Ocean. It has a catchment of about 3560 km^2 mainly drained by the Vouga River. The catchment is influenced by a humid and temperate climate, and largely covered by forest.

(2) The **Mar Menor** is located in Spain close to the Mediterranean Sea with a catchment of about 1380 km^2. Although the catchment is characterized by hot and dry summers, it is intensively used for irrigated agriculture. The largest river Albujon Wadi often dries up and is not a permanent stream.

(3) The **Tyligulskyi Liman** can be found in the Ukraine near the Black Sea with a catchment of about 5240 km^2. It is mainly drained by the Tyligul River and characterized by a warm temperate to continental climate. Due to very fertile soils in this region the catchment is mainly used for agriculture.

(4) The transboundary catchment of the **Vistula Lagoon** is located in Poland and Russia connected to the Baltic Sea. It covers an area of about 20.730 km^2 drained by several main rivers in a marine temperate climate. The drainage area is mainly used for agriculture with relatively numerous forested areas.

The locations of the four case study areas (CSA) within Europe as well as the digital elevation models (DEM) for the lagoon catchments can be found in Figure 1. Some catchment characteristics are listed in Table 1.

The four CSAs investigated in this study are subject to different degrees of human water management impacts on the natural river flow in the catchments. The water management measures in the Aveiro and **Vistula Lagoon** catchments are not substantial compared to two other cases studies. In relation to the total volume of freshwater inflow to the lagoons, the managed water flows amount to 2.3% in the **Ria de Aveiro** (abstraction and discharge for public water supply) and 8.5% in the **Vistula Lagoon** catchment (water transfer from the Vistula River). Additionally, 40% of the Pregolya River discharge does not reach the **Vistula Lagoon** directly but is instead transferred to another catchment beforehand. However, the management measures are more extensive in the other two cases. In the **Mar Menor** catchment, the total water inflow to the lagoon is influenced by water transfer from the Tagus catchment (used for irrigation) and the UWTP (urban water treatment plant) effluents (*i.e.*, 51% of the total discharge to the lagoon is related to management). The river discharge in the **Tyligulskyi Liman** catchment is highly influenced by the construction of ponds within the whole river network and by irretrievable water use (as a result, only 49% of the discharge reaches the lagoon). Therefore, even smaller absolute volumes of water transfer or abstraction can cause a high management degree in catchments with naturally low water discharges, as in the **Mar Menor** and **Tyligulskyi Liman** drainage basins, characterized by dry climate conditions.

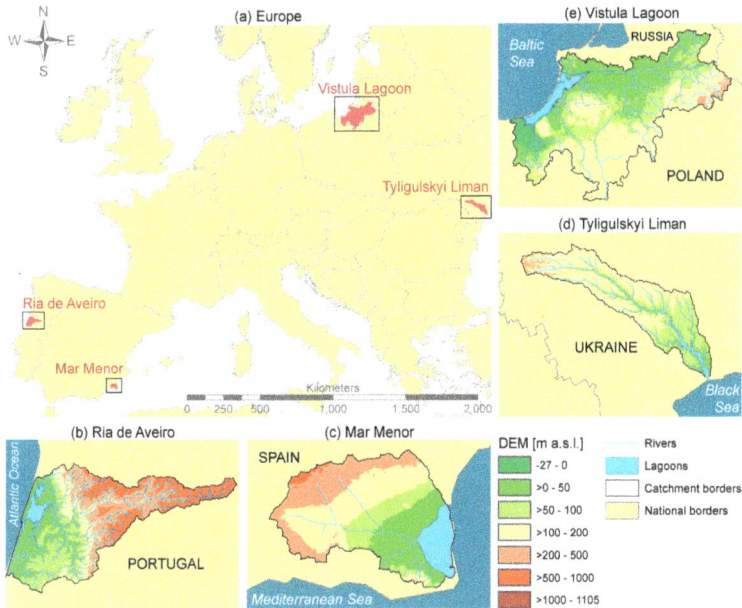

Figure 1. Location of the four case study areas (CSAs) within Europe (**a**) as well as the Digital Elevation Models (DEM), locations of lagoons, and main inflowing rivers per CSA (**b–e**).

Table 1. Characteristics of the four case study areas investigated in the study.

Parameter	Unit	Ria de Aveiro	Mar Menor	Tyligulskyi Liman	Vistula Lagoon
Lagoon area	km^2	75	135	160	322
Catchment area	km^2	3,556	1,380	5,240	20,730
Country(ies)	-	Portugal	Spain	Ukraine	Poland/Russia
Sea	-	Atlantic Ocean	Mediterranean	Black Sea	Baltic Sea
Total freshwater inflow	km$^3 \cdot$ year^{-1}	2.14	0.009	0.023	3.69
Main inflowing rivers	-	Vouga	Albujon	Tyligul	Pregolya / Pasleka / Elblag
Number of analysed inflowing rivers	-	10	7	6	12
Number of infl. rivers with avail. gauge data	-	1	0	1	5
Av. altitude (range)	m a.s.l.	363 (−10–1,105)	100 (−5–1,061)	102 (−6–254)	82 (−27–308)
Av. temperature	°C	14	25	9.7	7.7
Av. precipitation (range)	mm· year^{-1}	1,100 (600–2,100)	337 (300–370)	515 (470–570)	750 (670–860)
Major land uses					
Agriculture	%	29	82	80	67
Forest	%	56	1	4	25

89

3. Material and Methods

3.1. Soil and Water Integrated Model (SWIM)

The eco-hydrological model SWIM [28] was used for climate impact assessment in the four European lagoons. This process-based semi-distributed model was developed based on the models SWAT (Soil Water Assessment Tool) [30] and MATSALU [31] for river basins at the regional scale (200 up to 500,000 km^2). The hydrological, nutrient, sediment and vegetation processes are simulated with a daily time step at the hydrotope level. Hydrotopes are sets of the smallest spatial units in SWIM, and are defined as areas within one subbasin with unique land use class and soil type. It is assumed that such units behave similarly regarding water flows, nutrient cycling and vegetation growth (principle of similarity). After process calculation for hydrotopes, the components and flows are aggregated at the subbasin level, and then the lateral flows are routed to the basin's outlet (and enter the specific lagoon).

Soil hydrological processes in SWIM are based on the water balance equation, and consider surface, subsurface and groundwater flows as well as percolation and recharge of the aquifers. Surface flow is calculated by a non-linear function of precipitation and a retention coefficient which depends on land use, soil type, management, and the actual soil water content. Subsurface flow and percolation are calculated simultaneously, separately for each soil layer: subsurface flow occurs if percolation exceeds field capacity in a layer. The number of soil layers (up to ten can be considered) is defined depending on available soil parameterization for the catchments. Percolation from the bottom soil layer leads to a recharge of the shallow aquifer, from where water can rise again to the soil profile through capillaries, flow laterally to the river network, or percolate to a deep aquifer. From the deep aquifer water cannot rise up again.

Snow processes are simulated using the method of Gelfan *et al.* [32] as described in Huang [33]. Snow melting is calculated with a degree-day-factor. Water outflow from the snowpack depends on snow depth, content of ice and liquid water, and snow density. The processes of refreezing and snow metamorphism are also considered.

In the original SWIM model potential evapotranspiration is generally calculated based on solar radiation, daily mean temperature and elevation using the method of Priestley and Taylor [34]. However, in the case of the *Tyligulskyi Liman* catchment the Turc–Ivanov equation [35,36] was used to better reproduce observed values in this catchment. The actual evaporation/transpiration is calculated separately for soils and plants as functions of potential evapotranspiration and the leaf area index (LAI), while soil evaporation is reduced when its accumulated amount exceeds 6 mm. The limited soil water content leads to decreased plant transpiration [28].

Climate as well as land use and management practices are important external drivers for the processes represented in the model. The main external climatic drivers are precipitation and snow fall together with temperature and solar radiation influencing snow melt processes and the evapotranspiration potential of the vegetation and the landscape. Climate parameters are assumed to be homogeneous at the subbasin level. Measured (for model calibration) or projected (for impact assessment) climate data are interpolated to the subbasin centroids by using an inverse weighted distance method.

Due to its spatial resolution as well as climate and land use considered as boundary conditions, the SWIM model allows the analysis of impacts of climate and land use changes on the major model outputs. It is therefore a suitable tool for climate change impact assessments as planned in this study.

SWIM has been successfully applied in several catchments of different sizes firstly in Germany and later in other river basins in Europe, Africa, Asia and South America. It is still being developed further in accordance with the particular research needs or specific case characteristics. An overview of the main applications and implementations can be found in Krysanova *et al.* [37].

3.2. Input Data, Model Setup and Calibration Strategies

The SWIM model was applied to each of the four case study areas individually. The model setup for the drainage basins of the lagoons was a challenging task, as they mostly consist of several inflowing rivers and streams, with often no available data on water flows and discharges. A non-trivial modelling strategy was necessary to combine several river catchments within one SWIM modelling project for a specific lagoon.

The modelled river discharge could only partly be properly calibrated and validated at one or several gauges in four individual applications, and the sub-catchments of smaller rivers/streams flowing to the lagoons had to be modelled in an ungauged mode. Therefore, calibration and validation were performed for the largest river catchments within the lagoon drainage basins: two in the case of the *Vistula Lagoon*, and one in each of the three other cases. After that, SWIM was applied for the entire lagoon drainage areas transferring the same calibrated parameters to the ungauged parts, assuming similar geophysical and hydrological conditions in adjacent river catchments but under consideration of catchment-specific characteristics and management settings.

For model setup and hydrological calibration the SWIM model needs spatial data, time series and water management information as input. The spatial data include maps depicting DEM, land use classes and soil type distribution, as well as subbasin structure. The required time series for model setup especially include daily climate data (minimum, maximum and average temperature, precipitation, air humidity and solar radiation). Necessary management data for hydrological modelling include information on water abstraction, water transfer from and to adjacent catchments, major crops with their planting and harvesting dates, as well as irrigation practices on agricultural fields.

Furthermore, river discharge data at gauging stations are needed for model calibration and validation. In three of four catchments, discharge data from several gauges provided an opportunity for a multi-site calibration, which is more reliable in general and especially important in multi-river drainage basins as in our case.

Table 2 presents an overview of datasets used for SWIM application in the drainage basins of the four lagoons. Most of the data are case specific. In all four case study areas some data were missing, or data coverage in time and/or space was problematic. Therefore, in all four cases the model calibration was a very complicated task. The heterogeneity of spatial input data and inconsistent or missing time series of climate and water discharge required interpolation and empirical methods for data estimation.

The model setup and calibration strategies are described in more detail in LAGOONS (2013) as well as the *Ria de Aveiro* case study area in Stefanova *et al.* [38] and the *Vistula Lagoon* case study area in Hesse *et al.* [39].

During the calibration process, the model results were evaluated in regard to the ability of SWIM to adequately simulate observed water discharges in the lagoon catchments using two criteria of fit: Nash and Sutcliffe efficiency (NSE) and deviation in water balance (DB). The equations to calculate NSE and DB can be found in Hesse *et al.* [40].

The non-dimensional NSE [41] criterion describes the squared differences between the observed and the simulated values and is based on the dispersion of values around the line of equal values. The NSE can vary from minus infinity to 1 and should be as near as possible to 1. In hydrological modelling, NSE values above 0.5 are considered as corresponding to satisfactory, and above 0.65 as to good modelling results for a monthly time step [42].

Table 2. Data used for model setup and calibration in the catchments of the lagoons under study.

Data	Ria de Aveiro	Mar Menor	Tyligulskyi Liman	Vistula Lagoon
Observed climate	30 stations; large gaps in records; missing solar radiation was derived with the Hargreaves-Samani method	5 stations (4 in the basin), period: 2000–2011	4 climate and 2 precipitation stations outside the catchment available; poor correlation between precipitation and measured discharge → model data for 1979–2009 were used	Observed climate data with poor coverage in time and space → model data for 1979–2009 were used
	Sources: http://snirh.pt/ http://www.tutiempo.net/	Source: SIAM	Source: WFDEI climate data [43]	Source: WFDEI climate data [43]
DEM	SRTM (Shuttle Radar Topography Mission)	SRTM (Shuttle Radar Topography Mission)	SRTM (Shuttle Radar Topography Mission)	SRTM (Shuttle Radar Topography Mission)
	Source: http://srtm.csi.cgiar.org/	Source: http://srtm.csi.cgiar.org/	Source: http://srtm.csi.cgiar.org/	Source: http://srtm.csi.cgiar.org/
Land use	CORINE Land Cover 2006, Version 13	CORINE Land Cover 2006, Version 13	No digital data was available A paper map was scanned and digitized	CORINE (CLC2000), Kaliningrad oblast territorial planning scheme
	Source: EEA	Source: EEA	Source: DENR, RDIIM	Sources: EEA, PKO
Soil map and soil parameterization	Map: ESDB	Map: HWSD (1 km × 1 km)	Map: HWSD (1 km × 1 km)	Maps: HWSD and SGDBE
	Soil parameters from SGDBE and estimated using the German soil mapping guidelines [44]	Soil parameters from HWSD and estimated using the German soil mapping guidelines [44]	Soil parameters from HWSD and estimated using the German soil mapping guidelines [44]	Soil parameters from maps and estimated using the German soil mapping guidelines [44]
Spatial resolution in SWIM	90 m × 90 m raster maps	20 m × 20 m raster maps	100 m × 100 m raster maps	100 m × 100 m raster maps
	365 subbasins	215 subbasins	175 subbasins	442 subbasins
	2452 hydrotopes	744 hydrotopes	920 hydrotopes	4469 hydrotopes
Observed discharge	hourly/quarterly water levels and flow curve equations for 3 gauges for 2002–2005	no gauge data available, 24 survey measurements for period 09/2003–06/2006, estimated seasonal dynamics for 2003	one upstream gauge (1984–1988) and one downstream gauge (1984–1988 and 1998–2007)	10 discharge gauges for different sub-periods during 1995–2009 Main calibration gauges Lozy (Pasleka river) and Gvardeysk (Pregolya river)
	Source: http://snirh.pt/	Sources: UM, [45,46]	Source: CGO	Sources: IMGW-PIB, KCHEM
Main crops	corn	water melons, lettuce	winter wheat	winter wheat
	Source: [47]	Source: [48]	Source: [49]	Source: [50]

Table 2. *Cont.*

Data	Ria de Aveiro	Mar Menor	Tyligulskyi Liman	Vistula Lagoon
Water management	Water abstraction: from stream for public water supply with exact location	Irrigation with water from Tagus river (data on annual amounts and location of irrigated area)	Data on ponds and irretrievable water use provided by case study partners	Water inflow and outflow implemented according to literature
	Source: APA	Source: [51]	Sources: IWR-MR, PTR	Sources: [52,53]

Notes: APA—Portuguese Environment Agency (former ARH Centro). CGO—Central Geophysical Observatory http://cgo.kiev.ua. CORINE Land Cover—Coordination of Information on the Environment http://www.eea.europa.eu/data-and-maps/data. DENR—Department of Ecology and Natural Resources, Odessa Regional State Administration http://ecology.odessa.gov.ua/. EEA—European Environment Agency. ESDB/SGDBE—European Soil Database http://eusoils.jrc.ec.europa.eu/data.html. HWSD—Harmonized World Soil Database http://webarchive.iiasa.ac.at/Research/LUC/External-World-soil-database/HTML/. IMGW-PIB—Institute of Meterology and Water Management—National Research Institute. IWR-MR—Inventory of water resources in Mykolaiv Region. Mykolaiv, 2009, pp. 184 (in Ukrainian). KCHEM—Kaliningrad Centre for Meteorology and Environmental Monitoring. PKO—Government of the Kaliningrad Region http://old.gov39.ru/. PTR—Passport of Tyligul River. Odessa: UkrYuzhGIProVodChoz, 1994, pp. 148 (in Russian). RDILM—Odessa Research and Design Institute of Land Management http://www.landres.od.ua/kontakts.html. SIAM—Murcia Agriculture Information System http://siam.imida.es. UM—University Murcia. WFDEI—WATCH Forcing Data methodology applied to ERA-Interim data www.eu-watch.org/publications and ftp.iiasa.ac.at.

94

The DB criterion corresponds to the percent bias (PBIAS), but with the opposite algebraic sign, and shows the long-term percental difference of the observed values against the simulated ones. For best model results, DB should be as near as possible to 0. In hydrological model applications with a monthly time step, DB/PBIAS below $\pm25\%$ is considered satisfactory, and DB/PBIAS lower than $\pm10\%$ shows very good model results [42].

For an objective appraisal of the model results achieved in this study, it should be kept in mind that some criteria of fit presented here were calculated with the daily values, where good performance rates are generally more difficult to obtain than for simulations with the monthly time step.

3.3. Climate Scenario Description and Application

Climate impact assessment in the four lagoon catchments was performed by using a set of climate scenario data provided by the ENSEMBLES project [29]. In this project, a set of RCMs was run using the boundary conditions simulated by different GCMs. All scenarios were driven by the A1B emission scenario, which is an intermediate one concerning projections for increasing atmospheric CO2 concentrations. For the study presented here, only scenarios with the finer spatial resolution of 25 km and a simulation period until the end of the century (mainly until 2098) were selected from the ENSEMBLES climate scenario collection, resulting in a set of 15 different climate scenarios, later referred to as S1 to S15 (Table 3). Such a multi-model approach delivers several projections for the future climate, which have a higher reliability on average than each single scenario with its general uncertainty.

Table 3. Global (GCM) and Regional Climate Model (RCM) combinations and responsible institutes for the 15 ENSEMBLES climate scenarios (S1-S15) used for impact assessment in the case study areas.

Scenario	GCM	RCM	Institute	Country
S1	HadCM3Q3	RCA3	Swedish Meteorological and Hydrological Institute (SMHI)	Sweden
S2	HadCM3Q0	HadRM3Q0	Hadley Center for Climate Predictions and Research (HC)	Great Britain
S3	HadCM3Q3	HadRM3Q3	Hadley Center for Climate Predictions and Research (HC)	Great Britain
S4	HadCM3Q16	HadRM3Q16	Hadley Center for Climate Predictions and Research (HC)	Great Britain
S5	HadCM3Q16	RCA3	Community Climate Change Consortium for Ireland (4CI)	Northern Ireland
S6	HadCM3Q0	CLM	Swiss Federal Institute of Technology Zurich (ETHZ)	Switzerland
S7	ECHAM5-r3	RACMO2	Royal Netherlands Meteorological Institute (KNMI)	The Netherlands
S8	BCM	RCA3	Swedish Meteorological and Hydrological Institute (SMHI)	Sweden
S9	ECHAM5-r3	RCA3	Swedish Meteorological and Hydrological Institute (SMHI)	Sweden
S10	ECHAM5-r3	REMO	Max Planck Institute for Meteorology (MPI)	Germany
S11	ARPEGE	ALADIN RM5.1	National Center for Meteorological Research (CNRM)	France
S12	ARPEGE	HIRHAM5	Danish Meteorological Institute (DMI)	Denmark
S13	ECHAM5-r3	HIRHAM5	Danish Meteorological Institute (DMI)	Denmark
S14	BCM	HIRHAM5	Danish Meteorological Institute (DMI)	Denmark
S15	ECHAM5-r3	RegCM3	International Center for Theoretical Physics (ICTP)	Italy

As it is a common method in climate impact research to use 30-year-periods [54], scenarios S1 to S15 were divided into the reference period 1971–2000 (p0) and three scenario periods: near future 2011–2040 (p1), intermediate future 2041–2070 (p2), and far future 2071–2098 (p3) for climate impact assessment. They were analysed in advance by comparing the long-term monthly and annual averages of temperature and precipitation and calculating the according climate change signals per future period compared to the reference period of the same scenario (see results in Section 4.2).

The 15 climate scenarios were applied to the calibrated and validated SWIM for the four lagoon drainage basins under investigation. The climate parameters minimum, maximum and average temperature, precipitation, air humidity and solar radiation provided by the 15 individual scenarios were interpolated from the RCM grid points located within or at a maximum distance of 10 km around the catchments to the centroids of the subbasins in every case study area and then used to drive the SWIM model. The basin-specific land use and management input data of the reference period remained unchanged for the future periods in order to evaluate impacts of climate change only.

The climate impacts on runoff in the catchments were analysed considering (1) the long-term average daily, monthly and annual total water inflows to the lagoons; (2) the long-term average annual discharges of the main rivers per CSA; and (3) the long-term average runoff on a map with a hydrotope resolution, and all three analyses averaged over 30 years and 15 scenarios. The results of the three future scenario periods were always compared to the outputs simulated by SWIM under reference conditions driven by the same climate dataset (S1–S15), and not by observed climate. To get an impression of the uncertainty ranges of future projections, different percentiles as well as minimum and maximum values of the scenario results were also analyzed.

4. Results

This section summarizes the results of the study and describes model calibration and validation, presents the climate change signals and impacts on total water inflow to the lagoons and on inflows of separate rivers in the drainage areas with uncertainty estimates. The results section also deals with the impacts on spatial patterns of runoff in the catchment, and concludes with the general climate sensitivity of the total freshwater inflow to the four lagoons under study.

4.1. Model Calibration and Validation in the Four Case Study Areas

A detailed hydrological calibration and validation was performed for three gauges in the *Ria de Aveiro* catchment, for one gauge in the *Mar Menor* catchment (using an estimated hydrograph), for two gauges in the *Tyligulskyi Liman* catchment, and for two gauges in the *Vistula Lagoon* catchment. Figure 2 shows the observed and simulated long-term average daily and monthly discharges for the most downstream gauging stations of the largest rivers entering each of the four lagoons.

In addition, all available discharge data for smaller rivers within the entire lagoon catchments were used for a spatial validation of the calibration parameter set. All analysed gauges are listed in Table 4 together with the criteria of fit achieved there. If available, daily discharge data were used for evaluating model performance, but in some cases only biweekly or monthly values could be compared. The calibration and validation results are presented in more detail in a separate report [55].

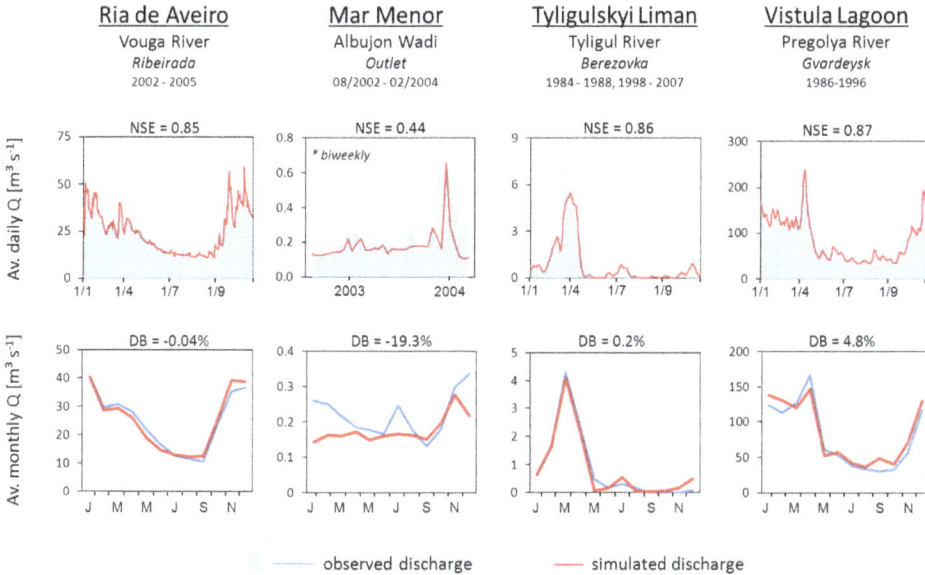

Figure 2. Comparison of the long-term average daily and monthly observed and simulated discharges (Q) at selected river gauges in the case study areas. Due to missing daily observed data in the *Mar Menor* case, the long-term average daily discharge could not be calculated.

In the *Ria de Aveiro* case, hydrological calibration was performed using data from three discharge gauges located within the Vouga catchment, the largest and most important river flowing to the lagoon, one on the main river, and two on tributaries. The observed daily discharges in the Vouga basin were estimated using water levels and unique flow curve equations for each gauge. Since the equations are valid only within certain ranges, some recorded values of very low and very high water levels could not be transformed into daily discharges, which contributed to uncertainty of model calibration in this CSA. The calibration was additionally complicated by human interventions and seasonal damming of water on one tributary, which masks the natural hydrograph. Another large uncertainty is related to missing or incomplete precipitation data in the catchment. Nevertheless, despite all uncertainties in input data, the hydrological calibration of the model was quite successful, reaching NSE values above 0.7 and a relative deviation in water balance of -0.04% at the most downstream gauge Ribeirada (Table 4). The long-term average dynamic of water discharge at this gauge is also reproduced quite well for the calibration period 2002–2005 (Figure 2).

Table 4. Criteria of fit achieved at all gauges with available water discharge data applying basin-wide calibration parameter sets in the lagoon case study areas.

Catchment	River	Gauge	Period	NSE	DB (%)
Ria de Aveiro	Águeda	Ponte Águeda	2002–2005	0.79	+5.6
	Cértima	Ponte Requeixo	2002–2005	0.72	−19.1
	Vouga	Ribeirada	2002–2005	0.7	−0.04
Mar Menor	Albujon	outlet	8/2002–2/2004 *	0.44	−19
Tyligulskyi Liman	Tyligul	Berezovka	1998–2007 °	0.36	+1.5
		Novoukrainka	1984–1988 °	−0.09	−0.1
			1984–1988 #	0.86	−0.4
Vistula Lagoon	Angrapa	Berestvo	1995–2000	0.63	−23.6
	Bauda	Nowe Sadulki	2009	0.55	−6.8
	Dzierzgon	Bagart	2009	0.34	−7.4
	Lava	Rodniki	1995–2000	0.70	−7.3
	Mamonovka	Mamonovo	2008–2009 °	0.62	−29.6
	Pasleka	Lozy	2007–2009	0.66	12.9
		Nowa Pasleka	1998–2000 °	0.72	−9.2
	Pissa	Zeleny Bor	1995–2000	0.73	0.4
	Pregolya	Gvardeysk	1983–1996	0.70	0.6
	Waska	Paslek	2009	0.48	−9.8

Notes: * biweekly values.° average monthly values.# long-term average monthly values.

In the **Mar Menor** catchment, hydrological calibration could not be performed as usual, as there were no observed time series of river discharge available. The calibration was undertaken using biweekly estimated data mainly for 2003 derived from Garcia-Pintado *et al.* (2007). An additional difficulty was that data on important water contribution to the Albujon Wadi from an urban water treatment plant (UWTP) were not available for 2003, and had to be estimated from the following years. The hydrological cycle in the catchment is additionally influenced by irrigation needs of highly water demanding crops, which are fulfilled by water transfer from adjacent catchments. The calibration results for the period 8/2002–2/2004 using the fortnightly estimated discharge are presented in Figure 2. The simulated curve shows acceptable results. In August 2003 the simulated discharge is missing one peak, which probably has its origin in the discharge coming from the UWTP since almost no rainfall was recorded for this month. Nevertheless, the average simulated discharge for 2003 is $0.20 \text{ m}^3 \cdot \text{s}^{-1}$, which is very close to the $0.24 \text{ m}^3 \cdot \text{s}^{-1}$ estimated by Garcia-Pintado *et al.* [45].

For the **Tyligulskyi Liman** catchment data from two gauges located at the Tyligul River, Novoukrainka and Berezovka, were used for hydrological calibration. The calibration was hampered by totally missing data on observed precipitation within the catchment, and poor correlation between observed discharge and precipitation from stations around the catchment. In addition, river discharge is highly influenced by numerous ponds in almost all reaches of the river network, and the high evaporation potential in this region often causing interruption of discharge in summer. The observed discharge could only be reproduced by the model after implementing ponds in the rivers using empirical data. Looking at the long-term average daily and monthly discharges of the Tyligul River (Figure 2) one can see that the model results with NSE of 0.86 and DB of 0.2% are acceptable, and match the average discharge dynamics quite well. However, the comparison of monthly averaged values for two stations (Table 4) shows that the discrepancies between the simulated and observed discharges are still high. The results are better for long-term average discharges, and the achieved DB values indicate that the simulated total water amount coming with the Tyligul River to the lagoon matches the observed values (Table 4). An acceptable balance could also be seen by comparing the simulated outputs with average discharge values from literature for four rivers flowing to the *Tyligulskyi Liman* as well as for the total inflow to the lagoon [55].

Hydrological calibration of the *Vistula Lagoon* catchment was a challenging task as well, mainly due to the heterogeneity of spatial input data in this transboundary catchment, and inconsistent time series of observed climate and discharge data with many gaps. The two largest sub-catchments Pregolya ($86 \, m^3 \cdot s^{-1}$ on average) and Pasleka ($14 \, m^3 \cdot s^{-1}$ on average), draining 82% of the total drainage area, were calibrated first, and then the calibrated parameters were used for the whole catchment. The observed and simulated average daily and monthly discharges of the Pregolya are depicted in Figure 2 and show good model performance. Most of the intermediate gauges in the catchment also show satisfactory or good NSE values (Table 4) driven by WFDEI climate data [43], except of the two small rivers Dzierzgon and Waska. These two rivers are located in the lowland area close to the former mouth of the Vistula River, which drains directly into the Baltic Sea now due to hydraulic engineering and rechanneling. The deviation in balance shows good to very good model results in the majority of cases, except two rivers (Angrapa and Mamonovka) with larger discrepancies (Table 4). As the last step in hydrological calibration and spatial validation, average annual data on inflowing water to the lagoon found in literature were compared with the corresponding simulated values at the outlets of the twelve most important rivers entering the *Vistula Lagoon* [39,55]. In general the results show a good comparison.

4.2. Climate Change Signals

Climate change signals were calculated as differences in long-term average (30 years) air temperature, precipitation and solar radiation between three future periods p1, p2 and p3 and the reference period p0 averaged over the whole case study areas, separately and averaged over 15 climate scenarios (Table 5).

Table 5. Annual climate change signals of the three future periods p1, p2, and p3 related to the reference period p0 for the parameters air temperature, precipitation and solar radiation averaged over 15 selected ENSEMBLES climate scenarios per lagoon catchment.

Catchment	Temperature (°C)			Precipitation (%)			Radiation (%)		
	p1	p2	p3	p1	p2	p3	p1	p2	p3
Ria de Aveiro	+1.03	+2.1	+3.1	−5.6	−7.5	−15.6	+2.2	+3.3	+3.9
Mar Menor	+0.9	+2.0	+3.0	−1.6	−10.7	−18.3	+0.03	+0.2	−0.3
Tyligulskyi Liman	+1.2	+2.4	+3.5	−3.3	+0.8	−4.1	+0.6	−0.1	−0.4
Vistula Lagoon	+1.1	+2.2	+3.1	+4.3	+10.5	+9.7	−0.8	−3.2	−4.3

The average climate change signals for air temperature show a continuous increase from period p1 to period p3 and are similar for all four case study areas. They amount to 1.06 °C for the period p1, 2.18 °C for p2 and 3.18 °C for p3 on average. For the *Tyligulskyi Liman* catchment the simulated raise in temperature is higher than for the other three catchments.

Regarding expected changes in precipitation, the change signals are more diverse among the catchments. Until the end of the 21st century, an increasing trend in precipitation is projected for the **Vistula Lagoon** catchment, while decreasing trends are projected for the **Ria de Aveiro** and **Mar Menor** catchments. The strongest relative decrease is simulated for the **Mar Menor** catchment. For the *Tyligulskyi Liman* catchment, only very small changes in precipitation are projected on average.

The average change signals for solar radiation show a continuous increasing trend for the **Ria de Aveiro** catchment, and a decreasing trend for the **Vistula Lagoon**. Average percental changes in projected solar radiation for the other two catchments are only minor.

However, if single scenarios are analysed, a wide range of possible climate change signals can be seen in the four case study areas (Figure 3). The cases with small changes and without clear trend direction, e.g., for precipitation in the *Tyligulskyi Liman* catchment or for solar radiation in the *Mar Menor* and *Tyligulskyi Liman* catchments, are visible. It can be observed that the projected changes in solar radiation show the opposite change direction than precipitation in the **Ria de Aveiro** and **Vistula Lagoon** catchments. Less precipitation means less cloudiness and higher radiation, and vice versa. This connection is less distinct for the other two catchments, which can be probably explained by the influence of some outlying scenarios.

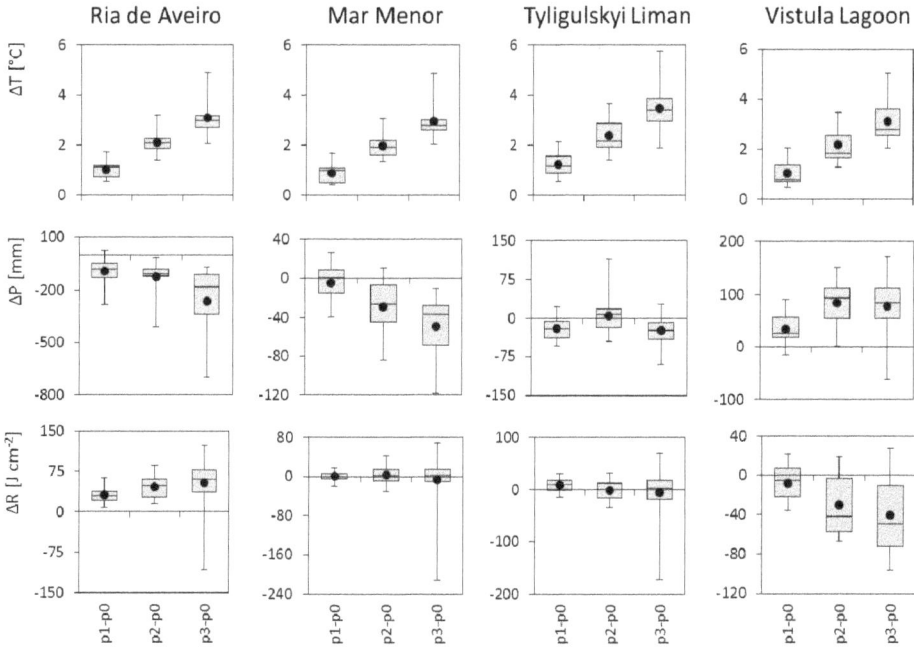

Figure 3. Ranges of climate change signals for 15 ENSEMBLES scenarios calculated as differences in the long-term average annual air temperature (T), precipitation (P) and solar radiation (R) between three future periods (p1, p2, p3) and the reference period (p0). The box plots visualize min/max (error bars), 25/75-percentiles (grey boxes), median (line) and average (dots) values.

In all cases the uncertainty between the 15 ENSEMBLES climate scenarios rises with time, from period p1 to p3. The wide ranges between the minimum and maximum changes are mostly due to outlying scenarios, but the 25/75-percentile boxes cover similar ranges in all three future periods in most cases.

4.3. Impacts of Climate Change on Total Water Discharge to the Lagoons

Figure 4 shows the climate change impacts on total freshwater inputs coming from the catchments to the four lagoons (sum of all inflowing rivers): the long-term average daily dynamics for period p3 (a), the mean monthly differences in total water inflow between periods p3 and p0 (b), and long-term average annual changes for all three future periods compared to the reference period (c). The black line means the average of 15 future scenario simulations, whereas the red line symbolizes the average of 15 simulations driven by climate model data for the reference period p0. The graphs include uncertainty bounds and ranges estimated based on all results driven by 15 climate scenarios and shown by grey shading or, respectively, error bars. The outer light grey uncertainty band is defined by the minimum and maximum outputs, and the dark grey inner band means the 25/75-percentiles of all results. The latter includes 50% of all simulation results around the average and means the range in which freshwater input can be expected in future with a high probability. The results of the future periods were compared with results of the reference period of the same scenario to calculate average monthly and long-term average changes.

Following the precipitation change signal, the simulated long-term average water discharge in the **Ria de Aveiro** show a moderate decrease in the 1st and 2nd future periods (−5% to −7%), which becomes higher by the end of the century (about −15% on average) (Figure 4, right). Although the decreasing trend is clear on average, there are some scenarios showing a slight increase in the first two scenario periods. The ranges between the 25/75-percentile values (meaning the level of uncertainty for the most probable scenarios) are increasing with time from period p1 to period p3. The projected river discharge to the lagoon in period p3 has a higher uncertainty in the winter months than in the summer season (Figure 4, left). A decrease in average water discharges can be detected during the whole year, while the absolute reduction in spring and autumn is stronger than in summer. Almost all scenarios show the negative trend.

The average climate impacts on freshwater inflow to the **Mar Menor** are similar to those of Aveiro, as the precipitation change signals have similar trends in these two regions. The results show a moderate decrease of the long-term average discharge to the lagoon by about 10% on average by the end of the century (Figure 4c). For the 1st future period the scenarios do not agree on a common trend and only a negligible decrease on average can be stated. The negative trend becomes more distinct but with a higher uncertainty towards the end of the century. Seasonal changes in water inflow to the **Mar Menor** show a decrease only in October-December, when generally the highest discharges can be observed (Figure 4b). The absolute uncertainty ranges are quite moderate, as usually water inflow to the lagoon is very low.

The total river discharge to the **Tyligulskyi Liman** is expected to decline in the scenario period p1, and increase in the two last periods p2 and p3 on average (Figure 4c). These changes do not follow the mean precipitation change signal of the last future period (Figure 3), and can probably be explained by the influence of a decreased solar radiation, which reduces evaporation and by that affects the total discharge. Besides, water inflow to the **Tyligulskyi Liman** is strongly influenced by water management (ponds), which was considered unchanged in the future in order to investigate "pure" impact of changing climate. The influence of the implemented ponds on the modelled river discharges could interfere with the impacts of a changing climate, and small climate induced changes could be masked. So, only in times when the available water volume exceeds the total effective pond volume in the catchment and evaporation effect on water cycle is low can changes in discharge be observed in future. Looking at the differences in seasonal dynamics of total water inflow to the Liman (Figure 4b), changes can be recognized in winter (an increase) and spring time (a decrease), when warmer winter temperatures influence snowfall and snowmelt processes, which lead to higher winter discharge and earlier and lower snowmelt peak.

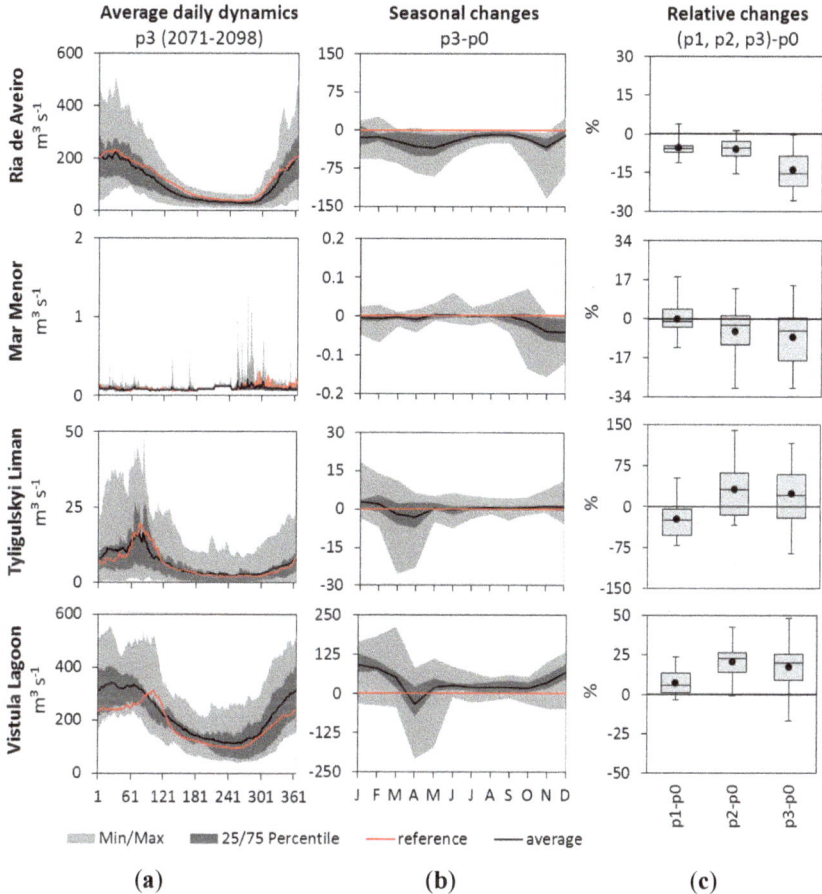

Figure 4. Climate change impacts on total freshwater inflow to the four lagoons with uncertainty ranges simulated by SWIM driven by 15 ENSEMBLES scenarios. (**a**): Long-term average daily dynamics in period p3 (black) with uncertainty ranges (grey) compared to that in the reference period (red); (**b**): Differences in long-term monthly dynamics between periods p3 and p0 (black) with uncertainty ranges (grey) and zero line (red); (**c**): Spread of 15 long-term average annual changes for three future periods compared to the reference period of the same scenario. Each box plot visualizes the min/max, 25/75-percentiles, median (line) and average (dot) of 15 average changes. The dynamics for *Tyligulskyi Liman* are shown without the outlying scenario S8.

103

The results of climate impact assessment on total inflow to the *Vistula Lagoon* show a notable increase of 7%, 21% and 18% on average in the three scenario periods due to higher precipitation (Figure 4c). The minimum and maximum ranges projected by SWIM driven by the 15 ENSEMBLES scenarios increase in time from period p1 to p3. The 25/75-percentile ranges (grey boxes) are quite narrow for all three periods, and are located above the zero line. Therefore, an overall increase in future water discharge as simulated by SWIM driven by 15 climate scenarios is quite certain. It can be seen in Figure 4b that positive absolute changes in river discharge are largest in winter and lower in summer. The only negative change in average discharge can be seen in April, probably due to increasing average temperatures causing a totally missing snow melt peak.

4.4. Impacts of Climate Change on Water Discharge of Single Rivers

Considering the spatial heterogeneity in the four lagoon catchments, it could be assumed that the different rivers and streams entering the lagoons may show different reactions to climate change, depending on the shares of specific land use classes and soil types within their catchments. Therefore, an analysis of climate change impacts was also done for discharges of the individual rivers flowing to the four lagoons. The same as before, percental changes were calculated by comparing the average discharge in the future period with the corresponding value simulated for the reference period of the same scenario. Some selected results comparing the intermediate period p2 with the reference period p0 are shown in Figure 5. This period was chosen, as the expectable future direction of changes should already be clearly visible, but the uncertainty ranges are quite moderate. The box plots show the average and median changes in discharge per river catchment together with their min/max-intervals and the 25/75-percentiles.

In Figure 5 it is evident that the region-specific average trends in total river discharge per CSAs presented in Figure 4 can be detected for the single rivers as well. For the *Ria de Aveiro* catchment, the majority of 15 climate scenarios project decreasing river discharges in period p2 for all rivers. The decreasing trend is less obvious for the *Mar Menor* rivers. Mostly an increasing trend with a high uncertainty originating in 15 climate scenarios can be seen for the *Tyligulskyi Liman* catchment. A distinct increasing trend projected by approximately 90% of the applied climate scenarios is obvious for all rivers flowing to the *Vistula Lagoon*.

However, some outliers still exist, for example the catchments with the smallest relative changes, e.g., the Albujon Wadi in the *Mar Menor* catchment, or the Nogat and Szkarpawa Rivers in the *Vistula Lagoon* catchment, resulting mainly from the water management measures. The effluents from an UWTP (Albujon Wadi) and water transfer from adjacent catchments via the small rivers (Vistula River) implemented in the model mask the climate change impacts in these cases, and result in only small relative changes with narrower uncertainty ranges. The wider uncertainty ranges can be detected in catchments with low or even no flow in drier summer months (e.g., in the smaller *Mar Menor* or *Tyligulskyi Liman* tributaries). Here, the relatively high percental changes mean only very low absolute differences in discharge values.

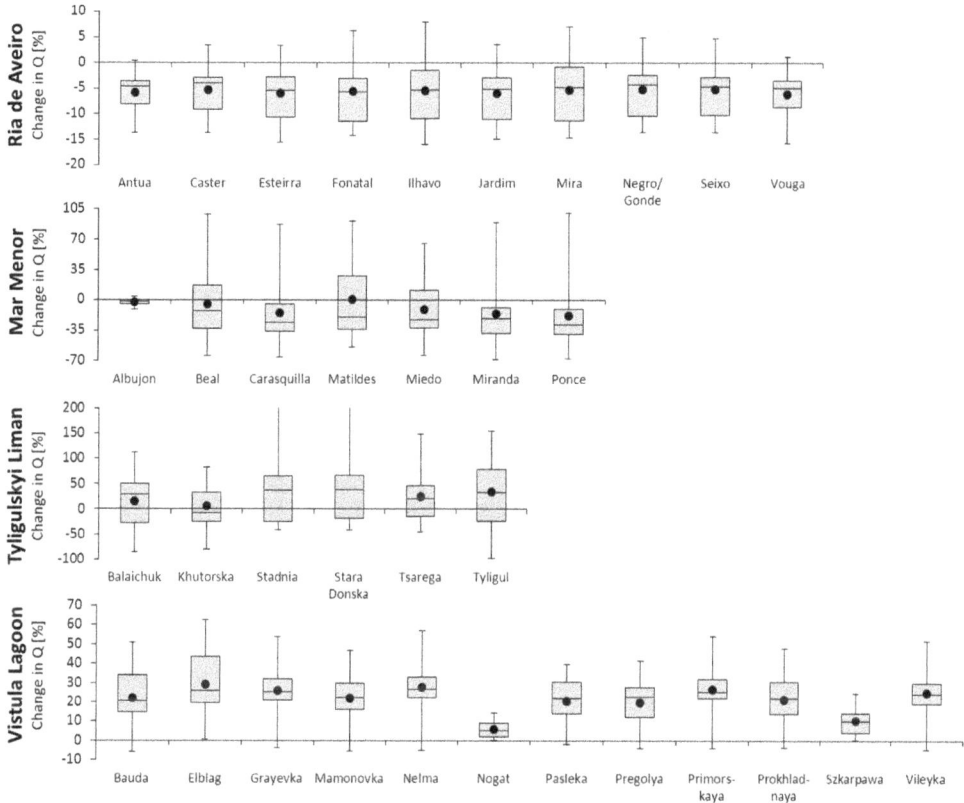

Figure 5. Percental changes in long-term average annual discharge of the main tributaries entering the four lagoons under study comparing the intermediate future period p2 (2041–2070) and the reference period p0 (1971–2000). The box plots visualize min/max, 25/75-percentiles, median (line) and average (dots) values based on simulated results driven by 15 ENSEMBLES scenarios.

The river-specific analysis of the scenario results shows that the differences in climate impacts on simulated water discharge between the rivers of one catchment can be mostly attributed to the degree of water management within the sub-catchments and less to climate variability or variability in land use and soil distribution. Water management makes the share of natural flow in the rivers considerably smaller and can mask the "pure" climate change impact. But temporally undifferentiated management measures (e.g., water transfer from other catchments, irrigation or waste water outflow from UWTP), which have to be implemented in the model using imprecise information, contribute to the general uncertainty of impact assessment. Additionally, the spatial variability within one (outlying) climate projection or between different climate scenarios can highly influence average results (e.g., for the *Tyligulskyi Liman* CSA). However, averaging the model outputs driven by a set of scenarios diminishes the influence of a single scenario, and provides more reliable results.

4.5. Impacts of Climate Change on Spatial Patterns of Runoff

Climate change impacts on water flows were additionally analysed in the four lagoon catchments on the hydrotope level to get an impression on the spatial distribution of changes within the catchments. Figure 6 shows the average differences in surface and subsurface runoff between the three future periods and the reference period. For the maps, the annual runoff values per hydrotope were averaged over 30-year-periods for every scenario and over 15 scenarios and then used for calculating and mapping the differences in runoff between periods.

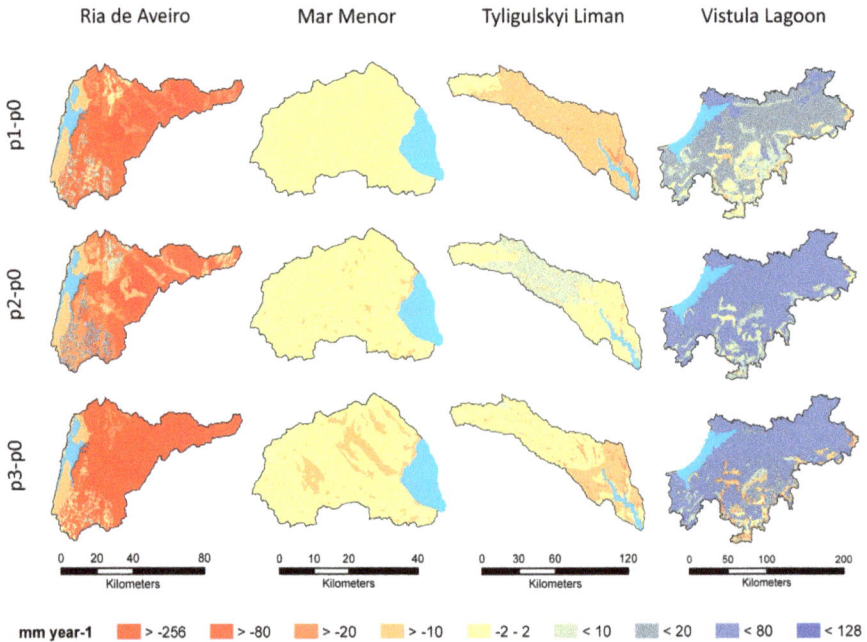

Figure 6. Spatial patterns of average annual changes in runoff (surface and subsurface flow) in the lagoon catchments under study simulated under the set of 15 ENSEMBLES climate change scenarios (average of 15 mean runoff maps for future periods p1, p2, p3 are compared to those of the reference period p0).

106

The largest average annual decrease in runoff in the **Ria de Aveiro** catchment is evident for period p3, especially in the eastern part of the catchment with higher elevation. In some areas the average reduction in runoff reaches -250 mm$\cdot$year^{-1}. However, during the 2nd future period, some areas (agricultural land) show a little increase (6 mm$\cdot$year^{-1}) due to lower evapotranspiration by affected crops leading to a slightly higher runoff. Besides the land use types, the runoff generation also depends on soil characteristics such as porosity and field capacity. Therefore, different soil types and land use classes can be clearly identified in the runoff maps.

As precipitation and water availability are generally low in the **Mar Menor** catchment, the average absolute changes in annual runoff are almost not visible in the near future period p1 ($< \pm 1$ mm$\cdot$year^{-1}), and show differences not larger than -10 mm$\cdot$year^{-1} in the periods p2 and p3. There are almost no changes visible in the driest western part of the catchment. The highest decrease in runoff can be observed in settlements, which had the highest runoff rates during the reference period, and thus a decrease in precipitation could result in runoff reduction.

Due to the fact that no continuous precipitation trend could be recognized in future in the **Tyligulskyi Liman** catchment, the spatial patterns of changes in runoff in the three scenario periods are quite diverse. The changes in surface and subsurface runoff more or less reflect the precipitation change signals of the future periods (Figure 3), and show decreasing runoff in the 1st and 3rd periods but practically unchanged or slightly increasing runoff in the 2nd period. They are also influenced by soil type and land use class distributions. The highest decreasing trend in runoff can be observed on areas covered by permanent vegetation (grassland) due to higher evapotranspiration values in these hydrotopes under warmer climate. The simulated decrease is highest in the currently drier south-eastern part of the catchment. It can be concluded that the increase in average discharge (sum of surface, subsurface and groundwater flow) to the **Tyligulskyi Liman** simulated for the 3rd future period (Figure 4) is probably caused by higher groundwater flow (which is not included in the maps) due to less evapotranspiration with decreased solar radiation.

Looking at the spatial distribution of changes in surface and subsurface runoff in the **Vistula Lagoon** catchment, the highest average increase is evident in period p2, and the heterogeneity of changes is highest in period p3. The simulated changes in runoff follow the changes in precipitation patterns, but are also influenced by soil characteristics and land use. The runoff increase is lower, or it is even decreasing, on highly permeable soil types, where additional precipitation may contribute to groundwater recharge instead. In general, the simulated changes in runoff are more diverse in the southern Polish part of the catchment, resulting from a soil map with better resolution used for this part of the catchment.

4.6. Climate Sensitivity of Freshwater Inflow to the Lagoons

In addition to the analysis of changes in the long-term average outputs under climate scenarios, the sensitivity of discharge to main climate parameters was analyzed. For that, the relative changes of the annual total water inflow for each future year of the time period 2011–2098 compared to the mean of the reference period 1971–2000 of the same scenario (ΔQ) were plotted against the corresponding relative changes in precipitation (ΔP) or absolute changes in temperature (ΔT) and radiation (ΔR). The latter were estimated the same: as the annual average climate parameters in the time period 2011–2098 related to the means in the reference period 1971–2000 of the same scenario (Figure 7).

Figure 7 illustrates the sensitivity of the total lagoon freshwater inflow (Q) to changes in climate parameters in the four lagoon catchments. The fitted regression curves are plotted as black lines together with their coefficient of determination (R^2) to illustrate the correlation between the simulated discharge and climate parameters.

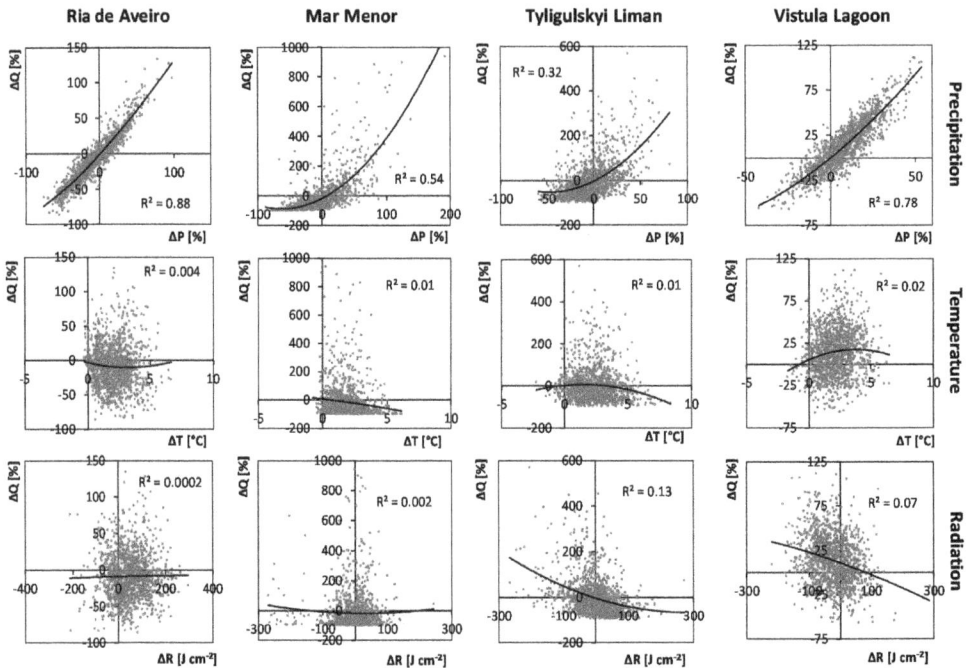

Figure 7. Climate sensitivity of the total freshwater inflow to the four lagoons for 15 ENSEMBLES climate scenarios: change of modelled annual discharge (ΔQ) per change of annual sum of precipitation (ΔP), average annual temperature (ΔT), and average annual solar radiation (ΔR) for the single years 2011–2098 compared to the mean of the reference period 1971–2000 for the same scenario.

It is evident from Figure 7 that the freshwater inflow is most sensitive to changes in precipitation than to changes in temperature or solar radiation. The highest correlation between Q and P can be seen for the **Ria de Aveiro** catchment, followed by the **Vistula Lagoon**, **Mar Menor** and **Tyligulskyi Liman** catchments. The degree of correlation is higher in catchments without extensive water management measures (**Ria de Aveiro**, **Vistula Lagoon**) and decreases with increasing human impacts on river discharge (**Mar Menor**, **Tyligulskyi Liman**).

According to the assumed regression lines, positive or negative changes in precipitation by 25% result in different relative changes in total freshwater inflows to the coastal water bodies. The discharge changes per catchment are listed in Table 6. In general, precipitation changes do not cause linear changes in river discharges. Rather, they can be potentiated, especially for catchments with very low natural river discharges and partially dry river beds (**Mar Menor**, **Tyligulskyi Liman**). In these catchments, a high number of years can be even detected, where the reduction of precipitation up to 50% causes a full absence of river discharges (-100%). It can be concluded from these experiments that especially water limited ecosystems are highly vulnerable to changing climate conditions.

Table 6. Response of annual modelled discharge to changes in annual sum of precipitation by $\pm$ 25%.

Change in Precipitation		+25%	−25%
	Ria de Aveiro	+30%	−25%
Resulting change in	Mar Menor	+50%	−62%
discharge	Tyligulskyi Liman	+61%	−41%
	Vistula Lagoon	+42%	−34%

The coefficients of determination are very low for the relations between discharge and temperature or solar radiation in all catchments (Figure 7). Although higher temperatures and/or radiation should cause an increase in potential evapotranspiration in the catchments and by that influence water inflows, such behaviour is not indicated in the graphs. The largest R^2 can be seen for the $\Delta Q/\Delta R$-ratio in the **Tyligulskyi Liman** catchment, where decreased solar radiation can affect the resulting water discharge in some years and even mask the average precipitation trend, as already mentioned above in Section 4.3.

5. Discussion

The results of climate change impact assessments for the four European lagoon catchments are in line with the projections described in the literature regarding expected climate change signals and their impacts in Europe. An overall trend in increasing precipitation (especially in winter) for Northern Europe, as well as decreasing precipitation (especially in summer) in the South is mentioned in several publications [4,6,7], and it was also detected in this study. Due to the fact that water limited ecosystems are highly vulnerable to changing climate conditions [56], the robust climate change signals produced by a range of global and regional models suggest that the Mediterranean region and the lagoons located there might be especially vulnerable to climate change [57,58].

The resulting changes in streamflow consequently show an increasing trend in the Northern European Baltic region [59,60], and a decreasing trend around the Mediterranean Sea and on the Iberian Peninsula [61,62]. Comparing with our results on impacts on the water availability in the four lagoon catchments under investigation, the same pattern could be found. Though expected changes in climate can be seen as more beneficial for the *Vistula Lagoon* catchment, one should not forget about water-related extreme events like floods and drought, which were not explored in this study.

Giorgi and Lionello [57] investigated climate change projections especially for the Mediterranean region. They detected an overall precipitation decrease in this area, but with an inter-seasonal spatial variability due to a shifting transition area between positive (in Northern Europe) and negative (in Southern Europe) precipitation change signals, which moves southward in winter. This could explain the high diversity in precipitation change signals between the 15 ENSEMBLES scenarios for the *Tyligulskyi Liman* CSA located exactly in this transition zone.

In general, climate change impact assessment performed for the *Tyligulskyi Liman* catchment reveals the limitations and constraints of the multi-model method for climate impact assessments. If climate input is highly uncertain (as in our case), the eco-hydrological catchment model driven by a set of climate scenarios, showing different directions of change in precipitation and radiation, generates quite heterogeneous model results with high uncertainty ranges, and a clear conclusion for future development is difficult to produce. In addition, as the modelled discharge is highly influenced by anthropogenic water management measures in this region, the ecosystem responses due to "pure" climate change are hardly detectable. Nevertheless, the scenario assessments undertaken in this study helped to identify the direction of potential changes in water quantity in the four lagoon catchments and delivered a first impression about a possible future also for the *Tyligulskyi Liman* catchment.

However, it cannot be assumed that the results of climate change impact assessment for the four lagoon catchments provide universally valid and strict results, and that impacts projected for a certain scenario (or average projections) will be realised in future. It is common knowledge that even multi-model climate change impact assessments come with some uncertainty [6,63], which should be kept in mind when the results are analysed and interpreted. The uncertainties are mainly related to (A) the ability of climate and eco-hydrological models to represent the interrelated processes in the atmosphere and in a landscape (so-called structural and parametrization uncertainties); (B) the reliability of climate scenarios applied for the impact assessment; and (C) data availability and quality for (eco-)hydrological model setup and calibration. These uncertainties are shortly discussed below.

(A) Models are always simplifications of reality and are characterised by some level of abstraction. Hydrological processes taking place in atmosphere, soils, water bodies and vegetation, as well as interrelations between them, are represented in models with a certain degree of accuracy. This is due to a restricted memory of computers and computation time, as well as due to a limited human knowledge and understanding of processes. Comparing simulated and observed climate data, several studies show the restricted ability of current climate models to satisfactorily reproduce the real local measurements [64,65]. Similar constraints can be found in the hydrological and eco-hydrological models as well. The SWIM model, for example, as a semi-distributed model simulating processes at a hydrotope-level resolution, tries to cover the heterogeneity within a catchment to a certain extent, but is not able to deliver locally exact projections.

(B) A further major uncertainty is connected to climate scenarios applied for impact assessments. Different models come along with different scenarios, and nobody knows the most probable future climate development in a region, as it is influenced by several unpredictable factors. A common method to overcome this problem is to use different scenarios from GCMs and/or RCMs in order to verify most probable projections and investigate ranges of uncertainty. Such method is preferable in comparison with a single climate scenario approach as recommend by many authors [24–27]. But this procedure still has limitations, and the uncertainty remains high, especially in case of a distinct diversity in climate projections, as detected for the *Tyligulskyi Liman* catchment in our study.

(C) A hydrological or eco-hydrological model used for an impact assessment should be calibrated and validated in advance. For that, appropriate homogeneous and complete spatial datasets (DEM, land use and soil maps) and time series (daily climate parameters and observed discharge) are necessary for a successful model setup. However, in our case in all four study areas some data were missing, or data coverage in time and/or space was problematic. Therefore, in all four CSAs, the model calibration was a very complicated task (as described in Sections 3.2 and 4.1), and the model outputs incorporate a certain degree of uncertainty.

Climate scenarios are often customized by applying at least one bias correction method in climate change impact assessment. The bias correction is aimed in avoiding unrealistic simulations of runoff by adjusting the climate scenario data in such a way that they better represent the observed climate. It is still argued whether the application of bias corrected climate scenarios provides better and more reliable results for hydrological impact analyses, especially when relative changes and a detection of trends in runoff are more of interest than the absolute values. While Teutschbein and Seibert [23] recommend the application of bias correction, Ehret *et al.* [66] state that the bias correction as it is currently used is "not a valid procedure" due to a general lack of the physical justification of corrections, and often missing, incomplete or deficient observed data. In our study it was decided we would not use bias correction, and we would compare simulations driven by 15 RCMs between periods.

6. Summary and Conclusions

The climate change impact assessments for the *Ria de Aveiro, Mar Menor, Tyligulskyi Liman* and *Vistula Lagoon* drainage basins were performed by applying a common technique for hydrological impact studies at the catchment scale using a set of 15 climate scenarios provided by the ENSEMBLES project to drive the eco-hydrological model SWIM. Despite some difficulties and uncertainties in model setup and calibration, satisfactory to good model results were achieved, delivering a sound basis for impact assessment. The scenarios covered three future 30-year periods until the end of the 21st century.

The analysis of the 15 scenarios indicates a continuously increasing trend in temperature for all four catchments under investigation. However, precipitation change signals are more diverse between the catchments as well as among the 15 scenarios within one catchment. The highest diversity could be seen for the *Tyligulskyi Liman* catchment, with only a small change in precipitation on average. A quite consistent increasing trend in precipitation was projected for the *Vistula Lagoon* catchment, while decreasing trends were projected for the *Ria de Aveiro* and *Mar Menor* catchments. The simulated river discharge more or less follows this precipitation pattern and is less sensitive to changes in air temperature or solar radiation. The sensitivity of river discharge to precipitation is more obvious in catchments less influenced by human water management.

The river discharge and flood risk will probably increase in the *Vistula Lagoon* catchment. In the South-European lagoon catchments, the water managers and stakeholders have to be ready for decreased water availability in the future, and adaptation measures should focus mainly on water-saving technologies. This is particularly important for the very dry *Mar Menor* catchment with its intensive irrigated agriculture and additional huge water demand in the tourist season. Water availability and freshwater inflow to the lagoon seem to be problematic in the *Tyligulskyi Liman* area as well. Water flows there are additionally hampered by numerous ponds in the river reaches. Pond management should definitely be adapted to climate change to allow some freshwater inflow to the Liman and prevent its salinization. More consistent and reliable climate scenarios would be desirable for this region in order to reduce uncertainty of projections.

Due to the quite high diversity of climate change impacts detected in the four lagoon catchments under study, it is a challenging task to list valid implications for other coastal areas or lagoon systems in Europe or even worldwide. A variety of lagoons, even in a similar geographical region like the Mediterranean, have a high hydrodynamic and/or saline variability, depending on their morphology and the location and number of river inflows and sea-water inlets influencing the water renewal time [67,68]. Therefore lagoons react differently to climate change with a trend to homogenize hydrodynamic and saline characteristics and to lose hydrodiversity, mainly due to increased inflow of sea water following the sea level rise [17,58]. As northern European lagoons are generally more influenced by river discharge due to the wet climate and the significantly higher runoff coefficients compared to the southern Mediterranean coastal areas [67], climate changes in the lagoon catchments and the resulting changes in river discharge will probably have greater effects on such lagoons fed by permanently flowing rivers than on lagoons located in dry climate.

The results provide some useful insights into possible future water availability and development of freshwater input to the lagoons under study. The model outputs were delivered to the lagoon modellers for evaluating resulting climate change impacts on the lagoon ecosystems. According to the obtained results, impacts are expected to be less pronounced in the near future but would increase in the middle of the century. Consequently, the implementation of adequate adaptation measures in the medium-term is recommended.

However, climate change is not the only factor to have remarkable effects on water resources in the study areas; human societies are important co-designers of the future conditions in these vulnerable coastal areas, too. It cannot be expected that future development will take place without any changes in human behaviour, land use pattern or economic conditions. Therefore, a subsequent combined assessment taking into account possible future climate and socio-economic changes is strongly recommended. This would help to better identify probable future risks and threats, and to virtually test possible adaptation measures as efforts to cope with probable future climate conditions and their impacts.

Acknowledgments: The research leading to these results has received funding from the European Community's Seventh Framework Programme FP7/2007–2013 through the collaborative research project LAGOONS (contract No. 283157). The ENSEMBLES data used in the study were produced in the EU FP6 Integrated Project ENSEMBLES (contract No. 505539). Hydrological data were provided by several local and national authorities for water management and environmental monitoring in Portugal, Spain, Ukraine, Poland and Russia and collected by the LAGOONS partners, whose support is gratefully acknowledged. The authors thank Vanessa Wörner and Lasse Scheele for their support in climate scenario data download, preparation and evaluation.

Author Contributions: All authors planned and designed the methods to study climate change impacts on water resources in the four case study areas. Cornelia Hesse set up, adjusted and calibrated the SWIM model for the *Tyligulskyi Liman* and *Vistula Lagoon* catchments and ran the scenarios for these two CSAs. Anastassi Stefanova set up, adjusted and calibrated the model for the *Ria de Aveiro* and *Mar Menor* catchments and ran the scenarios for these two CSAs. Cornelia Hesse and Anastassi Stefanova analyzed the model output data. Cornelia Hesse analyzed the climate sensitivity and impacts on single rivers, prepared all tables and figures for the publication, wrote the text and formatted the paper. Valentina Krysanova guided and supervised the whole process, discussed results during the modelling study, and edited the manuscript.

Conflicts of Interest: The authors declare no conflict of interest.

References

1. IPCC Summary for Policymakers. In *Climate Change 2007: Impacts, Adaptation and Vulnerability. Contribution of Working Group II to the Fourth Assessment Report of the Inter-governmental Panel on Climate Change*; Parry, M.L., Canziani, O.F., Palutikof, J.P., van der Linden, P.J., Hanson, C.E., Eds.; Cambridge University Press: Cambridge, UK, 2007; pp. 7–22.

2. IPCC Summary for Policymakers. In *Climate Change 2013: The Physical Science Basis. Contribution of Working Group I to the Fifth Assessment Report of the Intergovernmental Panel on Climate Change*; Stocker, T.F., Qin, D., Plattner, G.-K., Tignor, M., Allen, S.K., Boschung, J., Nauels, A., Xia, Y., Bex, V., Midgley, P.M., Eds.; Cambridge University Press: Cambridge, UK, 2013; pp. 3–29.

3. Kundzewicz, Z.W. Climate change impacts on the hydrological cycle. *Ecohydrol. Hydrobiol.* **2008**, *8*, 195–203.

4. Cassardo, C.; Jones, J.A.A. Managing water in a changing world. *Water* **2011**, *3*, 618–628.

5. Kløve, B.; Ala-Aho, P.; Bertrand, G.; Gurdak, J.J.; Kupfersberger, H.; Kværner, J.; Muotka, T.; Mykrä, H.; Preda, E.; Rossi, P.; *et al.* Climate change impacts on groundwater and dependent ecosystems. *J. Hydrol.* **2014**, *518*, 250–266.

6. European Environment Agency (EEA). *Climate Change, Impacts and Vulnerability in Europe 2012*; EEA-Report No 12/2012; European Environment Agency: Copenhagen, Denmark, 2012; p. 300.

7. Alcamo, J.; Moreno, J.M.; Nováky, B.; Bindi, M.; Corobov, R.; Devoy, R.J.N.; Giannakopoulos, C.; Martin, E.; Olesen, J.E.; Shvidenko, A. Europe. In *Climate Change 2007: Impacts, Adaptation and Vulnerability; Contribution of Working Group II to the Fourth Assessment Report of the Intergovernmental Panel on Climate Change*; Parry, M.L., Canziani, O.F., Palutikof, J.P., van der Linden, P.J., Hanson, C.E., Eds.; Cambridge University Press: Cambridge, UK, 2007; pp. 541–580.

8. Haylock, M.R.; Hofstra, N.; Klein Tank, A.M.G.; Klok, E.J.; Jones, P.D.; New, M. A European daily high-resolution gridded data set of surface temperature and precipitation for 1950–2006. *J. Geophys. Res.* **2008**, *113*.

9. Linderholm, H.W. Growing season changes in the last century. *Agric. For. Meteorol.* **2006**, *137*, 1–14.

10. Gabriel, K.M.A.; Endlicher, W.R. Urban and rural mortality rates during heat waves in Berlin and Brandenburg, Germany. *Environ. Pollut.* **2011**, *159*, 2044–2050.

11. Vittoz, P.; Cherix, D.; Gonseth, Y.; Lubini, V.; Maggini, R.; Zbinden, N.; Zumbach, S. Climate change impacts on biodiversity in Switzerland: A review. *J. Nat. Conserv.* **2013**, *21*, 154–162.

12. Wang, X.; Siegert, F.; Zhou, A.; Franke, J. Glacier and glacial lake changes and their relationship in the context of climate change, Central Tibetan Plateau 1972–2010. *Glob. Planet. Chang.* **2013**, *111*, 246–257.

13. Nicholls, R.J.; Wong, P.P.; Burkett, V.R.; Codignotto, J.O.; Hay, J.E.; McLean, R.F.; Ragoonaden, S.; Woodroffe, C.D. Coastal systems and low-lying areas. In *Climate Change 2007: Impacts, Adaptation and Vulnerability. Contribution of Working Group II to the Fourth Assessment Report of the Inter-Governmental Panel on Climate Change*; Parry, M.L., Canziani, O.F., Palutikof, J.P., van der Linden, P.J., Hanson, C.E., Eds.; Cambridge University Press: Cambridge, UK, 2007; pp. 315–356.

14. National Research Council (NRC). *Adapting to the Impacts of Climate Change*; The National Academies Press: Washington, DC, USA, 2010; p. 292.

15. Anthony, A.; Atwood, J.; August, P.; Byron, C.; Cobb, S.; Foster, C.; Fry, C.; Gold, A.; Hagos, K.; Heffner, L.; *et al.* Coastal lagoons and climate change: Ecological and social ramifications in U.S. Atlantic and Gulf coast ecosystems. *Ecol. Soc.* 2009, 14, p. 8. Available online: http://www.ecologyandsociety.org/vol14/iss1/art8/ (accessed on 3 December 2013).

16. Chapman, P.M. Management of coastal lagoons under climate change. *Estuar. Coast. Shelf Sci.* **2012**, *110*, 32–35.

17. De Pascalis, F.; Pérez-Ruzafa, A.; Gilabert, J.; Marcos, C.; Umgiesser, G. Climate change response of the *Mar Menor* coastal lagoon (Spain) using a hydrodynamic finite element model. *Estuar. Coast. Shelf Sci.* **2012**, *114*, 118–129.

18. Scavia, D.; Field, J.C.; Boesch, D.F.; Buddemeier, R.W.; Burkett, V.; Cayan, D.R.; Fogarty, M.; Harwell, M.A.; Howarth, R.W.; Mason, C.; *et al.* Climate change impacts on U.S. coastal and marine ecosystems. *Estuaries* **2002**, *25*, 149–164.

19. Jakimavicius, D.; Kriauciuniene, J. The Climate change impact on the water balance of the curonian lagoon. *Water Resour.* **2013**, *40*, 120–132.

20. Hirabayashi, Y.; Kanae, S.; Emori, S.; Oki, T.; Kimoto, M. Global projections of changing risk of floods and droughts in a changing climate. *Hydrolog. Sci. J.* **2008**, *53*, 754–772.

21. Arnell, N.W.; Gosling, S.N. The impacts of climate change on river flow regimes at the global scale. *J. Hydrol.* **2013**, *486*, 351–364.

22. Van Vliet, M.T.H.; Franssen, W.H.P.; Yearsley, J.R.; Ludwig, F.; Haddeland, I.; Lettenmaier, D.P.; Kabat, P. Global river discharge and water temperature under climate change. *Glob. Environ. Chang.* **2013**, *23*, 450–464.

23. Teutschbein, C.; Seibert, J. Regional climate models for hydrological impact studies at the catchment scale: A review of recent modeling strategies. *Geogr. Compass* **2010**, *4*, 834 860.

24. Giorgi, F.; Hewitson, B.; Christensen, J.H.; Hulme, M.; von Storch, H.; Whetton, P.; Jones, R.; Mearns, L.; Fu, C. Regional climate information—Evaluation and projections. In *Climate Change 2001: The Scientific Basis; Contribution of Working Group I to the Third Assessment Report of the Intergovernmental Panel on Climate Change*; Houghton, J.T., Ding, Y., Griggs, D.J., Noguer, M., van der Linden, P.J., Dai, X., Maskell, K., Johnson, C.A., Eds.; Cambridge University Press: Cambridge, UK, 2001; pp. 583–638.

25. Fowler, H.; Blenkinsop, S.; Tebaldi, C. Linking climate change to impact studies: Recent advances in downscaling techniques for hydrological modelling. *Int. J. Climatol.* **2007**, *27*, 1547–1578.

26. Graham, L.; Andreasson, J.; Carlsson, B. Assessing climate change impacts on hydrology from an ensemble of regional climate models, model scales and linking methods—A case study on the Lule river basin. *Clim. Chang.* **2007**, *81*, 293–307.

27. Tebaldi, C.; Knutti, R. The use of the multi-model ensemble in probabilistic climate projections. *Philos. Trans. R. Soc.* **2007**, *365*, 2053–2075.

28. Krysanova, V.; Wechsung, F.; Arnold, J.; Srinivasan, R.; Williams, J. *SWIM (Soil and Water Integrated Model): User Manual*; PIK Report No. 69; Potsdam Institute for Climate Impact Research (PIK): Potsdam, Germany, 2000; p. 239.

29. Van der Linden, P., Mitchell, J.F., Eds.; *ENSEMBLES: Climate Change and Its Impacts: Summary of Research and Results from the ENSEMBLES Project*; Met Office Hadley Centre: Exeter, UK, 2009; p. 160.

30. Arnold, J.; Allan, P.; Bernhardt, G. A comprehensive surface-groundwater flow model. *J. Hydrol.* **1993**, *142*, 47–69.

31. Krysanova, V.; Meiner, A.; Roosaare, J.; Vasilyev, A. Simulation modelling of the coastal waters pollution from agricultural watershed. *Ecol. Model.* **1989**, *49*, 7–29.

32. Gelfan, A.; Poneroy, J.; Kuchment, L. Modelling forest cover influences on snow accumulation, sublimation, and melt. *J. Hydrometeorol.* **2004**, *5*, 785–803.

33. Huang, S. Modelling of Environmental Change Impacts on Water Resources and Hydrological Extremes in Germany. Ph.D. Thesis, University Potsdam, Potsdam, Germany, November 2011. p. 206.

34. Priestley, C.; Taylor, R. On the assessment of surface heat flux and evaporation using large scale parameters. *Mon. Weather Rev.* **1972**, *100*, 81–92.

35. Turc, L. Évaluation des besoins en eau d'irrigation, évapotranspiration potentielle, formule simplifiée et mise à jour. *Ann. Agron.* **1961**, *12*, 13–49. (In French)

36. Wendling, U.; Schellin, H. Neue Ergebnisse zur Berechnung der potentiellen Evapotranspiration. *Z. Meteorol.* **1986**, *36*, 214–217. (In German)

37. Krysanova, V.; Hattermann, F.F.; Huang, S.; Hesse, C.; Vetter, T.; Liersch, S.; Koch, H.; Kundzewicz, Z. Modelling climate and land-use change impacts with SWIM: Lessons learnt from multiple applications. *Hydrol. Sci. J.* **2014**.

38. Stefanova, A.; Krysanova, V.; Hesse, C.; Lillebø, A. Climate change impact assessment on water inflow to a coastal lagoon: *Ria de Aveiro* watershed, Portugal. *Hydrol. Sci. J.* **2014**.

39. Hesse, C.; Krysanova, V.; Stefanova, A.; Bielecka, M.; Domnin, D. Assessment of climate change impacts on water quantity and quality of the multi-river *Vistula Lagoon* catchment. *Hydrol. Sci. J.* **2014**.

40. Hesse, C.; Krysanova, V.; Päzolt, J.; Hattermann, F.F. Eco-hydrological modelling in a highly regulated lowland catchment to find measures for improving water quality. *Ecol. Model.* **2008**, *218*, 135–148.

41. Nash, J.E.; Sutcliffe, J.V. River flow forecasting through conceptual models part I: A discussion of principles. *J. Hydrol.* **1970**, *10*, 282–290.

42. Moriasi, D.N.; Arnold, J.G.; van Liew, M.W.; Bingner, R.L.; Harmel, R.D.; Veith, T.L. Model Evaluation guidelines for systematic quantification of accuracy in watershed simulations. *Trans. ASABE* **2007**, *50*, 885–900.

43. Weedon, G.P.; Balsamo, G.; Bellouin, N.; Gomes, S.; Best, M.J.; Viterbo, P. The WFDEI meteorological forcing data set: Watch forcing data methodology applied to ERA-interim reanalysis data. *Water Resour. Res.* **2014**, *50*.

44. *Bodenkundliche Kartieranleitung*, 5th ed.; Boden AG: Hannover, Germany, 2005; p. 392.

45. García-Pintado, J.; Martínez-Mena, M.; Barberá, G.G.; Albaladejo, J.; Castillo, V.M. Anthropo-genic nutrient sources and loads from a Mediterranean catchment into a coastal lagoon: *Mar Menor*, Spain. *Sci. Total Environ.* **2007**, *373*, 220–239.

46. Velasco, J.; Lloret, J.; Millán, A.; Marín, A.; Barahona, J.; Abellán, P.; Sánchez-Fernández, D. Nutrient and particulate inputs into the *Mar Menor* lagoon (SE Spain) from an intensive agricultural watershed. *Water Air Soil Pollut.* **2006**, *176*, 37–56.

47. *The Ria de Aveiro Lagoon—Current Knowledge Base and Knowledge Gaps*; LAGOONS Report D2.1b; UA: Aveiro, Portugal, 2012; p. 52.

48. Jimenez-Martinez, J.; Skaggs, T.H.; van Genuchten, M.T.; Candela, L. A root zone modelling approach to estimating groundwater recharge from irrigated areas. *J. Hydrol.* **2009**, *367*, 138–149.

49. *The Tyligulskyi Lagoon—Current Knowledge Base and Knowledge Gaps*; LAGOONS Report D2.1d; OSENU: Odessa, Ukraine, 2012; p. 54.

50. Eriksson, H.; Pastuszak, M.; Löfgren, S.; Mörth, C.-M.; Humborg, C. Nitrogen budgets of the Polish agriculture 1960–2000: Implications for riverine nitrogen loads to the Baltic Sea from transitional countries. *Biogeochemistry* **2007**, *85*, 153–168.

51. *The Mar Menor Lagoon—Current Knowledge Base and Knowledge Gaps*; LAGOONS Report D2.1c; UM: Murcia, Spain, 2012; p. 65.

52. *The Vistula Lagoon—Current Knowledge Base and Knowledge Gaps*; LAGOONS Report D2.1a; IBW-PAN: Gdansk, Poland, 2012; p. 99.

53. Robakiewicz, M. Vistula River mouth—History and recent problems. *Arch. Hydro–Eng. Environ. Mech.* **2012**, *57*, 155–166.

54. *Guidelines on the Use of Scenario Data for Climate Impact and Adaptation Assessment*, 1st ed.; Carter, T.R., Hulme, M., Lal, M., Eds.; Intergovernmental Panel on Climate Change, Task Group on Scenarios for Climate Impact Assessment (IPCC-TGCIA): Helsinki, Finland, 1999.

55. *Results of Climate Impact Assessment—Application for Four Lagoon Catchments*; LAGOONS Report D5.1; PIK: Potsdam, Germany, 2013; p. 107.

56. Manfreda, S.; Caylor, K.K. On the vulnerability of water limited ecosystems to climate change. *Water* **2013**, *5*, 819–833.

57. Giorgi, F.; Lionello, P. Climate change projections for the Mediterranean region. *Glob. Planet. Chang.* **2008**, *63*, 90–104.

58. Ferrarin, C.; Bajo, M.; Bellafiore, D.; Cucco, A.; de Pascalis, F.; Ghezzo, M.; Umgiesser, G. Toward homogenization of Mediterranean lagoons and their loss of hydrodiversity. *Geophys. Res. Lett.* **2014**, *41*, 5935–5941.

59. Graham, L. Climate change effects on river flow to the Baltic Sea. *AMBIO* **2004**, *33*, 235–241.

60. Reihan, A.; Koltsova, T.; Kriauciuniene, J.; Lizuma, L.; Meilutyte-Barauskiene, D. Changes in water discharge of the Baltic States Rivers in the 20th century and its relation to climate change. *Nord. Hydrol.* **2007**, *38*, 401–412.

61. García-Ruiz, J.M.; López-Moreno, J.I.; Vicente-Serrano, S.M.; Lasanta-Martínez, T.; Beguería, S. Mediterranean water resources in a global change scenario. *Earth Sci. Rev.* **2011**, *105*, 121–139.

62. Arias, R.; Rodriguez-Blanco, M.L.; Taboada-Castro, M.M.; Nunes, J.P.; Keizer, J.J.; Taboada-Castro, M.T. Water resources response to changes in temperature, rainfall and CO_2 concentration: A first approach in NW Spain. *Water* **2014**, *6*, 3049–3067.

63. Koutsoyiannis, D.; Efstratiadis, A.; Mamassis, N.; Christofides, A. On the credibility of climate predictions. *Hydrol. Sci. J.* **2008**, *53*, 671–684.

64. Anagnostopoulos, G.G.; Koutsoyiannis, D.; Christofides, A.; Efstratiadis, A.; Mamassis, N. A comparison of local and aggregated climate model outputs with observed data. *Hydrol. Sci. J.* **2010**, *55*, 1094–1110.

65. Kundzewicz, Z.W.; Stakhiv, E.Z. Are climate models "ready for prime time" in water resources management applications, or is more research needed? *Hydrolog. Sci. J.* **2010**, *55*, 1085–1089.

66. Ehret, U.; Zehe, E.; Wulfmeyer, V.; Warrach-Sagi, K.; Liebert, J. HESS Opinions "Should we apply bias correction to global and regional climate model data? " *Hydrol. Earth Syst. Sci.* **2012**, *16*, 3391–3404.

67. Newton, A.; Icely, J.; Cristina, S.; Brito, A.; Cardoso, A.C.; Colijn, F.; Riva, S.D.; Gertz, F.; Hansen, J.W.; Holmer, M.; *et al.* An overview of ecological status, vulnerability and future perspectives of European large shallow, semi-enclosed coastal systems, lagoons and transitional waters. *Estuar. Coast. Shelf Sci.* **2014**, *140*, 95–122.

68. Umgiesser, G.; Ferrarin, C.; Cucco, A.; de Pascalis, F.; Bellafiore, D.; Ghezzo, M.; Bajo, M. Comparative hydrodynamics of 10 Mediterranean lagoons by means of numerical modelling. *J. Geophys. Res. Oceans* **2014**, *119*, 2212–2226.

A Structurally Simplified Hybrid Model of Genetic Algorithm and Support Vector Machine for Prediction of Chlorophyll *a* in Reservoirs

Jieqiong Su, Xuan Wang, Shouyan Zhao, Bin Chen, Chunhui Li and Zhifeng Yang

Abstract: With decreasing water availability as a result of climate change and human activities, analysis of the influential factors and variation trends of chlorophyll *a* has become important to prevent reservoir eutrophication and ensure water supply safety. In this paper, a structurally simplified hybrid model of the genetic algorithm (GA) and the support vector machine (SVM) was developed for the prediction of monthly concentration of chlorophyll *a* in the Miyun Reservoir of northern China over the period from 2000 to 2010. Based on the influence factor analysis, the four most relevant influence factors of chlorophyll *a* (*i.e.*, total phosphorus, total nitrogen, permanganate index, and reservoir storage) were extracted using the method of feature selection with the GA, which simplified the model structure, making it more practical and efficient for environmental management. The results showed that the developed simplified GA-SVM model could solve nonlinear problems of complex system, and was suitable for the simulation and prediction of chlorophyll *a* with better performance in accuracy and efficiency in the Miyun Reservoir.

Reprinted from *Water*. Cite as: Su, J.; Wang, X.; Zhao, S.; Chen, B.; Li, C.; Yang, Z. A Structurally Simplified Hybrid Model of Genetic Algorithm and Support Vector Machine for Prediction of Chlorophyll *a* in Reservoirs. *Water* **2015**, 7, 1610–1627.

1. Introduction

Water conflicts are key issues for sustainable water resources management. Under the dual effects of climate change and human activities, many water bodies are polluted to varying degrees, further exacerbating water conflicts [1,2]. Ecosystem studies such as water enhancement, water quality risk assessment, and early warnings have drawn much attention across the world [3]. As important engineering measures are developed to guarantee water supply, irrigation, electricity, and other functions, reservoirs can help solve these issues through the redistribution of runoff in both time and space; therefore, they are widely used throughout the world. Although water demands of each production department (e.g., industrial department, agricultural department, and so on) correspond to different water quality requirements, water quality always needs to be up to its appropriate standard in

119

different water usage [4]. Accordingly, it is important to forecast water quality accurately, which could provide a scientific decision basis for reservoir management.

Chlorophyll *a* is an important component of algae organisms, and its concentration in water bodies is closely related to the type and the quantity of algae [5]. Therefore, as an important symbol of phytoplankton stock, concentration of chlorophyll *a* can reflect water nutritional status, making chlorophyll *a* one of the indicators in controlling the eutrophication of lakes and reservoirs. The minimum threshold concentration of chlorophyll *a* for eutrophic lakes identified by the Organization for Economic Cooperation and Development (OECD) is 0.008 mg/L. Consequently, there is a need to control the concentration of chlorophyll *a* in water to prevent potential eutrophication. For this reason, accurate prediction of chlorophyll *a* is a worldwide concern.

The generation mechanism of chlorophyll *a* in water is accordingly complex, which is closely related to ecological, environmental, and societal activities. Therefore, the elements involved in the prediction for chlorophyll *a* in water are complex accordingly. In the existing literature, prediction models for chlorophyll *a* mainly included two categories: statistical regression models [6] and mechanism models [7]. Statistical regression models were established with the applications of statistical correlation analysis theory and methods. This means that the sample size had a major influence on prediction accuracy. Moreover, these models usually applied a linear relationship to simplify complex problems, leading to unsatisfactory prediction results under the situation when the limiting factors of chlorophyll *a* changed. Mechanism models mainly included the nutrient model, phytoplankton model, and ecological dynamic models such as CE-QUAL-ICM, WASP, CAEDYM, AQUATOX, and ECOPATH [8–10]. Based on the principle of hydrodynamics and ecosystem dynamics, these models comprehensively considered the interaction mechanisms among indicators of water resources system and ecosystem, and then predicted the future development of the system accurately. However, these models also had a high demand for data quantity, which was inconvenient for model calibration and verification, leading to a decline in reliability and applicability. Furthermore, due to uncertainty factors such as the concentration of phosphorus input from non-point source pollution, the prediction of chlorophyll *a* based on deterministic differential equations was not reasonable [11]. For this reason, the uncertainty of input parameters and the nonlinearity of the system required further consideration when constructing models.

To improve the accuracy and efficiency of nonlinear system simulations, intelligent algorithms have been applied in recent years [12]. Widely used intelligent algorithms include the artificial neural network (ANN), the genetic algorithm (GA), the particle swarm algorithm, the wavelet theory, and the projection pursuit algorithm, *etc.*; these intelligent algorithms overcome the uncertainty to a certain extent with high simulation precision. In recent years, the support vector machine (SVM) algorithm, which is a new type of machine learning tool based on statistical learning theory, has drawn more attention [13]. This intelligent algorithm can solve nonlinear system problems and has reasonable generalization ability when using small samples, ameliorating the weaknesses of the above intelligent algorithms, e.g., large sample size requirements and being susceptible to underfitting and overfitting the data for the ANN. The SVM has demonstrated promise for applied studies of water environments, especially for the prediction of hydrologic factors, such as wave height [14], inflow [15], and water levels [16]. Previous studies of chlorophyll *a* based on the SVM algorithm often focused on the retrieval of chlorophyll *a* in water [17], although very few results of chlorophyll *a* prediction have been reported [18]. Furthermore, chlorophyll *a* is affected by many factors, and irrelevant and redundant information is often hidden in the time series of high dimensional feature vectors, leading to structurally complex models and a decrease of analysis precision and application efficiency of the SVM model when using conventional modeling processes [19]. To simplify the model structure and avoid the interference of redundant information in chlorophyll *a* forecasts, it is desirable to obtain more accurate and reliable prediction results by using SVM models with the most relevant influence factors as input vectors and simple structures. Feature selection is an important approach for getting structurally simplified model by removing those redundant parameters. Cho *et al.* [20] used principal component analysis (PCA) to extract variables for the prediction model of chlorophyll *a*. Compared to conventional parameter extraction approaches such as PCA, the GA has a distinct advantage in fast random search. Thus, the SVM model can be expected to get satisfactory prediction results through combing the GA to extract feature variables and simplify the model structure in such complex water bodies as reservoirs, whose water quality variations likely result from a combination of multiple factors. However, the GA-SVM hybrid model needs to be further developed for nonlinear water resources system.

This study developed a hybrid model of GA and SVM algorithms to predict chlorophyll *a* in the Miyun Reservoir of northern China. Based on the feature selection with the GA, we extracted appropriate input vectors, so that the redundant information was effectively eliminated with the simplified model structure. This model could analyze water quality and its change trend with reliable results, and was of great practical significance in preventing water pollution accidents.

2. Study Area and Data Description

The Miyun Reservoir is located in the Miyun County of Beijing City. Built in 1960, it is the largest reservoir and is a unique surface source of drinking water in Beijing City (Figure 1). In addition to functioning as a water supply, the Miyun Reservoir also provides irrigation, flood control, power generation, aquaculture, tourism, and other comprehensive benefits. The Miyun Reservoir's surface area is 183.6 km^2, its maximum depth is 153.93 m, and the maximum volume of reservoir storage is 3.349×10^{10} m^3. Monitoring data shows that in recent years, the total phosphorus concentration in the reservoir fluctuated between 0.010 and 0.025 mg/L, which means the nutrition status of the water is at a mesotrophication to oligotrophication level. The total nitrogen concentration ranges between 0.62 and 1.43 mg/L, indicating that the nutrition status is at a mild or moderate eutrophication level. Planktonic algae have rich diversities, and the dominant population in various periods is different in the Miyun Reservoir. As for cyanobacterium, from 2001 to 2003 it was the dominant algae from September to October [21]; from 2008 to 2010, it was the dominant algae from June to September [22,23]. Considering the current water quality situation, we should take effective measures to alleviate adverse influences resulting from climate change and human activities on the reservoir.

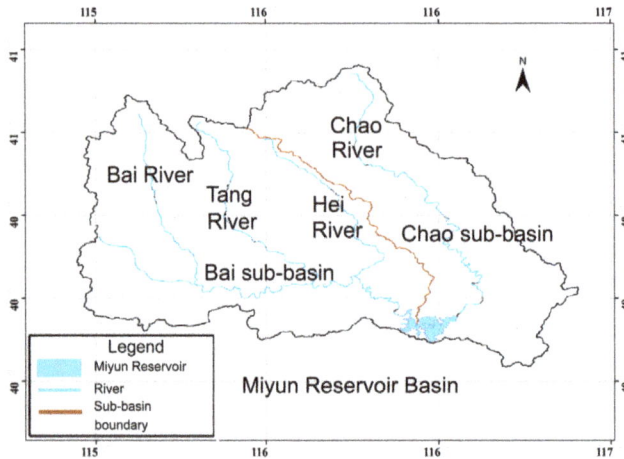

Figure 1. The Miyun Reservoir in northern China.

In this paper, the Baihe Key Dam in the Miyun Reservoir is taken as the research area. Data for the model establishment and calibration are from the monitored data, including water quality indicators (*i.e.*, chlorophyll *a* in water, total nitrogen, total phosphorus, permanganate index, and dissolved oxygen), hydrological indicators (*i.e.*, water temperature, pH, transparency, flow, reservoir storage, inflow, outflow, and water level), and meteorological indicators (*i.e.*, precipitation and temperature). The monthly data from 2000 to 2010 were obtained from the Miyun Reservoir Management Office. Because the Miyun Reservoir is frozen for the period from December to March, the prediction of chlorophyll *a* focused on the period from April to November in each year, and other indicators in the SVM model corresponded to these months. The average monthly variations of parts of these indicators are shown in Figure 2.

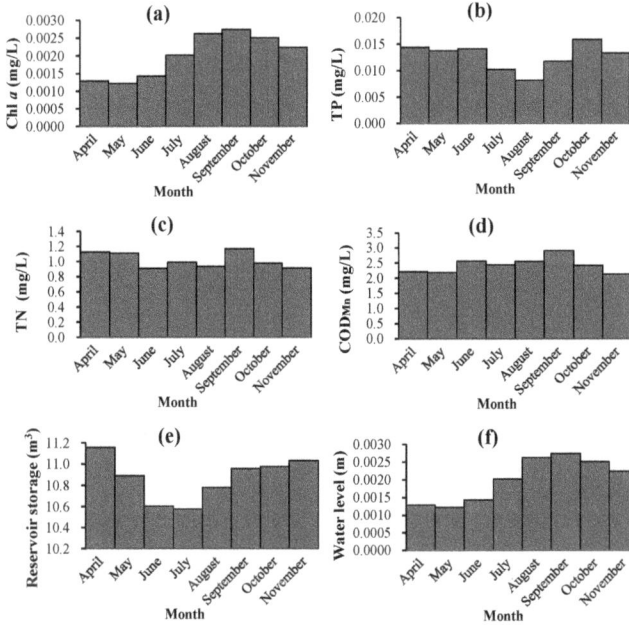

Figure 2. Water environmental situation in Miyun Reservoir: (**a**) Concentration of chlorophyll *a*; (**b**) TP concentration; (**c**) TN concentration; (**d**) COD$_{Mn}$ concentration; (**e**) reservoir storage; and (**f**) water level.

3. Methods

Originally proposed in 1985 by Cortes and Vapnik, the SVM algorithm was widely used to solve highly nonlinear classification and regression problems with good generalizability [24]. The SVM algorithm can be easily applied to other machine learning problems such as function fitting. It is based on the VC dimension theory and the structure risk minimum principle of the statistical learning theory. By seeking the best compromise among the complexities in the model with a limited sample (*i.e.*, learning accuracy of particular training samples) and learning ability (*i.e.*, learning ability to identify random sample), the SVM algorithm can achieve the best generalization ability. There are many meteorological and hydrological parameters that influence chlorophyll *a*. To avoid blindness in selecting the input vector during the process of chlorophyll *a* prediction, this study firstly took feature selection to determine the best input vectors of prediction model with the GA, and then constructed the SVM model with a simplified model structure to achieve the purpose of improving prediction accuracy.

123

3.1. The Flow Chart for Developing a Simplified Structural GA-SVM Hybrid Model

The penalty factor and kernel function parameters of the SVM model may directly influence simulation results. This study developed a SVM model by using the GA to optimize the input parameters in the SVM model and extract feature parameters with the aim of simplifying the model structure. The flow chart of the simplified structural SVM model based on the GA is shown in Figure 3. After data pre-treatment, the input and output vectors were determined, and the sample set was divided into a training data set and a testing data set. The GA was applied to optimizing the parameters of the SVM model and extracting input vectors. The SVM model was trained and calibrated with optimal parameters, then used to predict chlorophyll *a* in the water. This study applies the LIBSVM software package developed by Lin Chih-Jen *et al.*, of Taiwan University to run the program on the MATLAB platform [25].

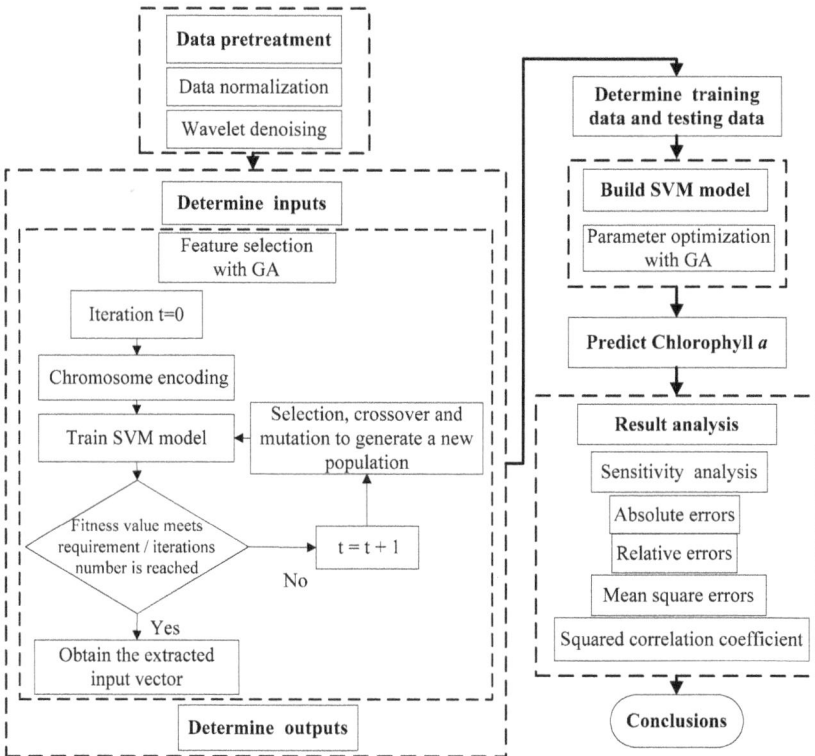

Figure 3. The flow chart for developing simplified structural GA-SVM hybrid model for chlorophyll *a* prediction.

3.2. Construction of Chlorophyll a Prediction Model Based on the SVM Algorithm

This paper applied the principle of the SVM algorithm to establish a prediction model for chlorophyll a in the Miyun Reservoir. The basic principle of the SVM algorithm was to first select a nonlinear mapping algorithm as a kernel function, through which the input vectors were mapped into a high dimensional feature space, and in this space simpler linear regressions can replace complex nonlinear regressions of the original input space [26]. Then an optimal decision function was produced in the feature space to realize the nonlinear decision function of the original input space, and finally the linear learning method can be applied to solve the classification and regression problems in the input space. This process can be expressed as:

$$y = f(x, w) = w \bullet \varnothing(x) + b,$$ (1)

where y is the output, $y \in R$; x is the input vector, $x \in R^n$; w is the matrix of the regression weight vector; $\varnothing$ is a non-linear function by which x is mapped into a high dimensional feature space; b is a bias; and b and w can be obtained with Equation (3). In the mapping process, a kernel function $k(*, *)$ can be constructed by $k(x, x') = (\varnothing(x) \bullet \varnothing(x'))$. Therefore, we only need to replace the x or x_i of the original space with $\varnothing(x)$ or $\varnothing(x_i)$, while it is not necessary to know the explicit expression of nonlinear mapping $\varnothing$. In this study, we selected radial basis function (RBF) as the kernel function:

$$K(x, x_i) = exp\left(-\gamma \| x - x_i \|^2\right),$$ (2)

where x_i is the input vector, $x \in R^n$; and γ is the parameter of the RBF kernel function.

In Equation (1), the concentration of chlorophyll a in reservoir water was selected as y in the SVM model, whereas other water quality factors, hydrological factors, and meteorological factors were selected as x in the SVM model. In this way, the concentration of chlorophyll a was predicted based on the other factors. To solve Equation (1), the following regularized risk function (*i.e.*, Equation (3)) was used. These constraints ensured the regression errors of the samples being within the area that was delineated by the error tolerance and the slack variables. Equation (3) can be solved with the Lagrange technique.

Minimize

$$\frac{1}{2} \| w \|^2 + C \sum_{n=1}^{N} (\xi_n + \xi_n^*)$$

subject to

$$\left\{ \begin{array}{l} y_i - W^T \varnothing(X_i) - b \leqslant \varepsilon + \xi_i \\ W^T \varnothing(X_i) + b - y_i \leqslant \varepsilon + \xi_i^* \\ \xi_i, \xi_i^* \geqslant 0 \end{array} \right\} i = 1, 2, \ldots, N,$$ (3)

where C is a penalty parameter that determines the penalty degree for the sample classification errors in the optimization problem; N is the number of the samples of $\{x_n, y_n\}_{n=1}^{N}$; ξ_n and ξ_n^* are slack variables that penalize training errors by the loss function over the error tolerance (ε); ξ_n represents the upper training errors subject to ε; ξ_n^* represents the lower training errors subject to ε; ξ_n and ξ_n^* can be calculated with Equation (3); and was normally set to 0.001.

Overall, the chlorophyll a simulation with the SVM model depended on the ability to learn the nonlinear causality between the historical data of the concentration of chlorophyll a and its influencing factors (i.e., other water quality factors, hydrological factors, and meteorological factors). The modeling process for the SVM model was introduced below.

Step 1. Determine chlorophyll a as the output value of the prediction model, with the other indicators as input values:

$$\rho_{Chla} = F(S, W_I, W_O, L, TP, TN, COD_{Mn}, DO, T_W, pH, SD, T_A, P), \qquad (4)$$

where S is reservoir storage, W_I is inflow, W_O is outflow, L is water level, TP is the concentration of total phosphorus in water, TN is total nitrogen in water, COD_{Mn} is permanganate index in water, DO is the dissolved oxygen concentration in water, TW is water temperature, pH is hydrogenion concentration of water, SD is water transparency, T_A is temperature, and P is precipitation.

Step 2. Before establishing the SVM model, to extract useful information in the original data and determine the most reasonable and relevant input vectors of the prediction model, this study applied data normalization, wavelet denoising, and feature selection for the data pre-treatment in the MATLAB software. To test the prediction ability of the SVM model, the sample set was divided into separate training and testing sets. Data between 2000 and 2004 were used as the training set, and those between 2005 and 2010 as the testing set. Thus, the testing set data were independent and not used to train the model. In the parameter optimization, the initial conditions of the GA were set: the biggest evolution generation was 100, the largest population was 20, the gap of genetic algorithm was 0.9, and the k-fold cross-validation number was 5.

Step 3. To determine the effect of each input indicator to the prediction model, we carried out sensitivity analysis of the chlorophyll a prediction model. The analysis method was to change a particular input variable (increase or decrease by 10%) while the other input variables remained fixed and then applied the established SVM model to re-predict; the variable of the sensitivity model was obtained by calculating the relative changes in chlorophyll a with the output value.

Step 4. To eliminate the irrelevant and redundant information hidden in the time series of high dimensional feature vectors, and reveal the more representative features that influenced the concentration of chlorophyll *a* in the Miyun Reservoir, we applied feature selection to the input vectors of the SVM model for chlorophyll *a* simulation by using the GA. Subsequently, we established the hybrid SVM model using the extracted feature vectors and improved prediction accuracy, generalization ability, and efficiency.

3.3. Feature Selection and Parameter Optimization Based on Genetic Algorithm Optimization

The feature means each attribute of the data set. Too many features will increase the complexity of the work, while the accuracy of data mining may not be improved. To pick out the most representative and effective feature vectors of the chlorophyll *a* prediction model, we used the GA for feature selection. In addition, considering that the SVM model did not provide a method for selecting the parameter in the RBF kernel function and the penalty parameter (C), we used the GA for optimizing these two parameters in the SVM model. The principle of the GA is based on a specific operation for the structure of objectives, according to a predefined criteria function, to improve the new population by comparing it with the original one. In the process of generation, proper coding was used and the operator was applied to imitate the path of natural selection. Reproduction, crossover, and mutation were taken to operators in the current population [27]. Procedures of feature selection and parameter optimization with applications of the GA were as follows.

Step 1. Chromosome encoding. In the selection and optimization process, the iterations were set to zero. Chromosomes were encoded with binary coding. Each operator was composed of N codes, where N is the number of characteristic vectors or SVM parameters that need to be optimized. When a number in the operator was 1, it represented the characteristic vector and the parameter was selected; otherwise it was not selected, and the initial population was generated randomly.

Step 2. Evaluation of the fitness function value. Determine the square of the root mean square error in the training phase as the objective function for the fitness value. Then, calculate the fitness function value of the current generation. Choose a certain adaptation level, retaining the individuals whose fitness function value is greater than the adaptation level; these individuals compose the next generation.

Step 3. Selection, crossover and mutation of operators. Apply genetic operation of selection, crossover, and mutation to individuals in the group; the next generation was produced after the genetic optimization.

Step 4. Termination judgment. If the iteration number was greater than the set value, or the accuracy of the fitness function value reached the expected value, then terminate the iteration [28]. The extracted features and the optimal model parameters were then determined.

3.4. Model Calibration

In order to analyze the performance of the model, four indicators—The absolute error (AE), relative error (RE), root mean square error (RMSE), and square of correlation coefficient (R^2)—Were selected to evaluate the fit and prediction effect of the model. AE represented the deviation between monitoring and prediction values, and RE was the ratio of AE and monitoring values, reflecting the objective accuracy of measurement results. RMSE reflected the performance of the prediction model, i.e., generally, the smaller the RMSE the better the performance. R^2 represented the degree of linear relevance among the variables, i.e., the closer R^2 was to 1 the higher the relevance. The expressions of these four indicators were as follows:

$$AE = |y_i - \hat{y}_i| \tag{5}$$

$$RE = \frac{AE}{y_i} \times 100\% \tag{6}$$

$$RMSE = \sqrt{\frac{1}{n} \sum_{i=1}^{n} (y_i - \hat{y}_i)^2} \tag{7}$$

$$R^2 = 1 - \frac{\sum_{i=1}^{n} (y_i - \hat{y}_i)^2}{\sum_{i=1}^{n} (y_i - \bar{y})^2} \tag{8}$$

where y_i is the real value of the data set, $\hat{y}_i$ is prediction value, $\bar{y}$ is the average of the original data, and n is the amount of data for the testing set.

4. Results and Discussion

4.1. Wavelet Denoising

The results of applying the wavelet denoising method to the original data of chlorophyll *a* are shown in Figure 4. The upper and lower figures represented the time series data before and after the denoising, respectively. It was clear to see that after wavelet decomposition and reconstruction the low frequency characteristics of the original data were preserved, while eliminating the high frequency data. After this process, the abrupt change points of the time series data were smoothed, and the main information regarding the concentration of chlorophyll *a* was well preserved.

(a)

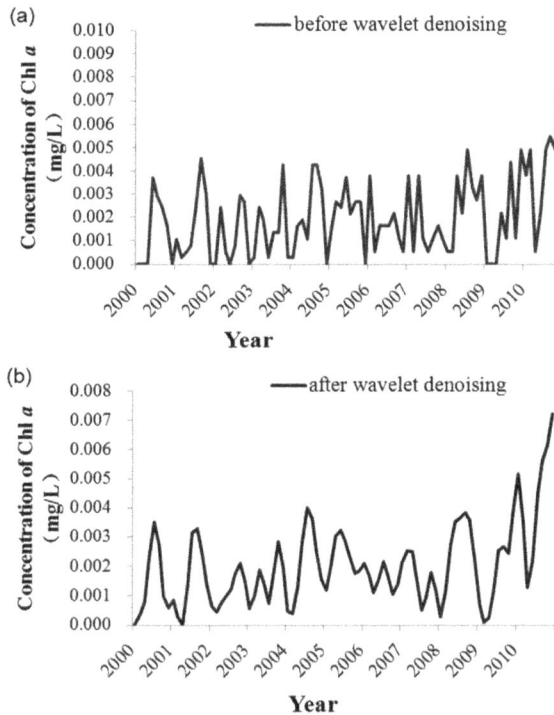

(b)

Figure 4. Comparisons of before and after the wavelet denoising.

The noise of the original data about chlorophyll *a* was mainly caused by the errors in sampling and deviations in experiments due to unsuitable experimental conditions and improper operation caused by human errors. The wavelet denoising was realized through multi-scale decomposition of sequence data and reconstruction. The original signal was decomposed into a series of low frequency and high frequency components by using the wavelet decomposition, and the noise of chlorophyll *a*'s raw data was concentrated in the high frequency components. The high frequency components were processed with threshold, and the low frequency components were reconstructed to obtain the smooth data series. Because the low frequency components could preserve the details of the original data, excessive deviations can be avoided in data applications. Therefore, it was reasonable and concise to use the time series data of chlorophyll *a* after the wavelet denoising as input variables of the SVM model, without the loss of important information.

4.2. Results of Sensitivity Analysis and Feature Selection

The use of the GA aimed to automate and enhance SVM designing process. The results of parameter optimization indicated that the optimal penalty factor C was 1.0737, and the optimal parameter γ in RBF kernel function was 1.0005.

The final SVM model was established based on these results. Figure 5 shows sensitivity analysis results for input vectors of the SVM model. It can be seen that when the values of eight parameters (*i.e.*, rainfall, water level, inflow, pH, water temperature, permanganate index, total nitrogen, and total phosphorus) increased by 10%, the sensitivity degree was greater than zero, which meant the prediction of chlorophyll *a* showed a positive correlation. Alternatively, when the values of the other five parameters (*i.e.*, temperature, outflow, reservoir storage, dissolved oxygen, and transparency) increased by 10%, the sensitivity degree was less than zero, which meant the prediction of chlorophyll *a* showed a negative correlation. These correlations coincided with the mechanism of action for hydrology and water quality. Compared with other feature vectors, the prediction model for chlorophyll *a* in the Miyun Reservoir water had greater sensitivity to dissolved oxygen, transparency, permanganate index, pH, temperature, total nitrogen, and total phosphorus. It should be noted that the variations of transparency and dissolved oxygen were the results of the change in chlorophyll *a*.

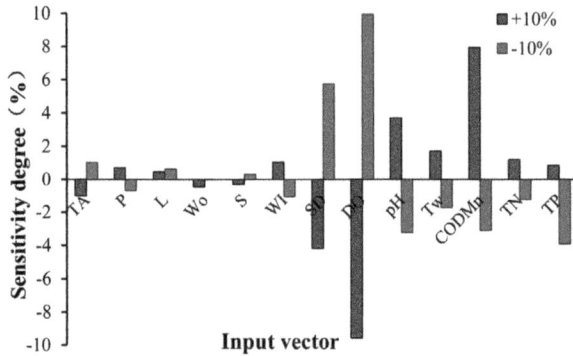

Figure 5. Sensitivity analysis for input vectors of the SVM model.

Four parameters, *i.e.*, total phosphorus, total nitrogen, permanganate index, and reservoir storage, were extracted through feature selection. The results of the feature selection were consistent with those of the sensitivity analysis. Compared with other studies, although the influence factors of chlorophyll *a* were different for various research objects, two factors including the concentration of TN and TP were always main factors. For example, Canfield [29] applied statistical analysis to pick out the total phosphorus concentration and the total nitrogen concentration as the explanatory variables in their developed prediction model of chlorophyll *a*.

With these four parameters as the input vectors, the simplified SVM model was constructed. Figure 6a shows the results of the model training process. It can be seen that the simulation values are perfectly consistent with the monitored values, with the exception of a bias in the extreme points. Accordingly, the RMSE was only 0.00017, and the R^2 was 97.33%. After the SVM prediction model was trained, chlorophyll *a* in the Miyun Reservoir was predicted for the period between 2005 and 2010. In Figure 6b, we can see that from 2005 to 2009, the simulation effect was passable. However, in 2010 the simulation effect was not satisfactory. Calculations showed that the RMSE of the testing set was 0.000641 and the R^2 was 81.97%. From 2005 to 2009, the RMSE was 0.0004 and the R^2 was 85.96%; however, in 2010, the RMSE was 0.0013 and the R^2 was only 79.00%. This was primarily related to the fluctuations and periodicity of the monitored data. For the training data set, from 2000 to 2004, the concentration of chlorophyll *a* generally showed a peak in the middle of the year, but there was no obvious periodic trend for the concentration of chlorophyll *a* in the testing data set. In addition, in 2010 the concentration of chlorophyll *a* in the Miyun Reservoir was relatively higher compared with previous years. During April and for the period from August to November, the concentration of chlorophyll *a* was anomalously high, exceeding 0.004 mg/L, whereas in the training data set, the concentration of chlorophyll *a* had never achieved this level. Therefore, the SVM model was sensitive to the data. To explore which indicators were most relevant to chlorophyll *a*, sensitivity analysis for each input vector of the model was conducted.

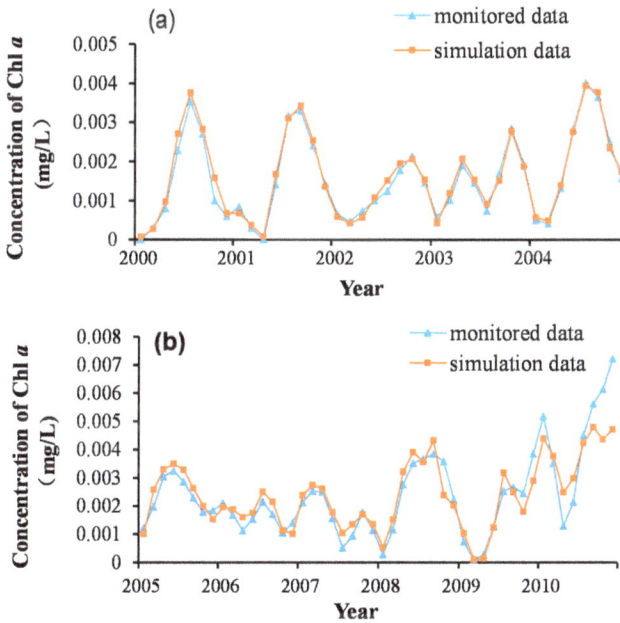

Figure 6. Training and prediction results of the SVM model: (**a**) training results and (**b**) prediction results.

131

4.3. Relative Errors of the SVM Model

Figure 7 shows the relative errors in each testing year of the SVM prediction model. Overall, the relative error between 2005 and 2009 was smaller than that in 2010. This indicated that the model had improved prediction accuracy during the early part of the testing data. This result was related to the variations in the testing data. The first five years of the testing data were consistent with the training data with regard to the amplitude, cycle, and peak values. However, data in the last year did not exhibit regularity, as evidenced by the notable fluctuations in each month. Given the obvious fluctuation in the overall trend, the average relative error of the prediction in 2010 was the largest.

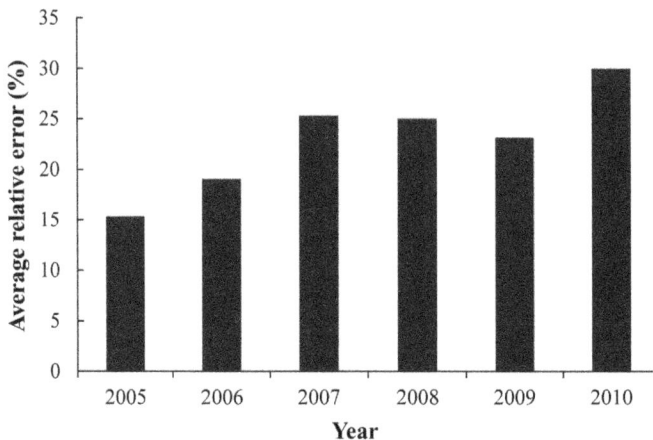

Figure 7. Average relative errors for each year.

To analyze the errors of chlorophyll *a* prediction in different months, the monthly average relative error of the SVM model was compared between April and November (Figure 8). Prediction results showed that the average relative errors from April to November were 29.10%, 15.90%, 35.84%, 14.64%, 27.53%, 19.41%, 19.15%, and 22.24%, respectively. It can be seen that in these six years, the biggest relative error occurred in June, followed by April, August, November, September, October, May, and July. Differences among these months may be due to the precipitation during summer. Inflow to the reservoir increased with rainfall, augmenting the frequency of soil and water erosion and leading to more nutrients being deposited into the reservoir. It required a significant amount of time for these nutrients to be used by the microorganisms in the water. Considering the cumulative effect, in autumn the concentration of chlorophyll *a* fluctuated markedly. Therefore, due to the synergy of rainfall and inflow, as well as the cumulative effect of the nutrients, the prediction error of chlorophyll *a* in summer and autumn was larger.

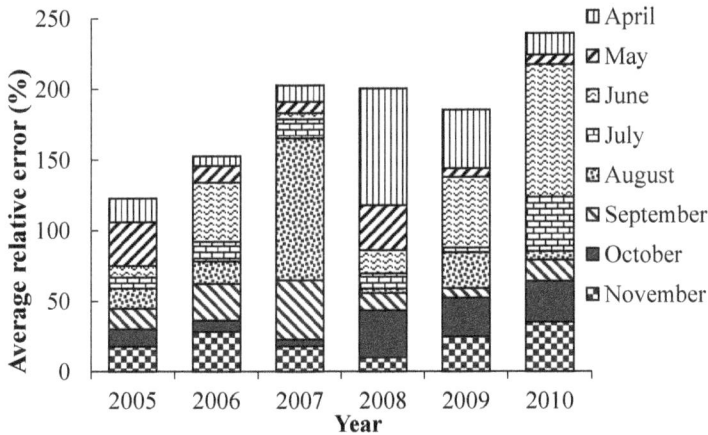

Figure 8. Average relative errors for each month.

4.4. Comparisons of Model with Feature Selection and Model without Feature Selection

To determine the effect of feature selection on the model, we also established a prediction model without feature selection. Table 1 shows the comparisons of model structure and prediction results between these two models. With feature selection, the input vectors were simplified from 13 to 4, and the extracted features were consistent with the sensitivity analysis. For the prediction results, in the training process, the mean AE, mean RE and R^2 of the model with feature selection were slightly larger, and the RMSE was slightly smaller; in the testing process, the mean AE, mean RE, and RMSE of the model with feature selection were smaller, and the R^2 was significantly higher than that of model without feature selection.

Table 1. Comparisons of model with feature selection and model without feature selection.

Description		Model with Feature Selection	Model without Feature Selection
Number of input vectors		4	13
Input vectors		TP, TN, COD_{Mn}, S	$TP, TN, COD_{Mn}, S, W_I, W_O, L, DO, T_W,$ pH, SD, T_A, P
Training process	Mean AE	0.00014244	0.00013824
	Mean RE	12.35%	10.64%
	RMSE	0.00017	0.00018
	R^2	97.33%	97.19%
Testing process	Mean AE	0.00045199	0.00057325
	Mean RE	22.98%	26.16%
	RMSE	0. 000641	0.000836
	R^2	81.97%	69.36%

Notes: S is reservoir storage; W_I is inflow; W_O is outflow; L is water level; TP is the concentration of total phosphorus in water; TN is total nitrogen in water; COD_{Mn} is permanganate index in water; DO is the dissolved oxygen concentration in water; T_W is water temperature; pH is hydrogenion concentration of water; SD is water transparency; T_A is the temperature; and P is precipitation.

133

It can be seen that feature selection picked out the representative and effective feature vectors from the original features that were more related with the chlorophyll *a*, so the dimensions of feature space were reduced. When the redundant or irrelevant information was deleted and the data set was simplified, the model was more concise and understandable [30]. Though the simulation accuracy in the training process was similar or even smaller than that prior to feature selection, relevant input vectors and reasonable model structure led to better prediction results in the testing process. As a whole, the SVM model with feature selection showed better performance both in model structure and prediction effect, and this indicated the model with feature selection had great potential in prediction ability, which had close relation with the internal structure of the model. In brief, feature selection with the GA in this study played important roles in three specific aspects. Firstly, it determined the feature vectors that were most relevant to chlorophyll *a* concentration. Their information was preserved accordingly in the simplified model, leading to the accuracy improvement with the decreased amount of calculation. It was a rather feasible way to improve calculation efficiency, especially for large-scale computing with multiple parameters. Secondly, it reduced the dimensions of the input vectors to avoid dimension disaster (*i.e.*, with the increase of the input vectors' dimensions, the complexity of the calculation would greatly increase), while revealing the representative factors that influenced the chlorophyll *a* in the Miyun Reservoir. Thirdly, it was easy to combine with other algorithms (e.g., SVM) to improve the generalization ability of the SVM prediction model, reflecting good convergence and robustness.

Besides feature selection, the proposed SVM model showed good performance mainly due to the proper initial settings of parameters, which would affect the computational complexity and convergence rate directly. Although the GA used in the prediction model effectively avoided falling into a local optimal solution and producing a low convergence speed, the initial values of the GA's parameters were determined through the trial method in this study. Recent research mainly used two advanced approaches to optimize the initial settings for the GA's parameters: one approach was to optimize the initial population's characteristics and quantity by combining other approaches, such as the heuristic algorithm [31]; another approach was to improve the crossover and mutation rates with adaptive GA [32], such as clustering-based adaptive GA [33]. In future research, the determination of reasonable initial values of the GA's parameters would combine with these approaches.

5. Conclusions

A GA-based SVM model for predicting the monthly concentration of chlorophyll *a* of the Miyun Reservoir was constructed. We firstly carried out a sensitivity analysis of the prediction model, and identified that the concentration of chlorophyll *a* had great sensitivity to seven input indicators. With the GA being used for the removal of redundant features and the feature selection of input vectors, the four most relevant influence factors of chlorophyll *a* (*i.e.*, total phosphorus, total nitrogen, permanganate index, and reservoir storage) were screened as new input vectors, which were consistent with the results of the sensitivity analysis. With these new input vectors, the prediction model had simpler structure and better prediction accuracy than the model without the feature selection. Due to the stronger correlation of the input vector structure, the simplified GA-SVM model showed improved calibration and prediction ability. This proved that the SVM prediction model was sensitive to the structure of the input variables. In brief, this study proposed an intelligent algorithm for predicting the concentration of chlorophyll *a* of the reservoir water, which provided an effective tool for the management of reservoirs, especially for an early warning of eutrophication. Besides, this model could solve practical problems with different nutritional load conditions, and its applications can be extended to other reservoirs. In future research, the interaction mechanism of influence factors should be further considered to optimize the parameters used in the developed hybrid model of GA-SVM algorithm to get more reliable results for the prediction of chlorophyll *a*, and empirical models will be explored to get better application performance in chlorophyll *a* prediction.

Acknowledgments: This study was supported by the Fund for Innovative Research Group of National Natural Science Foundation of China (No. 51421065), the National Science and Technology Support Program (No.2011BAC12B02) and the key project of National Natural Science Foundation of China (No. 51439001).We would like to extend special thanks to the editors and anonymous reviewers for all their detailed comments and valuable suggestions in greatly improving the quality of this paper.

Author Contributions: Xuan Wang determined the focus of this work; Jieqiong Su performed the numerical simulations and drafted the initial manuscript under the guidance of Xuan Wang; Shouyan Zhao provided the data of the Miyun Reservoir; Bin Chen and Chunhui Li reviewed the work and helped bring it to its final form; and Zhifeng Yang was supporting this work as an expert in model building and fault detection.

Conflicts of Interest: The authors declare no conflict of interest.

References

1. Scholz, M. Sustainable water systems. *Water* **2013**, *5*, 239–242.
2. Cai, Y.P.; Huang, G.H.; Tan, Q.; Yang, Z.F. An integrated approach for climate-change impact analysis and adaptation planning under multi-level uncertainties. Part I: Methodology. *Renew. Sustain. Energy Rev.* **2011**, *15*, 2779–2790.
3. Cai, Y.P.; Huang, G.H.; Yang, Z.F.; Tan, Q. Identification of optimal strategies for energy management systems planning under multiple uncertainties. *Appl. Energy* **2009**, *86*, 480–495.

4.   Tan, Q.; Huang, G.H.; Cai, Y.P. Radial interval chance-constrained programming for agricultural non-point source water pollution control under uncertainty. *Agric. Water Manag.* **2011**, *98*, 1595–1606.

5.   Mulia, I.E.; Tay, H.; Roopsekhar, K.; Tkalich, P. Hybrid ANN-GA model for predicting turbidity and chlorophyll-*a* concentrations. *J. Hydro-Environ. Res.* **2013**, *7*, 279–299.

6.   Liu, Y.; Guo, H.; Yang, P. Exploring the influence of lake water chemistry on chlorophyll *a*: A multivariate statistical model analysis. *Ecol. Model.* **2010**, *221*, 681–688.

7.   Cerco, C.F.; Noel, M.R. Twenty-one-year simulation of Chesapeake Bay water quality using the CE-QUAL-ICM eutrophication model. *J. Am. Water. Resour. Assoc.* **2013**, *49*, 1119–1133.

8.   Blancher, E.C. Modeling nutrients and multiple algal groups using AQUATOX: Watershed management implications for the Braden River Reservoir, Bradenton Florida. *Proc. Water Environ. Feder.* **2010**, *10*, 6393–6410.

9.   Rinke, K.; Yeates, P.; Rothhaupt, K.O. A simulation study of the feedback of phytoplankton on thermal structure via light extinction. *Freshw. Biol.* **2010**, *55*, 1674–1693.

10. Seo, D.G.; Ahn, J.H. Prediction of chlorophyll-a changes due to weir constructions in the Nakdong River using EFDC-WASP modelling. *Environ. Eng. Res.* **2012**, *17*, 95–102.

11. Chen, Q.; Han, R.; Ye, F.; Li, W. Spatio-temporal ecological models. *Ecol. Inform.* **2011**, *6*, 37–43.

12. Gandomi, A.H.; Yun, G.J.; Yang, X.S.; Talatahari, S. Chaos-enhanced accelerated particle swarm optimization. *Commun. Noulinear Sci. Numer. Simul.* **2013**, *18*, 327–340.

13. Maity, R.; Bhagwat, P.P.; Bhatnagar, A. Potential of support vector regression for prediction of monthly streamflow using endogenous property. *Hydrol. Process.* **2010**, *24*, 917–923.

14. Malekmohamadi, I.; Bazargan-Lari, M.R.; Kerachian, R.; Nikoo, M.R.; Fallahnia, M. Evaluating the efficacy of SVMs, BNs, ANNs and ANFIS in wave height prediction. *Ocean. Eng.* **2011**, *38*, 487–497.

15. Karamouz, M.; Ahmadi, A.; Moridi, A. Probabilistic Reservoir Operation Using Bayesian Stochastic Model and Support Vector Machine. *Adv. Water Resour.* **2009**, *32*, 1588–1600.

16. Çimen, M.; Kisi, O. Comparison of Two Different Data-driven Techniques in Modelling Lake Level Fluctuations in Turkey. *J. Hydrol.* **2009**, *378*, 253–262.

17. Zhang, Y.C.; Qian, X.; Qian, Y.; Liu, J.P.; Kong, F.X. Application of SVM on Chl-*a* concentration retrievals in Taihu Lake. *China Environ. Sci.* **2009**, *29*, 78–83. (In Chinese)

18. Xiang, X.Q.; Tao, J.H. Eutrophication Model of Bohai Bay Based on GA-SVM. *J. Tianjin Univ.* **2011**, *44*, 215–220. (In Chinese)

19. Liu, C.; Tang, D. Spatial and temporal variations in algal blooms in the coastal waters of the western South China Sea. *J. Hydro-Environ. Res.* **2012**, *6*, 239–247.

20. Cho, K.H.; Kang, J.H.; Ki, S.J.; Park, Y.; Kim, J.H. Determination of the optimal parameters in regression models for the prediction of chlorophyll-*a*: A case study of the Yeongsan Reservoir, Korea. *Sci. Total Environ.* **2009**, *407*, 2536–2545.

21. Wang, L.; Yang, M.; Guo, Z.H.; Zhang, Y.; Jiang, Y.; Fan, K.P. Study on water quality transformation in Miyun Reservoir. *China Water Wastewater* **2006**, *22*, 45–48. (In Chinese)

22. Jia, D.M.; Wang, J.S.; Xue, X.J.; Qi, Z.Y. Research on phytoplankton characteristics of Miyun Reservoir. *Beijing Water* **2013**, *1*, 12–15. (In Chinese)

23. Pan, K.M.; Wang, J.M. Control and management of eutrophication of the Miyun reservoir. *Beijing Water* **2010**, *6*, 25–27. (In Chinese)

24. Cortes, C.; Vapnik, V. Support-vector networks. *Mach. Learn.* **1995**, *20*, 273–297.

25. Chang, C.C.; Lin, C.J. LIBSVM: A library for support vector machines. *ACM Trans. Intell. Syst. Technol.* **2011**, *2*, 1–39.

26. Maus, A.; Sprott, C. Neural network method for determining embedding dimension of a time series. *Commun. Nonlinear Sci. Numer. Simul.* **2011**, *16*, 3294–3302.

27. Pourbasheer, E.; Riahi, S.; Ganjali, M.R.; Norouzi, P. Application of genetic algorithm-support vector machine (GA-SVM) for prediction of BK-channels activity. *Eur. J. Med. Chem.* **2009**, *44*, 5023–5028.

28. Fernandez, M.; Caballero, J.; Fernandez, L.; Sarai, A. Genetic algorithm optimization in drug design QSAR: Bayesian-regularized genetic neural networks (BRGNN) and genetic algorithm-optimized support vectors machines (GA-SVM). *Mol. Divers.* **2011**, *15*, 269–289.

29. Canfield, D.E., Jr. Prediction of chlorophyll *a* concentrations in Florida lakes: The importance of phosphorus and nitrogen. *J. Am. Water Resour. Assoc.* **1983**, *19*, 255–262.

30. Noori, R.; Karbassi, A.R.; Moghaddamnia, A.; Han, D.; Zokaei-Ashtiani, M.H.; Farokhnia, A.; Gousheh, M.G. Assessment of input variables determination on the SVM model performance using PCA, Gamma test, and forward selection techniques for monthly stream flow prediction. *J. Hydrol.* **2001**, *401*, 177–189.

31. Besalatpour, A.A.; Ayoubi, S.; Hajabbasi, M.A. Feature selection using parallel genetic algorithm for the prediction of geometric mean diameter of soil aggregates by machine learning methods. *Arid Land Res. Manag.* **2014**, *28*, 383–394.

32. Zandieh, M.; Karimi, N. An adaptive multi-population genetic algorithm to solve the multi-objective group scheduling problem in hybrid flexible flowshop with sequence-dependent setup times. *J. Intell. Manuf.* **2011**, *22*, 979–989.

33. Halder, U.; Das, S.; Maity, D. A cluster-based differential evolution algorithm with external archive for optimization in dynamic environments. *IEEE Trans. Cybern.* **2013**, *43*, 881–897.

Experimenting with Coupled Hydro-Ecological Models to Explore Measure Plans and Water Quality Goals in a Semi-Enclosed Swedish Bay

Berit Arheimer, Johanna Nilsson and Göran Lindström

Abstract: Measure plans are currently being developed for the Water Framework Directive (WFD) by European water authorities. In Sweden, such plans include measures for good ecological status in the coastal ecosystem. However, the effect of suggested measures is not yet known. We therefore experimented with different nutrient reduction measures on land and in the sea, using a model system of two coupled dynamic models for a semi-enclosed bay and its catchment. The science question was whether it is worthwhile to implement measures in the local catchment area to reach local environmental goals, or if the status of the Bay is more governed by the water exchange with the Sea. The results indicate that by combining several measures in the catchment, the nutrient load can be reduced by 15%–20%. To reach the same effect on nutrient concentrations in the Bay, the concentrations of the sea must be reduced by 80%. Hence, in this case, local measures have a stronger impact on coastal water quality. The experiment also show that the present targets for good ecological status set up by the Swedish water authorities may be unrealistic for this Bay. Finally, we discuss when and how to use hydro-ecological models for societal needs.

Reprinted from *Water*. Cite as: Arheimer, B.; Nilsson, J.; Lindström, G. Experimenting with Coupled Hydro-Ecological Models to Explore Measure Plans and Water Quality Goals in a Semi-Enclosed Swedish Bay. *Water* **2015**, *7*, 3906–3924.

1. Introduction

Water pollution is widespread and at the same time highly complex, as it originates from the interaction between natural processes and human activities. Many cause–effect relationships are affecting the water simultaneously, with various extensions in time and space e.g., [1–3]. Soluble chemical components are being caught and carried by water, an aggressive liquid on continuous move, following shallow, intermediate or deep pathways from the land surface to the sea. Nutrients and oxygen are the best-documented water quality determinants e.g., [4,5], with major impacts downstream and on the coast [6–8]. During the last decades, this evidence base from monitoring has supported the development of a large flora of dynamic models that numerically describe the coupling between water and nutrients during the movement through the landscape. The models include interactions with sources and sinks, both within single catchments e.g., [9] and across continents in multi-basins e.g., [10]. Such models are normally developed by scientists but when used in practical applications, together with stakeholders, they may serve as platforms for communication and bridge the gap between science and practice e.g., [11–13].

The problems with water pollution have led to great efforts worldwide to detect deterioration and to achieve more sustainable, holistic and integrated water management. In Europe, the Water Framework Directive (WFD, 2000/60/EC) for integrated river basin management was adopted in year 2000 [14]. The environmental objectives are the core of this directive and the definition of "good ecological status" is essential. Each member state decides how to implement the WFD and in Sweden five water districts are responsible for the characterization of ecological status, setting up objectives for each water body, providing management and measure plans, and the implementation and continuous monitoring. In 2008, the European Union also adopted the Marine Strategy Framework Directive (MSFD, 2008/56/EC) which, just as the WFD, applies an adaptive management approach in a six years cycle. The MFSD addresses four marine regions surrounding Europe, where Sweden is part of the Baltic Sea region and the North-East Atlantic Ocean.

The Baltic Sea is one of the largest brackish water systems in the world and it is enclosed except from the narrow connection with the North-east Atlantic Ocean at the Danish straights and Öresund. The drainage basin is home to 85 million people and the sea suffers from pressures like eutrophication, overfishing, industrial waste, and heavy traffic by ships. International agreements, also addressing nutrient load reductions from each country surrounding the sea, have been made within the Helsinki Commission (HELCOM) Baltic Sea Action Plan, BSAP [15–17]. The BSAP is a substantial part of the implementation of the MSFD in the region, although local marine targets have been suggested as well [18]. The coastal zone along the Baltic Sea is highly affected by the increased load of nutrients, and just as for the sea as a whole, the problems are especially severe in semi-enclosed basins, where the water exchange is reduced. Such coasts are more affected by riverine nutrient loads from land and suffer from eutrophication and oxygen depletion. Hence, the ecological status in the coastal zone is affected by decisions taken both with reference to the WFD and the MSFD.

In the present study, we coupled a hydrological and an oceanographic model to simulate the effects of suggested measures and management plans from the two different directives for a semi-enclosed Bay in South-Eastern Sweden, called Slätbaken Bay (Figure 1). The catchment model Hydrological Predictions for the Environment, HYPE [19], which calculates water and nutrients on a daily time-step, was used to estimate effects of land-based measures. The effects in the bay itself were estimated using the Coastal Zone Model, CZM [20], which is an oceanographic ecosystem model with nine biogeochemical variables calculated with a ten minutes time-step. The hypothesis is that measures both in the coastal drainage basin and in the Baltic Proper will affect the nutrient status of the Bay, as it is semi-enclosed.

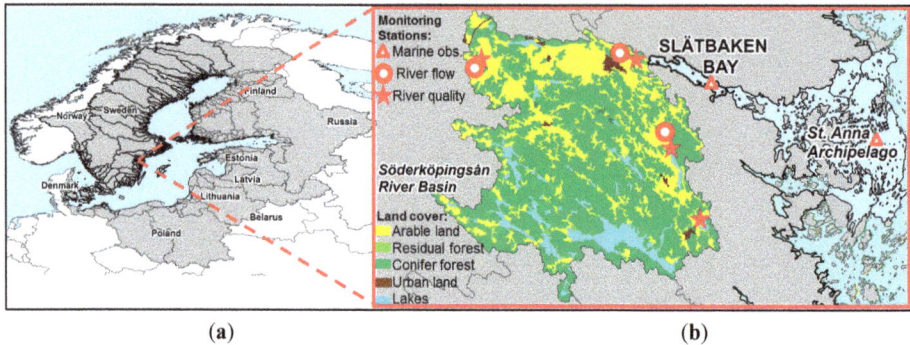

(a)   (b)

Figure 1. The Baltic Sea catchment area and the location of the semi-enclosed Slätbaken Bay (**a**), catchment and land cover of the major river, and monitoring sites (**b**).

An experiment was set up to quantify the effects from various combat measures in relation to the environmental targets set up by the water authorities for Slätbaken Bay. Local land-based measures in the catchment were compared with remote measures for the total sea, to evaluate the influence of the two different EU directives on this specific coastal environment. The aim was to explore whether it is worthwhile to implement measures in the local catchment area to reach local environmental targets, or if the status of the Bay is more governed by the water exchange with the Sea. The paper thus shows how coupled hydro-ecological models can assist water authorities in practical water management issues.

2. Materials and Methods

2.1. The Study Site and Environmental Quality Targets

This study explores different ways to improve the nutrient status in Slätbaken Bay (Figure 1). The bay has an area of 15.5 km^2, is of fjord-like character and further enclosed by the St. Anna archipelago, which is part of the Baltic Proper. The major river inflow to the bay is the Söderköpingsån River, which has a catchment area of 880 km^2, where 9% are lakes, 64% forests, 26% agricultural land and 1% urban areas. About 16,000 inhabitants live permanently in the catchment area and, in addition, there are 900 summer cottages.

140

In accordance with the WFD, the local water authorities have defined targets for good environmental status in Slätbaken Bay. These are winter concentrations of total nitrogen (N) < 29 μmolN/L and phosphorus (P) < 0.61 μmolP/L. Targets for summer concentrations are <30 μmolN/L and <0.46 μmolP/L.

The MSFD, on the other hand, prescribes reductions, relative to current conditions, in summer concentrations of −3% for N and −27% for P for the Baltic Proper as a whole, according to the BSAP [15]. The status of the Slätbaken Bay is thus affected by implementation of both EU directives.

2.2. Catchment Modelling with HYPE

The HYPE model [19] is a process-based hydrological model that simulates the flow and turnover of water and nutrients in the soil, rivers and lakes. Catchments are divided into subcatchments, which are further divided into hydrological response units, *i.e.*, combinations of soil type and land use. Agricultural land is further divided into main crop types. The soil profile is normally divided into three layers. The turn-over and flow of water, N and P is simulated daily for each computational element. The model simulates concentrations of inorganic and organic N, as well as dissolved and particulate P. The HYPE model code is continuously developed and released in new versions at http://hype.sourceforge.net/. For this study, the version HYPE3.5.3 was used.

For Sweden, the HYPE model is set up according to the resolution decided by the water authorities to support the WFD work across the whole country e.g., [21,22]. This set-up is called S-HYPE and the latest version can be found for inspection and be downloaded at http://vattenwebb.smhi.se/. The national model system covers more than 450,000 km^2 and produces daily values of nutrient concentration and water discharge in 37,000 catchments since 1961. The latest version has an average Nash and Sutcliffe [23] Efficiency (NSE) for water discharge of 0.7 and an average relative error of 5%, including both regulated and unregulated rivers with catchments from 1 to 50,000 km^2 and various land-uses across the country. The modelled long-term flow weighted concentrations of nutrients generally fall within ±10% for N and ±20% for P compared to observations (http://vattenwebb.smhi.se/).

For this study, the requested input data [22] such as land cover, topography, soil types, emissions, and forcing data (precipitation and temperature) were obtained from the S-HYPE version 2010. The Söderköpingsån River catchment was then divided into 43 subcatchments and calibrated separately for this experiment. In addition to the river, calculations of water and nutrient inflow from adjacent coastal land near the Bay were included in the simulation. Observations of N- and P-concentrations were available at 4 sites and water discharge at 3 sites (Figure 2). The model was calibrated simultaneously for all observation sites in the specific catchment, using a step-wise iterative procedure [22]. The parameters in the model are linked to soil type and land use; hence the set of parameters for a subcatchment is determined by the distribution of these characteristics.

2.3. Coastal Zone Modelling with CZM

The Coastal Zone Model (CZM) [20] is a coupled one-dimensional hydrodynamic and biogeochemical model where the hydrodynamic part is based on the Program for Boundary Layers in the Environment, PROBE model [24]. PROBE is a general equation solver including, among other things, heat exchange, a two-equation turbulence model and a parameterization of deep sea mixing. The water exchange in the model is driven by baroclinic pressure gradients, *i.e.*, density differences. The model has been set up for the whole Baltic Sea divided into thirteen subbasins in an application called PROBE-Baltic [25], for examining salinity and temperature variations.

The CZM also solves for nine biogeochemical variables described within the Swedish Coastal and Ocean Biogeochemical model, SCOBI model [26]. The variables solved for are nitrate, ammonium, phosphate, oxygen, phytoplankton, zooplankton, detritus and benthic detritus as N and as P. For every time step of 10 min, the CZM generates vertical profiles of both the hydrodynamic and the biogeochemical variables. Every subbasin is considered as horizontally homogenous and thus the horizontal resolution is determined by the division into subbasins. The vertical resolution is 0.5 m in the upper 4 m, then 1 m down to 70 m and then sparser [20]. The model is applied along the whole Swedish coast in approximately 630 subbasins, with water exchange between neighboring subbasins.

The CZM for Slätbaken Bay comprises 12 coupled basins and 15 sounds (represented by double-headed arrows in Figure 2). The Slätbaken Bay has a maximum depth of 47 m and a water exchange time of about 6 months, while the basin Trännofjärden, which is less enclosed, has an exchange time of only 20–30 days. The mean surface salinity in the bay is 4 practical salinity units (psu) compared to the Baltic Proper which has a mean surface salinity of 6–8 psu and a bottom salinity of 11–13 psu.

The CZM is coupled to S-HYPE so that is receives input of freshwater and nutrients generated from land to each grid connected to the shoreline. The daily values from HYPE are linearly interpolated to fit with the higher temporal resolution of CZM. In addition, the CZM receives input regarding the state of the Baltic Proper, which is calculated with a data-assimilated version of the PROBE-Baltic model [25]. The boundary conditions from land and the Baltic Proper are then combined with meteorological forcing (temperature, wind velocity, cloud cover, relative humidity and precipitation) at every third hour to drive the CZM model.

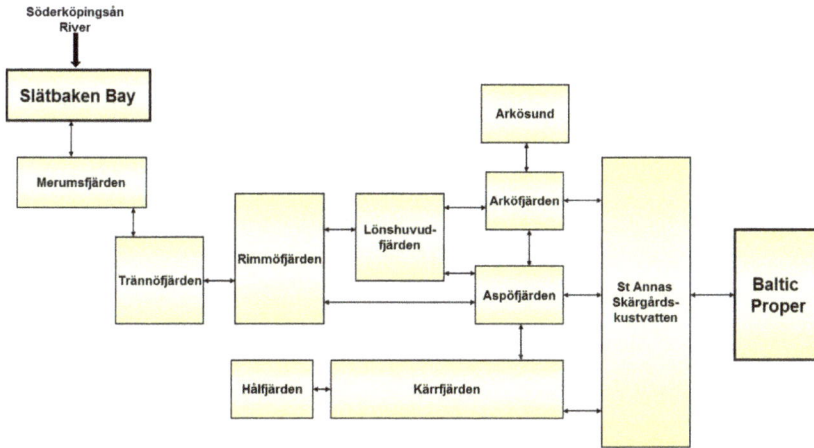

Figure 2. The CZM set-up in the St Anna archipelago, showing the Slätbaken Bay and the water exchange with surrounding basins and eventually the Baltic Proper.

The CZM has been calibrated against available oceanographical data in the area, received from the SHARK database http://vattenwebb.smhi.se/, which is the national host of marine environmental monitoring data. The calibration was mostly done by studying the correlation between observed and simulated salinity, temperature and oxygen conditions since these reflect if the transports and mixing in the model are described correctly.

2.4. Model Experiment

After the calibration, the models were run for the time period 2000–2009 with consideration to various measures for nutrient reduction. The models were then used as laboratories, experimenting by changing model input while keeping all other variables constant. The model response was explored by changing one factor at a time and then with combined factors. In all, 10 different scenarios of land-based reduction measures were simulated. The measures were suggested by the local water authorities at the County Board as means to fulfil the environmental status targets of the Slätbaken Bay. Remedial measures to reduce nutrient emissions were addressing several societal sectors (Table 1).

Table 1. Scenarios of reduced nutrient contribution from land-based sources and sea concentrations, respectively, which were used in the experiment on effects on water quality in the Skälderviken Bay.

Societal Sector	Scenario No.	Description
Waste Water Treatment Plants (WWTP)	1	Removal of N by 80% and maximum concentration of P by 0.2 mg/L
	2	Removal of N by 80% and maximum concentration of P by 0.1 mg/L
	3	Complete removal of the two largest WWTP by using a pipe to a nearby town with discharge to another bay
Rural households	4	200 rural households connected to WWTP
	5	All present (1882) rural households connected to WWTP
Wetlands	6	56 constructed wetlands of 1 ha each
	7	56 constructed wetlands of 0.5 ha each
Agriculture	8	Protection zones along the river wherever possible
	9	Reduced total load from arable land by 10% P and 20% N
Combined measures	10	Combination of the most efficient land-based measures in (*i.e.*, scenario No: 3 + 5 + 6 + 8 + 9) and the BSAP targets for the Baltic Proper (−3% N and −27% P).
International agreements on measures for the Baltic Proper	11	Reduction in sea boundary conditions by −10% of N and P
	12	Reduction in sea boundary conditions by −20% of N and P
	13	Reduction in sea boundary conditions by −40% of N and P
	14	Reduction in sea boundary conditions by −80% of N and P

A combined scenario was tried, using the most efficient land-based measures for each sector, *i.e.*, complete removal of waste water treatment plants (WWTP), connection of all rural households, 56 wetlands of 1 ha each, reduced load from agriculture, and BSAP [15] target levels for the Baltic Proper. The latter corresponds to reduced concentrations in the sea boundary conditions by −3% N and −27% P, respectively. Finally, the impact from only reducing concentrations in the boundary conditions of the sea was tested by reductions of concentrations in the Baltic Proper (*i.e.*, scenario 11–14).

3. Results and Discussion

The modelled spatial pattern of nutrient concentrations in surface water clearly reflects the influence of land cover in the catchment and dilution by sea water in the Archipelago (Figure 3). River water from forested areas show much lower concentrations than the agricultural rivers (*cf.* Figure 1) and the influence of lakes as nutrient traps is significant. The more dark green areas are all found down-stream of lakes, which illustrates the low travel-times and rapid transport from land to sea from these areas. The semi-enclosed Slätbaken Bay is greatly affected by the inflow from the Söderköpingsån River, showing much higher concentrations than the nearby more open coastal basins to the right in each figure.

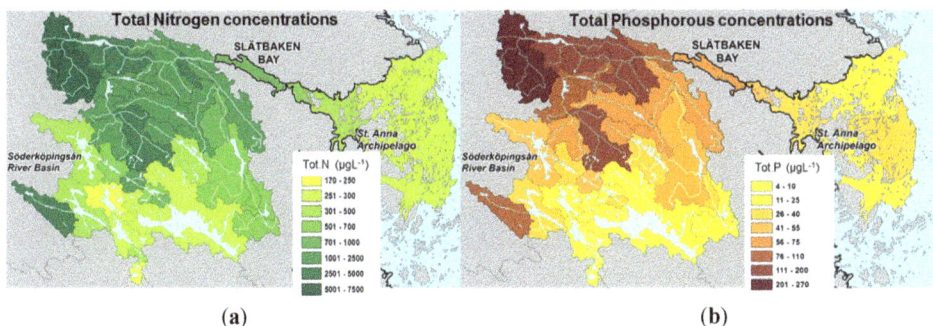

Figure 3. Spatial distribution of modelled surface-water concentrations of total N (**a**) and Total P (**b**) in the Söderköpingsån catchment, the Slätbaken Bay and St. Anna Archipelago (average values for all depths are shown).

The coupled models thus clearly illustrates going from low concentrations in the head waters to higher concentration in agricultural and more populated lowlands, leading to high coastal concentrations, which are being rapidly decreased with distance from the shore towards the sea.

3.1. Model Performance

The model performance of water discharge in the three sites, using the HYPE model, gave an average NSE of 0.72 (best = 0.83 and poorest = 0.62). Model results for river flow and nutrient concentrations showed similar level of agreement with observations during the calibration and validation periods (Table 2). Although the same set of parameters was used in the whole catchment, the model captured the very different characteristics between the small agricultural catchment (Ryttarbacken) and the larger catchment close to the outlet with a high lake percentage (Figure 4). At all sites, however, the model underestimated the highest peaks. The simulations for N agreed better with the observations than those for P. The model underestimates the highest P peaks implying that the description of processes concerning P needs to be improved. To address these problems, the HYPE model has been further developed and the next model version includes a new process description of organic nutrient turnover, which results in particulate P also in soil water. This will give P concentrations of particulate P that are more in agreement with observations.

Table 2. HYPE model performance in the Söderköpingsån catchment (cc = correlation coefficient, re = relative error; all in %). Cal. = calibration period: 2000–2009 and Val. = validation period: 1990–1999.

Variable, Station	cc, Cal.	re, Cal.	cc, Val.	re, Val.
Q, Hälla	85	8	90	10
Q, Ryttarbacken	83	−8	79	−9
Q, Söderköping	91	−1	86	−10
Mean, Q	*86*	*0*	*85*	*−3*
N, outlet Byngaren	46	−6	48	−15
N, outlet Strolången	62	−1	56	−2
N, inlet Hällerstadsjön	45	−8	56	5
N, Söderköping	59	12	51	0
Mean, N	*53*	*−1*	*53*	*−3*
P, outlet Byngaren	25	14	20	−15
P, outlet Strolången	32	9	31	−1
P, inlet Hällerstadsjön	65	−16	53	−4
P, Söderköping	37	9	40	10
Mean, P	*40*	*4*	*36*	*−2*

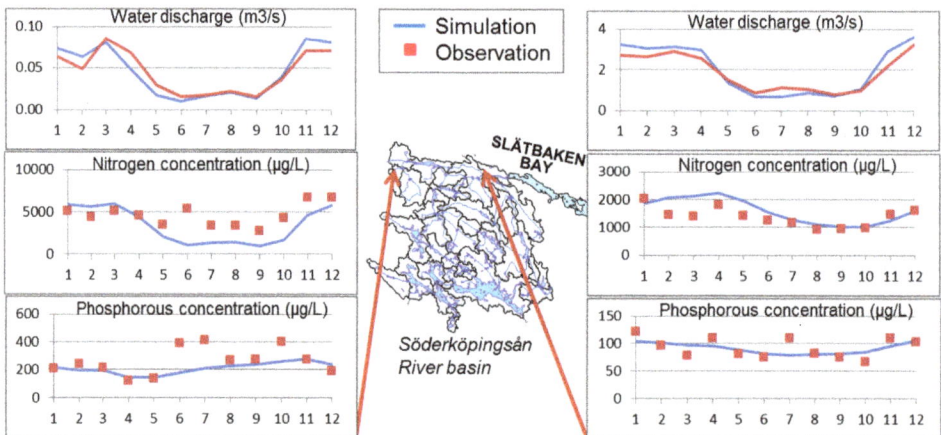

Figure 4. Daily averages of HYPE simulations *vs.* monthly averages of observations for water discharge, total N and P concentrations for Ryttarbacken, a small subcatchment dominated by arable land, and in Storån River, close to the outlet (1–12 in *x*-axis indicates the months January to December).

The coastal zone model was evaluated in the surface layer in two subbasins; Slätbaken Bay and Kärrfjärden. The model performance shows a good correlation for salinity in both subbasins (Figure 5). For Slätbaken Bay, the simulations for N and P are well correlated although the highest peaks are, most prominently for P, underestimated which is a reflection of the underestimation in the modelled load from land. In the outer basin, Kärrfjärden, the simulations match observation rather well concerning the range, although the dynamics are not totally described by the CZM. However, the relative error is less in this basin. The CZM in the Archipelago is more influenced by boundary conditions from the outer sea, which are determined by using a three dimensional sea model including data assimilation. Hence, the model performance at this observation point is biased by data in the model results.

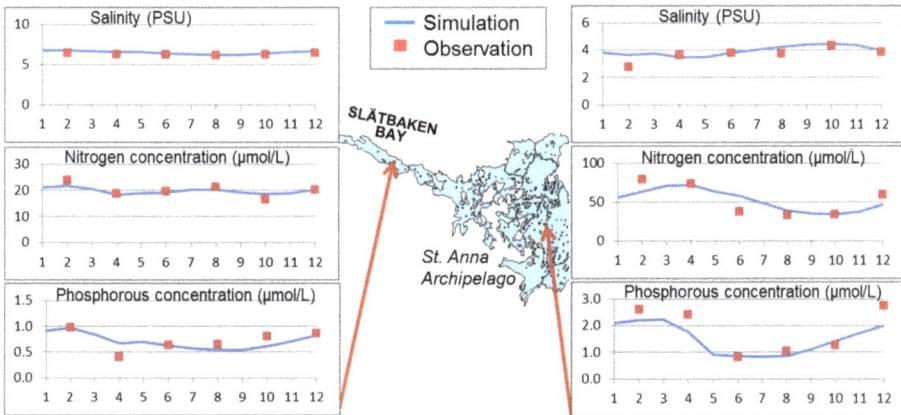

Figure 5. Daily averages of CZM simulations *vs.* bi-monthly observations of salinity, total N and P concentrations in the surface layer for Slätbaken Bay and Kärrfjärden (1–12 in *x*-axis indicates the months January to December).

Overall, the model shows the same seasonal dynamic as the observations, which reflect the biogeochemical cycle in the ecosystem; phytoplankton consume nutrients during spring and summer, they grow and are grazed by zooplanktons, until they die during autumn and are decomposed to nutrients again and create the large winter storage of soluble nutrients, which is to be digested during next spring by phytoplankton growth again. This is described in the model and confirmed by observations.

3.2. Source Apportionment and Mass Balance

Long-term mass-balance of all sources and sinks in the model are compiled along the flow paths to estimate the contribution from each sector and catchment at the river outlet. This is done by book keeping of all emissions and transformations in the model storages, assuming that equal part from all sources is transformed in each storage and time step. When knowing the rate of transformation in each sink, the origin of the load at the outlet can be traced through the flow paths upstream. Modelled contributions from various inland sources to the Slätbaken Bay indicate that 75% or more of the riverine nutrients originates from arable land in the catchment

147

(Figure 6). Waste Water Treatment Plants (WWTP) emissions correspond to 8% of the total N and 2% of the total P river inflow to the coast.

Simulated long-term annual means in the Slätbaken Bay show that about equal nutrient contributions were received from land and sea (Figure 7). The removed amounts of nutrients refer to sedimentation processes of P and sedimentation or denitrification of N. Approximately 90% of the N removal in the Slätbaken Bay refers to denitrification and thus leaves the water as atmospheric N gas. In return, the atmospheric deposition to the water surface is about 10 tons of N/year and 0.1 tons P /year. During winter, almost no biomass remains in the water storage.

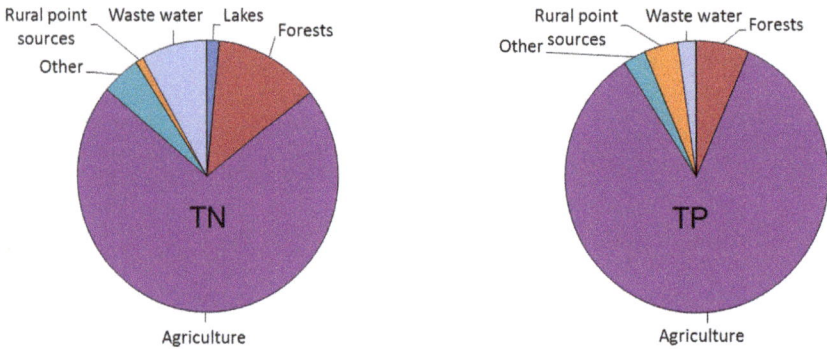

Figure 6. Long-term source apportionment for total N and total P in the Söderköpingsån River calculated with the HYPE-model.

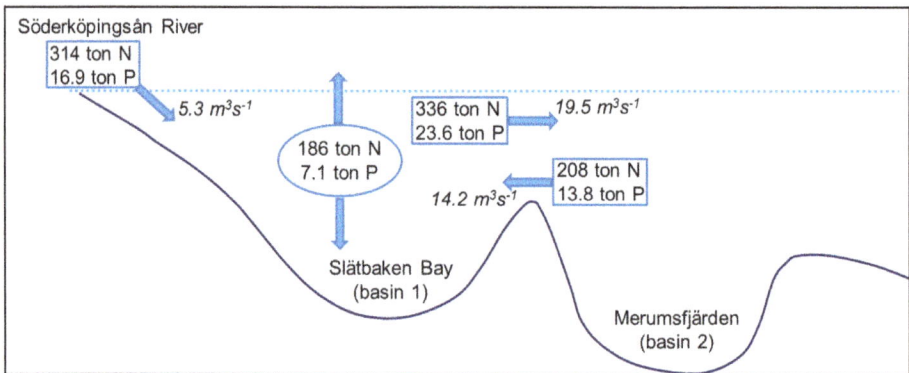

Figure 7. Long term annual means for total N and total P fluxes in the Slätbaken Bay calculated with the coastal zone model.

3.3. Scenario Results

The numerical experiments for reduced emissions from WWTP show that the suggested reduction levels reduced the total load by 3%–6% for N and 2% for P (bars

148

1–3 in Figure 8). However, reduced emissions from rural households gave higher impact on P, up to 4% (bar 5 in Figure 8). Protection zones did not give any significant impact, while reducing arable land leaching by 20% gave 15% reduction on riverine N load and 8% reduction on P load at the outlet (bar 9 in Figure 8). The lower effect at the outlet than at the sources is due to processes taking place within the river system and on the land surface, such as N denitrification in discharge areas and surface water, and sedimentation and absorption processes in bottom sediments. The combination of reductions of both WWTP emissions and arable land leaching gave the highest effect in land-based reductions and is illustrated in bar 10 in Figure 8. The resulting concentrations in the Slätbaken Bay was then reduced by 8% for P and almost 15% for N.

The corresponding reductions in Slätbaken Bay of either land-based measures or measures in the Baltic Proper were lower due to dilution and biological processes taking place within the coastal zone (lower panel of Figure 8). It was notable that 80% load reduction in the Baltic Proper sources corresponded to 15%–20% load reduction from land-based sources, to achieve about the same effect on nutrient concentrations in Slätbaken Bay. Hence, it is much more efficient to reduce the load of the river to improve the status of the Slätbaken Bay.

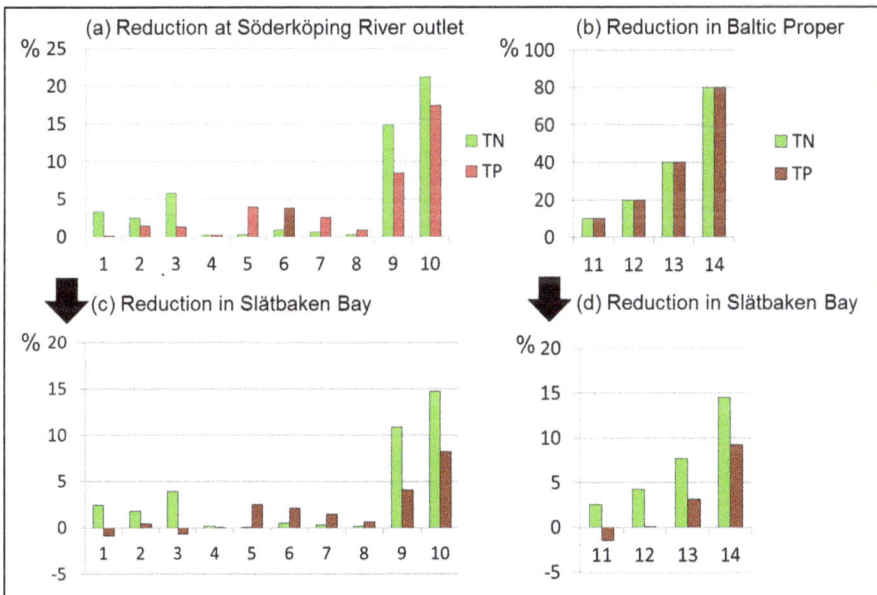

Figure 8. Simulated effect of various remedial measures in total nutrient load from land-based sources (**a**) and reduction in the Baltic Proper (**b**) and their corresponding effect on concentrations in the Slätbaken Bay (**c**) and (**d**). Bars correspond to Scenarios No. 1–14 in Table 1.

Comparing the modelled reductions with the reduction targets set-up by the local water authorities to fulfil the WFD, it was found that no measures reached the goals, neither for N nor for P (Figure 9). The achieved effect reached only half the summer target value for the concentration of N and one fourth for the concentration of P, whereas for the winter target values, they were only reached by 1/3 for N and 1/5 for P. It should be noted that some scenarios (Nos. 1, 3 and 11) actually resulted in increased P concentrations due to a shift in the internal biogeocycling of nutrients in the Bay (Figure 8). This can happen if one nutrient is reduced more than the other so that the algae up-take is restricted or if the bottom sediments start to leach due to changes in redox potentials.

The combined scenario (No. 10) was used to explore the integrated effect of the BSAP target and the best composite of land-based actions. This resulted in about 30%–50% achievement of the desired reduction of nutrients in the Slätbaken Bay (Figure 10). The land-based measures accounted for the largest part of this reduction. The reduction was still only about 1/3 of the goal for P, although all the best land-based actions were combined. It seems unrealistic to achieve roughly three times as much reduction on the contribution from land than what is already included in the combination of best land-based actions. It can thus be questioned whether the selected local environmental goal for Slätbaken Bay is realistic and if this definition of "good ecological status" for this Bay will ever be reached.

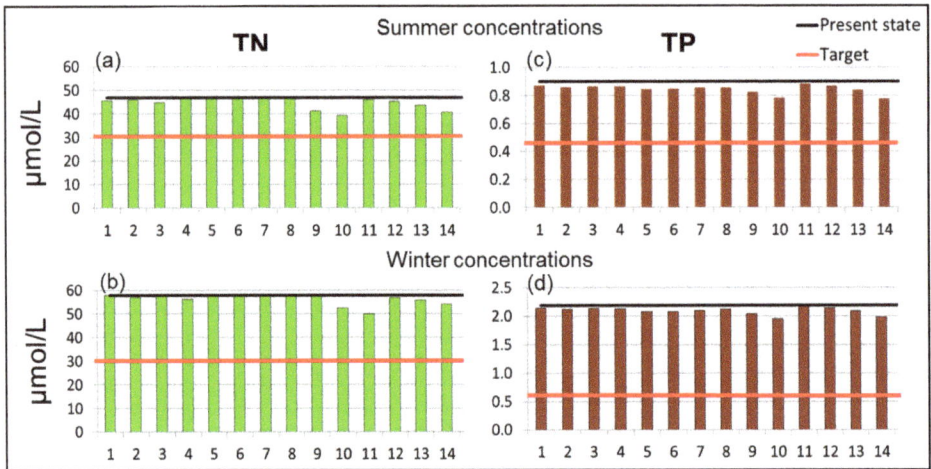

Figure 9. The effect of remedial measures on summer and winter concentrations of total N (**a, b**) and total P (**c, d**) in the surface water of Slätbaken Bay during summer (June to August) and winter (December to February). The x-axis corresponds to scenario No. 1–14 in Table 1. Horizontal lines indicate present conditions and the target for good ecological status according to the WFD.

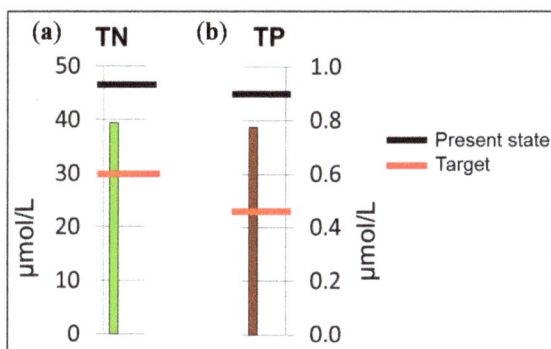

Figure 10. The effect when combining the BSAP target for the Baltic Proper and the most efficient land-based actions as found in this study. The simulated concentrations of total N (**a**) and P (**b**) are compared to the present state and the target for good ecological status (horizontal lines) according to the WFD.

3.4. Implications of Findings and Uncertainties

The numerical experiment clearly shows the importance of good decision-support before implementing remedial measures and environmental quality targets. The coupled hydro-ecological models indicate that the present targets may actually be unrealistic to be achieved for the Slätbaken Bay. It also shows that local land-based measures can be important for semi-enclosed bays. However, exact values from the experiment results should be taken with caution. All models are based on assumptions, both in the model structure itself, in input data generation, and in estimation of model coefficients, constants and parameters. For instance, Arheimer *et al.* [21] examined some major uncertainties when using HYPE at various scales. The most uncertain part of the catchment modelling is processes concerning soil leakage, how agricultural practices vary in time, and the assumptions on soil and sediment storages. The overall results mainly reflect the water budget of the Slätbaken Bay, for which the results are considered to be rather robust. Just analysing the mass balance of water gives a clear picture of dominating fluxes; in this case the exchange with the Baltic Proper is relatively small as there are many basins to cross with potential for denitrification and sedimentation before the water and nutrients from the sea reaches the Slätbaken Bay. Thereby, locally generated load from land will have a larger impact.

According to the experiment, local measures for the arable land gave the largest effect on nutrient reduction. Nevertheless, it should be recognised that 20% reduction of total load from the arable land may exceed the potential for reduction. A large part of the soil leaching is natural and the agriculture in the region is not very intensive. In previous studies of southern Sweden with much more intensive agricultural production, the potential reduction for the anthropogenic load from this sector was estimated to at most 15% [27,28]. It can thus be suspected that the suggested reduction of 20% by the regional authorities is too optimistic. According to HELCOM, most effort ought to be put on the agricultural sector as it is supposed to increase with the recent expansion of the European Union in the South-Eastern part of the Baltic basin. On the other hand, Humborg *et al.* [29], showed that an improvement of sewage treatment in these countries may compensate for the larger nutrient load from agriculture, at least as far as P is concerned.

When simulating the scenarios with reduced load from the Baltic Proper, it was not taken into account by which measures this could be achieved. Gren *et al.* [30] showed that a reduction in nutrient load to the Baltic Sea by 50% would bring its status back to what it was 1960 before the high increment in loads. The 80% reduction scenario thus seems unrealistic but was included to test the sensitivity of sea concentration on the semi-enclosed bay. For credibility, it is also important to recognise additional on-going changes, for instance climate change. Arheimer *et al.* [31] found that the BSAP targets on reductions from land-based sources may be reached in a future climate for the Baltic Proper if emissions from both agriculture and WWTP are reduced. In fact, impact from climate change may be beneficial for N reduction due to increase in denitrification by higher temperatures and longer residence times of water in the southern part of the region. However, there were large differences between the climate projections in this respect, which show the large uncertainties in climate impact assessments.

Even though there are uncertainties involved when experimenting with hydro-ecological models, it clearly shows that local analyses of mass-balance are important when improving the status of semi-enclosed bays. Although uncertain, the models are capable of separating between large and small fluxes. According to the results for the Slätbaken Bay, neither the efforts according to WFD nor the MSFD will improve the water quality as much as the local authorities wish to achieve "good ecological status". This conclusion is probably valid in spite of model uncertainties, and can be justified by examining the mass balance and exchange between different parts of the system. The modelling should thus serve as a platform for knowledge transfer and understanding of dominant pressures for a specific site, more than providing exact numbers for planning.

3.5. On the Practical Use of Hydro-Ecological Models for Societal Needs

Water authorities are currently asking for hydro-ecological information across administrative borders e.g., [32] while most model developers claim that their models should be used in practice, also for water pollution management e.g., [33–35]. There is a contemporary movement in the hydrological research community to emphasise societal needs and practical applications, pointing out "Science in Practice" as one major target for the up-coming scientific decade e.g., [11,36]. However, it is not yet clear which role the scientists aim at towards societal actors, and as Pielke [37] argued, it can be as a (i) pure scientist out of decision context; (ii) science arbiter answering inquiries without interfering with the questions; (iii) issue advocate providing solutions; or (iv) honest broker trying to explain several cause-effect in relation to the broader context. The latter is probably appealing for many scientists but it is not easy to achieve as it requires good understanding of the stakeholders knowledge, setting and commission. There is a need for a common learning process, which may take long time and much efforts, before the practitioner and the scientist fully understand the vocabulary and actual needs.

For instance, the coupled model system HYPE-CZM described in this study was in an early version [38] embedded in a graphical user interface and provided to the Swedish water authorities. They were asking for a model tool to help them plan nutrient reduction measures on a catchment scale. However, although the tool fulfilled all the user requirements and regular training courses were arranged, they did not use the model as it was far too complex and slow for everyday work at the authorities. It was an expert system delivered to infrequent users in a multi-task environment. Ten years later, they received a simple web-based tool at http://vattenwebb.smhi.se/ with an emulated model for catchment analysis that is not as precise and with very little functionality. Nevertheless, it gives a rough estimate, it is easy and quick to use, and it is much appreciated. The results differ only by some 10% compared to the more complex role-model (S-HYPE). So, for practical use, a simple model was preferred and the more complex model is nowadays used only by experts to provide new data, assessments and for research.

The HYPE-CZM model system (and its earlier versions) has also been applied to facilitate dialogues and participatory processes among various stakeholders to elaborate on management plans [12,13,39]. The model then served as a platform for the establishment of a common view of present conditions and the causes behind these conditions. The benefits were found to be twofold: it increased the willingness to carry out remedies or necessary adaptations to a changing environment, and it increased the level of understanding between the various stakeholder groups and therefore ameliorated the potential for future conflicts. Compared to traditional use of model results in environmental decision-making, the experts' role was transformed from a one-way communication of final results to assistance in the various steps of the participatory process. Hence, to use the vocabulary from Pielke [37], the scientists evolved from the role of science arbiters to honest brokers. The participatory process, however, is time- (and cost-) consuming and may thus not be feasible to implement at the large scale.

It was interesting to note that model performance, which is so much in focus in the discussions among scientists, was less important for model credibility among the stakeholders. The use of local input data was essential for confidence [40] and giving explanations for discrepancies was more convincing than the "best fit" between models and measurements [39]. Another way to improve confidence among practitioners is to present results from an ensemble of models. This is the common procedure at operational forecast institutes, where results from several models are considered before warnings of floods and droughts are issued e.g., [41–43]. To sum up, based on our experience, we recommend the following for practical use of hydro-ecological models when addressing societal needs:

1. Simple scenario tools (based on emulated hydro-ecological models) for rough and quick results on the web for water authorities working with WFD and MSFD;
2. Participatory modelling involving experts and stakeholders in critical conflict areas;
3. Hydro-ecological models should primarily be used for research to provide new knowledge on process interactions and dominant drivers under specific conditions, which is the basis for simplified models;
4. An ensemble of different models should be used for more reliable decision support.

4. Conclusions

This paper illustrates the societal relevance of hydro-ecological modelling as the study shows how coupled models can be used to evaluate environmental targets and improve policy making for complex systems. Two hydro-ecological models were coupled in a case study to experiment with various nutrient reduction measures and to explore the effect in a semi-enclosed Bay *vs.* local environmental targets set-up by water authorities. In addition, the practical use of such models was discussed based on previous experience. The study shows that:

1. For nutrient concentrations in the Slätbaken Bay, the reduction of land-based load by 15%–20% corresponded to 80% reduction of concentrations in the Baltic Proper. Local measures are thus recommended for this semi-enclosed Bay, as they have the largest potential to being implemented.
2. The best effect was achieved when combining measures for WWTP and agriculture, both locally (in the river catchment) and internationally (for the Baltic Proper). However, implementation of both the MSFD (BSAP targets) and the most efficient combined land-based actions suggested by the WFD, is not enough to achieve the locally established environmental quality targets for this bay. The present water quality targets thus seem unrealistic.
3. To overcome problems when using hydro-ecological models in practice for societal needs, we recommend the following: (i) emulate hydro-ecological

models into simple scenario tools for water authorities working with WFD and MSFD; (ii) involve both experts and stakeholders in participatory modelling of critical conflict areas; (iii) use hydro-ecological models by experts only; (iv) use an ensemble of different models for more reliable decision support.

Acknowledgments: We would like to thank Jörgen Sahlberg, SMHI, for valuable insights in coastal zone modelling and Kristina Isberg for helping with GIS illustrations. We also thank two anonymous reviewers for valuable comments on the manuscript. The present work was (partially) developed within the framework of the Panta Rhei Research Initiative of the International Association of Hydrological Sciences (IAHS). The work is further a demonstration of how to apply tools from the Water Management Programme of SMHI Core Services. Data from S-HYPE and CZM can be inspected and downloaded for free at http://vattenwebb.smhi.se/.

Author Contributions: Berit Arheimer contributed with designing the experiment, putting it into a context, result compilation and analyses, figures and maps, and writing of the final manuscript. Johanna Nilsson contributed with coupled modelling using both HYPE and CZM, computational runs of the experiment, result compilation, figures and writing of results. Göran Lindström contributed with data collection, calibrating the catchment model, result compilation and commenting the manuscript.

Conflicts of Interest: The authors declare no conflict of interest.

References

1. Meybeck, M. Riverine quality of the Anthropocene: Propositions for global space and time analysis, illustrated by the Seine River. *Aquat. Sci.* **2002**, *64*, 376–393.

2. Falkenmark, M. Water—A Reflection of Land Use: Understanding of Water Pathways and Quality Genesis. *Int. J. Water Resour. Dev.* **2011**, *27*, 13–32.

3. Vörösmarty, C.J.; McIntyre, P.B.; Gessner, M.O.; Dudgeon, D.; Prusevich, A.; Green, P.; Glidden, S.; Bunn, S.E.; Sullivan, C.A.; Reidy Liermann, C.; *et al.* Global threats to human water security and river biodiversity. *Nature* **2010**, *467*, 555–561.

4. Billen, G.; Garnier, J.; Ficht, A.; Cun, C. Modeling the response of water quality in the Seine River estuary to human activities over the last 50 years. *Estuaries* **2001**, *24*, 977–993.

5. Neal, C.; Davies, H. A summary of river water quality data collected within the Land-Ocean Interaction Study: Core data for Eastern UK rivers draining to the North Sea. *Sci. Total Environ.* **2000**, *251*, 585–665.

6. Rabalais, N.N.; Turner, R.E.; Justic, D.; Dortch, Q.; Wiseman, W.J., Jr.; Sen Gupta, B.K. Nutrient changes in the Mississippi River and system responses on the adjacent continental shelf. *Estuaries* **1996**, *19*, 386–407.

7. Humborg, C.; Conley, D.J.; Rahm, L. Silicon retention in River basins: Far reaching effects on biogeochemistry and aquatic foodwebs in coastal marine environments. *Ambio* **2000**, *29*, 45–50.

8. Seitzinger, S.P.; Kroeze, C.; Bouwman, A.F.; Caraco, N.; Dentener, F.; Styles, R.V. Global patterns of dissolved inorganic nitrogen and particulate nitrogen inputs to coastal systems: Recent conditions and future projections. *Estuaries* **2002**, *25*, 640–665.

9.   Silgram, M.; Schoumans, O.F.; Walvoort, D.J.; Anthony, S.G.; Groenendijk, P.; Strömqvist, J.; Bouraoui, F.; Arheimer, B.; Kapetanaki, M; Lo Porto, A.; *et al.* Subannual models for catchment management: Evaluating model performance on three European catchments. *J. Environ. Monit.* **2009**, *11*, 526–539.

10.   Donnelly, C.; Arheimer, B.; Capell, R.; Dahné, J.; Strömqvist, J. Regional overview of nutrient load in Europe—Challenges when using a large-scale model approach, E-HYPE. In Proceedings of the H04, IAHS-IAPSO-IASPEI Assembly, Gothenburg, Sweden, 22–26 July 2013; pp. 49–58.

11.   Hipsey, M.R.; Arheimer, B. Challenges for water-quality research in the new IAHS decade on: Hydrology under societal and environmental change. In Proceedings of the H04, IAHS-IAPSO-IASPEI Assembly, Gothenburg, Sweden, 22–26 July 2013; pp. 17–30.

12.   Alkan Olsson, J.; Jonsson, A.; Andersson, L.; Arheimer, B. A model supported participatory process: A socio-legal analysis of a bottom up implementation of the EU Water Framework Directive. *Int. J. Agric. Sustain.* **2011**, *9*, 379–389.

13.   Andersson, L.; Alkan-Olsson, J.; Arheimer, B.; Johnsson, A. Use of participatory scenario modelling as platforms in stakeholder dialogues. HELP special edition. *Water SA* **2008**, *34*, 439–447.

14.   Directive 2000/60/EC of the European Parliament and of the Council of 23 October 2000 Establishing a Framework for Community Action in the Field of Water Policy (Water Framework Directive). *Off. J. Eur. Commun.* **2000**, *43*, 1–73.

15.   Helsinki Commission (HELCOM). Towards a Baltic Sea unaffected by eutrophication. In Proceedings of the HELCOM Ministerial Meeting, Krakow, Poland, 15 November 2007.

16.   Baltic Marine Environment Protection Commission. Overview of implementation of the HELCOM Baltic Sea Action Plan (BSAP). 2013. Available online: http://helcom.fi/Documents/Ministerial2013/Associated%20documents/ Supporting/BSAP_Overview_with%20cover.pdf (accessed on 9 June 2015).

17.   Helsinki Commission Baltic Marine Environment Protection Commission. Approaches and Methods for Eutrophication Target Setting in the Baltic Sea Region. Baltic Sea Environment Proceedings No. 133. Available online: http://www.helcom.fi/Lists/ Publications/BSEP133.pdf (accessed on 9 June 2015).

18.   Schernewski, G.; Friedland, R.; Carstens, M.; Hirt, U.; Leujak, W.; Nausch, G.; Neumann, T.; Petenati, T.; Sagert, S.; Wasmund, N.; *et al.* Implementation of European marine policy: New water quality targets for German Baltic waters. *Mar. Policy* **2015**, *51*, 305–321.

19.   Lindström, G.; Pers, C.; Rosberg, J.; Strömqvist, J.; Arheimer, B. Development and testing of the HYPE (Hydrological Predictions for the Environment) water quality model for different spatial scales. *Hydrol. Res.* **2010**, *41*, 295–319.

20.   Sahlberg, J. *The Coastal Zone Model*; SMHI Reports Oceanography No. 98; Swedish Meteorological and Hydrological Institute: Norrköping, Sweden, 2009.

21.   Arheimer, B.; Dahné, J.; Donnelly, C.; Lindström, G.; Strömqvist, J. Water and nutrient simulations using the HYPE model for Sweden *vs.* the Baltic Sea basin—Influence of input-data quality and scale. *Hydrol. Res.* **2012**, *43*, 315–329.

22. Strömqvist, J.; Arheimer, B.; Dahné, J.; Donnelly, C.; Lindström, G. Water and nutrient predictions in ungauged basins—Set-up and evaluation of a model at the national scale. *Hydrol. Sci. J.* **2012**, *57*, 229–247.

23. Nash, J.E.; Sutcliffe, J.V. River flow forecasting through conceptual models part I—A discussion of principles. *J. Hydrol.* **1970**, *10*, 282–290.

24. Svensson, U. *Program for Boundary Layers in the Environment—System Description and Manual*; SMHI Reports Oceanography No. 24; Swedish Meteorological and Hydrological Institute: Norrköping, Sweden, 1998.

25. Omstedt, A.; Axel, L.B. Modelling the variations of salinity and temperature in the large Gulfs of the Baltic Sea. *Cont. Shelf Res.* **2003**, *23*, 265–294.

26. Eilola, K.; Meier, H.E.M.; Almroth, E. On the dynamics of oxygen and phosphorus and cyanobacteria in the Baltic Sea: A model study. *J. Mar. Syst.* **2009**, *75*, 163–184.

27. Arheimer, B.; Löwgren, M.; Pers, B.C.; Rosberg, J. Integrated catchment modeling for nutrient reduction: Scenarios showing impacts, potential and cost of measures. *Ambio* **2005**, *34*, 513–520.

28. Arheimer, B.; Torstensson, G.; Wittgren, H.B. Landscape planning to reduce coastal eutrophication: Constructed Wetlands and Agricultural Practices. *Landsc. Urban Plan.* **2004**, *67*, 205–215.

29. Humborg, C.; Mörth, C.-M.; Sundbom, M.; Wulff, F. Riverine transport of biogenic elements to the Baltic Sea—Past and possible future perspectives. *Hydrol. Earth Syst. Sci.* **2007**, *11*, 1593–1607.

30. Gren, I.-M.; Söderqvist, T.; Wulff, F. Nutrient Reductions to the Baltic Sea: Ecology, Costs and Benefits. *J. Environ. Manag.* **1997**, *51*, 123–143.

31. Arheimer, B.; Dahné, J.; Donnelly, C. Climate change impact on riverine nutrient load and land-based remedial measures of the Baltic Sea Action Plan. *Ambio* **2012**, *41*, 600–612.

32. Künitzer, A. Pan-European information needs on quality of freshwater. In *Understanding Freshwater Quality Problems in a Changing World*; Arheimer, B., Krysanova, V., Lakshmanan, E., Meybeck, M., Stone, M., Eds.; IAHS Press: Wallingford, UK, 2013; pp. 39–48.

33. Collins, A.L.; Strömqvist, J.; Davison, P.S.; Lord, E.I. Appraisal of phosphorus and sediment transfer in three pilot areas for the catchment sensitive farming initiative in England: Application of the prototype PSYCHIC model. *Soil Use Manag.* **2007**, *23*, 117–132.

34. Silgram, M.; Anthony, S.G.; Fawcett, L.; Stromqvist, J. Evaluating catchment-scale models for diffuse pollution policy support: Some results from the EUROHARP project. *Environ. Sci. Policy* **2008**, *11*, 153–162.

35. Crossman, J.; Whitehead, P.G.; Futter, M.N.; Jin, L.; Shahgedanova, M.; Castellazzi, M.; Wade, A.J. The interactive responses of water quality and hydrology to changes in multiple stressors, and implications for the long-term effective management of phosphorus. *Sci. Total Environ.* **2013**, *454*, 230–244.

36. Montanari, A.; Young, G.; Savenije, H.H.G.; Hughes, D.; Wagener, T.; Ren, L.L.; Koutsoyiannis, D.; Cudennec, C.; Grimaldi, S.; Bloschl, G.; *et al.* "Panta Rhei—Everything Flows": Change in hydrology and society—The IAHS Scientific Decade 2013–2022. *Hydrol. Sci. J.* **2013**, *58*, 1256–1275.

37. Pielke, R.A., Jr. *The Honest Broker: Making Sense of Science in Policy and Politics*; Cambridge University Press: Cambridge, UK, 2007.

38. Marmefelt, E.; Arheimer, B.; Langner, J. An integrated biogeochemical model system for the Baltic Sea. *Hydrobiologia* **1999**, *393*, 45–56.

39. Andersson, L.; Arheimer, B.; Gustafsson, D.; Lexén, K.; Glaumann, K. Potentials for numerical models in water management—Recommendations for local water management with stakeholder involvement. *VATTEN J. Water Manag. Res.* **2013**, *69*, 163–171.

40. Arheimer, B.; Andersson, L.; Alkan-Olsson, J.; Jonsson, A. Using catchment models for establishment of measure plans according to the WFD. *Water Sci. Technol.* **2007**, *56*, 21–28.

41. Cloke, H.L.; Pappenberger, F. Ensemble flood forecasting: A review. *J. Hydrol.* **2009**, *375*, 613–626.

42. Arheimer, B.; Lindström, G.; Olsson, J. A systematic review of sensitivities in the Swedish flood-forecasting system. *Atmos. Res.* **2011**, *100*, 275–284.

43. Pechlivanidis, I.G.; Bosshard, T.; Spångmyr, H.; Lindström, G.; Gustafsson, G.; Arheimer, B. Uncertainty in the Swedish Operational Hydrological Forecasting Systems. In Proeedings of the Second International Conference on Vulnerability and Risk Analysis and Management (ICVRAM2014), University of Liverpool, Liverpool, UK, 13–16 July 2014.

Chapter 2:
Water Flux Impact on Ecosystem Functions

The Stability of Revegetated Ecosystems in Sandy Areas: An Assessment and Prediction Index

Lei Huang and Zhishan Zhang

Abstract: The stability and sustainability of revegetated ecosystems is a central topic in ecological research. In this study, long-term monitoring and focused research on vegetation, soil and soil moisture from 2006 to 2012 were used to develop a model for evaluating indices of ecosystem stability using the analytical hierarchy process method. The results demonstrated that rainfall (R), vegetation coverage (C), and surface soil moisture (S) were the three most influential factors among the 14 indicators considered in a revegetated desert area in the Tengger Desert, China. A stability index (SI) was defined as $SI = VAR\ (R) \times VAR\ (C)/VAR\ (S)$, and a comparative study was conducted to examine the stability index of the natural vegetation community. The SI was divided into three regimes: $SI < 0.006$ was stable, $0.006 \leqslant SI < 0.015$ was semi-stable, and $0.015 \leqslant SI$ was unstable. The stable, semi-stable and unstable periods of revegetated ecosystems in our simulations were 191, 17 and 11 years, respectively, within the total modeling period of 219 years. These results indicated that the revegetated desert ecosystem would be stable in most years during the vegetation succession, and this study presents new ideas for future artificial vegetation management in arid desert regions.

Reprinted from *Water*. Cite as: Huang, L.; Zhang, Z. The Stability of Revegetated Ecosystems in Sandy Areas: An Assessment and Prediction Index. *Water* **2015**, *7*, 1969–1990.

1. Introduction

Stability has frequently been considered an important characteristic of ecological systems because of its theoretical and practical significance [1–4]. However, the concept of ecosystem stability is complex and has often been discussed in vague terms because of the complex physical and biological structures or integrated functions [5]. This complexity also occurs in artificial ecosystems [6]. Grimm and Wissel (1997) [7] presented a review and analyzed ecological stability, addressing 163 definitions of 70 different ecological stability concepts. Constancy, resilience, persistence, resistance, elasticity, mathematical stability and other concepts have been discussed in the literature in relation to specific problems in various ecosystems [8,9]. In these studies, debates on complexity-stability and diversity-stability have provided points of contention [1,10]. Early studies suggested that simple ecosystems were less stable

161

than complex ecosystems, but later studies came to the opposite conclusion [11]. Until recently, questions on the relationship between stability and complexity have not been answered [12–14]. However, for the diversity-stability debate, evidence from multiple ecosystems at a variety of temporal and spatial scales suggests that biological diversity acts to stabilize ecosystem functioning when presented with environmental fluctuations [1]. Moreover, variation among species in their response to such fluctuations is an essential requirement for ecosystem stability [15,16]. Thus, we can conclude that stability is a multi-dimensional concept that has scale-dependent features. Generally, ecosystem stability refers to the capacity of a natural system to apply self-regulating mechanisms and return to a steady state after an outside disturbance.

Ecologists have developed a variety of approaches to measure ecosystem stability [2,17–19]. In most cases, mathematical models or empirical methods are involved, with the former developed and expanded by Robert May in 1973 [20] using linear stability analysis on models constructed from a statistical universe (randomly constructed communities with randomly assigned interaction strengths). However, mathematically derived models are only suitable for characterizing the dynamic behavior of simple dynamic systems, whereas ecological systems are not usually uniform [21]. Most ecosystems operate in a variable environment that includes events at a wide range of frequencies and intensities, and it is often difficult to determine the degree of changes or disturbances. Empirical methods involve the development of a stability index that incorporates the main characteristics of ecosystem structure or environmental factors; the stability of an ecosystem can then be determined from these indicators [22,23]. However, this solution must still manage the problem of parameter selection [24]. For instance, in a forest ecosystem, stability may be described with biomass, diversity, dominant species density, nutrient cycling and soil characteristics, *etc.* Thus, developing a method that can combine quantitative measurements and qualitative descriptions is particularly pertinent when evaluating analyses of ecosystem stability. The best method for evaluating problems that involve a number of uncertain indices is the analytic hierarchy process (AHP) method [25–27]. The AHP is a multiple-criteria decision-making tool that has been widely applied in diverse fields, such as resource allocation, project design, maintenance management and policy evaluation. This method is particularly useful because it enables the decomposition of a given problem into a hierarchy of more easily comprehended sub-problems that can each be analyzed independently. The elements of the hierarchy can be related to any aspect of the qualitative and quantitative problems, including aspects that are tangible or intangible and carefully measured or roughly estimated. Once the hierarchy is built, the AHP systematically evaluates its various elements and derives numerical priorities for each of the decision alternatives [28,29]. Thus, one may easily apply the AHP to select promising technologies, and it provides a simple method for making decisions or performing environmental impact assessments [30–32]. However, relatively few studies have assessed the application of AHP to ecosystem stability [27,33,34], especially in revegetated desert areas.

Sand-binding vegetation has been widely used in arid desert regions and is considered one of the most effective methods of mitigating desertification [35]. However, in recent years, climate change and harsh natural environments have produced a series of problems, such as declining vegetation cover, poor plant growth and widespread water stress, which have led us to investigate the stability of artificial vegetation [6]. In this research, we hypothesized that the natural vegetation community is stable because natural vegetation in the study area is the result of long-term (*i.e.*, over thousands of years) evolution [36]. Therefore, this hypothesis is reasonable, and several studies have also illustrated that ecosystems with artificial vegetation would become stable over plant succession, with these stabilizing changes mainly reflected in the increased vegetation cover and soil texture improvements compared with that of the adjacent natural vegetation communities [37–39]. However, these improvements are only concomitant with the individual process of vegetation succession, and a comprehensive system for evaluating stability indices, including the integration of vegetation, soil and water factors, is still lacking. Furthermore, previous studies primarily focused on qualitative concept models, and a quantitative model is not available. In the present study, over 50 years of vegetation succession and long-term monitoring data (2006–2012) from the Shapotou Desert Research and Experimental Station on the southeastern edge of the Tengger Desert were used to investigate soil and vegetation characteristics at different stages of plant succession. The key influencing factors of the ecosystem stability were then determined with the AHP method, and a stability index was defined. Finally, the ecosystem's stability was predicted with a dynamical model of vegetation cover and soil moisture. Our results provide basic suggestions for sustainable ecosystem management and new hypotheses regarding vegetation succession models.

2. Materials and Methods

2.1. Study Site

The study was conducted at Shapotou Desert Research and Experimental Station of the Chinese Academy of Sciences, located in the Shapotou region at the southeastern margin of the Tengger Desert (37°32' N, 105°02' E). The climate at the site is characterized by abundant sunshine and low relative humidity. The average minimum monthly relative humidity is 33% in April, and the average maximum monthly relative humidity is 54.9% in August. The elevation of the area is 1330 m, and the mean annual precipitation is 188.2 mm according to meteorological records from 1956 to 2009, with rainfall occurring primarily between June and September [37]. The mean annual temperature is 9.6 °C, and the mean monthly temperatures are −6.9 °C in January and 24.3 °C in July. The evapotranspiration potential during the growing season (May to September) is 2300 to 2500 mm.

To ensure the smooth operation of the desert section of the Baotou-Lanzhou railway, a system involving sand-binding vegetation was established by the Chinese Academy of Sciences and other related departments in 1956. First, mechanical sand fences were installed at right angles to the prevailing winds, and then 1 m × 1 m straw sand barriers were erected in a checkerboard pattern behind the mechanical sand fences. Under non-irrigated conditions, xerophytic shrubs dominated by *Caragana korshinskii*, *Artemisia ordosica*, *Hedysarum scoparium*, *Caragana intermedia*, *Calligonum arborescens* and *Atraphaxis bracteata* were planted at a spacing of 1 m × 2 m or 2 m × 3 m using the checkerboard of straw barriers as a protective screen. This ecological shelter was extended in 1964, 1981 and 1987. As shown in Figure 1, a 16 km long protective system of vegetation was eventually established, and our research site was part of this protective system, which was 500 m wide on the north side and 200 m wide on the south side of the railway. Over the 50 years since the establishment of the vegetation, the environment in the area has improved, and the stabilized sand surface has created conditions that support the colonization of a number of species. The mass propagation of psammophytes has transformed the original moving sand into a complex man-made and natural desert vegetation landscape [37]. This ecological engineering project was viewed as a successful model for desertification control and ecological restoration along the transport line in the arid desert region of China. Because sites with different ages were stabilized using similar approaches, including the planting of shrub seedlings of the same species with the same density in similar straw checkerboards (see Table 1), they can represent the different successional stages of sand-binding vegetation.

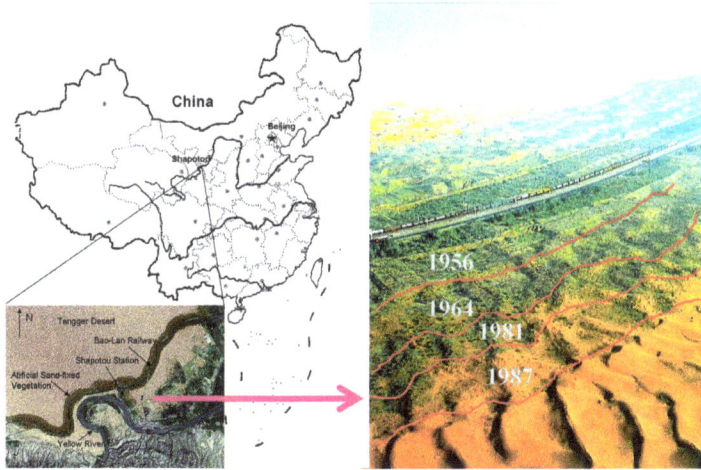

Figure 1. Schematic description of the study site (revegetated in 1956a, 1964a, 1981a and 1987a) in the Shapotou region of the Tengger Desert, northwestern China.

Table 1. Description of four revegetation sites with different ages and a natural community located in the southeastern fringe of the Tengger Desert, Northern China.

Year of Revegetation	Approaches to Sand Stabilization and Revegetation	Remaining Shrub Species of Revegetation	Native/Invasion Dominant Plant Species
1956	Straw-checkerboard of 1 m^2 planted with 10 xerophytic shrubs at a density of 16 individuals per 100 m^2	*Artemisia ordosica, Caragana korshinskii, Hedysarum scoparium*	*Artemisia ordosica, Scorzonera mongolica, Sonchus arvensis, Chloris virgata, Aristida adscensionis, Setaria viridis, Bassia dasyphylla, Chenopodium aristatum*
1964	Straw-checkerboard of 1 m^2 planted with 10 xerophytic shrubs at a density of 16 individuals per 100 m^2	*Artemisia ordosica, Caragana korshinskii, Hedysarum scoparium*	*Artemisia ordosica, Bassia dasyphylla, Eragrostis poaeoides, Sonchus arvensis, Scorzonera mongolica, Euphorbia humifusa*
1981	Straw-checkerboard of 1 m^2 planted with 10 xerophytic shrubs at a density of 16 individuals per 100 m^2	*Artemisia ordosica, Caragana korshinskii, C. microphylla, Hedysarum scoparium*	*Artemisia ordosica, Hedysarum scoparium, Bassia dasyphylla, Eragrostis poaeoides, Corispermum patelliforme*
1987	Straw-checkerboard of 1 m^2 planted with 10 xerophytic shrubs at a density of 16 individuals per 100 m^2	*Amorpha fruticosa, Artemisia ordosica, A. sphaerocephala, Caragana korshinskii, C. microphylla, Calligonum arborescens, Hedysarum scoparium*	*Hedysarum scoparium, Agriophyllum squarrosum, Bassia dasyphylla, Echinos gmelinii, Eragrostis poaeoides*
Natural	No	No	*Artemisia ordosica, Caragana korshinskii, Lespedeza davurica, Ceratoides latens, Oxytropis aciphylla, Stipa breviflora, Carex stenophylloides, Cleistogenes sogorica, Allium mongolicum, Oxytropis myriophylla, Enneapogon brachystachyus, Asparagus gobicus*

165

2.2. Methods

2.2.1. Sampling Method and Data Collection

Three 10 m × 10 m quadrats were established in each of the fixed observation plots in the sand-binding vegetation districts established in previous years (1956, 1964, 1981 and 1987) as well as in the adjacent natural vegetation zones for a total of 15 quadrats. The plant species number, height, and coverage for each species in the sand stabilization areas of different years were recorded or measured monthly from 2006 to 2012. Plant diversity (H') was estimated with the Shannon–Wiener index for each region and month according to the formula: $H' = -\sum p_i ln p_i$, where p is the proportion of each species i. Precipitation was recorded every 30 min using tipping bucket-type rain gauges (Casella) and Campbell CR30X data-loggers (Campbell Scientific, Logan, UT, USA). Samples were collected monthly with a soil auger, and the soil moisture of the samples was determined using the oven-drying method (0–40 cm) and neutron moisture probe method (40–300 cm). To avoid confusion between the surface and deep soil moisture, the gravimetric moisture content of the surface soil layer and bulk density measurements were used to calculate the volumetric moisture content.

Soil parameters were measured at depths of 0–20 cm at each site during the growing season in 2006 and 2010. In each plot, 100 soil sampling points were mechanically arranged in 10 m × 10 m vegetation plots of different ages. The transverse and longitudinal spacing were both 1 m. The surface of the sampling plots was flat, and composite samples were sieved through a 2 mm mesh screen and used for further analysis. Particle size was analyzed using the pipette method [40], and soil bulk density was measured using the ring-cutting method [41]. Soil organic carbon (SOC) was determined according to the dichromate oxidation method of Walkley–Black [42]. Total nitrogen was measured with a Kjeltec System 1026 Distilling Unit (Tecator AB, Hoganas, Sweden), and electrical conductivity (EC) was determined by preparing a suspension that consisted of a soil–water mixture in a ratio of 1:5 and was measured using a portable conductivity meter (Cole-Parmer Instrument Company, Vernon Hills, IL, USA). Topographic parameters (elevation, slope angle and aspect) were determined with a Real-Time Kinematic (RTK) global positioning system (GPS) (S86T, Southern Technology, Guangzhou, China) in 2006.

In this study, we chose the commonly applied method of space-for-time substitution, which assumes that simultaneous sampling of different sites of different ages is equivalent to resampling the same site through time [36]. The clay percentage was selected to represent the soil texture, and soil organic matter and total N content were used to reflect the soil nutrient regime. According to the depth distribution of the mass root systems of herbaceous plants and shrubs, the soil moisture content at 0–40 cm and 40–300 cm, respectively, were determined. The clay percentage was measured because soil texture is considered an important factor that determines the vegetation structure and composition under uniform climatic conditions, whereas soil moisture was measured because it is considered a driving force for ecological processes in arid zones. Thus, these indices can reflect the overall stability of the ecosystem [38].

2.2.2. Analytic Hierarchy Process Methodology

The AHP method was applied to select the major influencing factors from the above datasets. The AHP procedure involves three basic steps: (i) Design of the decision hierarchy; (ii) Pair-wise comparison of elements of the hierarchical structure; and (iii) Construction of an overall priority rating. For more details, refer to Appendix 5. In this study, fourteen important criteria were selected to evaluate the stability of the revegetated desert ecosystem (Figure 2). The top level of the diagram shows the overall goal of the hierarchy, "stability of revegetated desert ecosystems"; the second level lists the most influential factors obtained from other literature [36,37], such as soil moisture, soil characteristics, plant cover and topography; and the third level describes the attributes of each factor. After defining the criteria for selecting the evaluating index, five comparison matrices were developed: A–B, B1–C, B2–C, B3–C and B4–C. At each level, the criteria were compared pairwise according to their levels of influence and according to the specified criteria at the higher level. In AHP, multiple pairwise comparisons are based on a standardized comparison scale of nine levels, and ten experts were asked to perform pair-wise comparisons using a 1–9 preference scale that indicates the importance or dominance of one element over another. On this preference scale, 1 indicates equal preference and 9 indicates absolute preference. Intermediate values are used to express increasing preference/performance for one weight/alternative [43]. For example, if the criteria for soil moisture (B1) were judged as essential or of moderate importance, then the soil criteria (B2) with respect to the preservation of revegetated desert ecosystem stability would be given a score of 3. In addition, for other pairwise comparisons matrices, such as B2–C, the degree of importance was determined by the number of years of recovery required to reach the level of native ecosystem, such as desert steppe [36]. All of the important factors were then assigned appropriate weights, and a standardization index was calculated with the Z-score method. Finally, the integrated index was calculated, and the most influential factors in the revegetated ecosystem were determined.

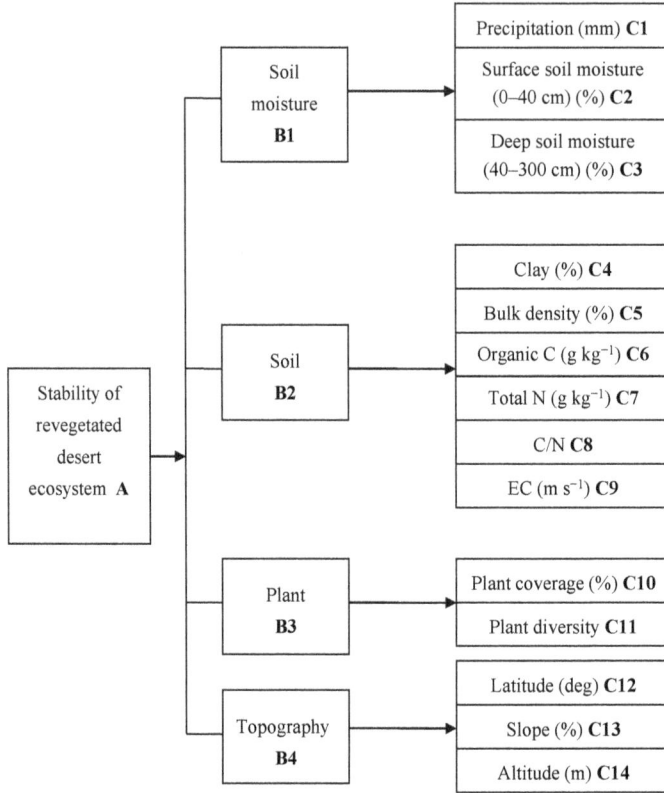

Figure 2. A hierarchy for the stability evaluation of revegetated desert ecosystems.

2.2.3. Coupled Dynamics of Soil Moisture and Vegetation

Soil water is the major driving force shaping vegetation patterns and processes in desert areas; however, plant growth, vegetation succession or landscape variability may also generate temporal and spatial heterogeneity of soil moisture [6]. Thus, to predict dynamic changes in soil water and vegetation succession in arid desert regions, a simplified ecohydrological box model that contains the coupled dynamics of vegetation and soil moisture was selected [44]. The model is as follows:

$$\frac{ds}{dt} = I\left(s, t\right) - \left[bx_b\left(s\right) + \left(1 - b\right)x_0\left(s\right)\right] \tag{1}$$

$$\frac{db}{dt} = g\left(s\right)b\left(1 - b\right) - u\left(s\right) \tag{2}$$

In Equation (1), s is the relative soil moisture averaged over the root zone $(0 \leqslant s \leqslant 1)$ and b is the fraction of vegetated sites, where $0 < b < 1$. Variations in site occupancy depend on the rate at which empty sites are colonized and the rate at which vegetated sites become vacant as a result of mortality. $I(s, t)$ is the infiltration rate (mm day^{-1}); X_b and X_0 denote evapotranspiration (mm day^{-1}), which is distinct between vegetated and bare sites; $g(s)$ denotes the colonization rate; and $u(s)$ denotes the local extinction rate (year^{-1}). All of the aforementioned rates were dependent on soil moisture. Each item in Equation (1) and the above parameter values are detailed in Appendix 5. Matlab 7.0 (The MathWorks, Natick, MA, USA) and Origin 7.0 (OriginLab, Northampton, MA, USA) software were used for data simulation and analysis.

3. Results

3.1. Results of the Analytic Hierarchy Process (AHP) Application

Following the input of each factor and its importance into the expert choice and quantitative approach functions, the comparison results and weights of the four main criteria were calculated, and they are shown in Table 2. The results (principal vectors) show that the attributes have the following approximate priority weights: Soil moisture (0.91), soil (0.22), plant cover (0.23), and topography (0.28). The consistency ratio (CR) for this comparison was 0.086 < 0.10, which indicated that the weighted results were valid and consistent. In contrast, if the CR value were larger than the acceptable value of 0.10, the matrix results would be inconsistent and exempt from further analysis. Similarly, pair-wise comparisons of the sub-criteria indices with respect to the four criteria in the middle level were conducted. The comparison and weight results for this level are shown in Tables 3–6. The final stage of the AHP was to compute the contribution of each index to the overall goal, and the global weights were calculated by multiplying the local weights with criteria and sub-criteria. The final weights and ranking of the indices were then obtained as shown in Table 7. The ranking of critical ecosystem stability factors showed that soil moisture and plant factors are the most influential. The highest-ranked factor was precipitation (0.86), followed by plant coverage (0.25) and surface soil moisture (0.23). The CR values of all matrices were less than 0.10 and were therefore accepted. The largest value in the priority weight was the most important criterion, which means that precipitation, surface soil moisture and plant coverage were the three most influential factors determining the stability of the revegetated desert ecosystem.

Table 2. Judgment matrix of the objective hierarchy and the criterion hierarchy.

A	B1	B2	B3	B4	Priorities	AHP Criteria
B1	1.00	3.00	5.00	4.00	0.91	$\lambda_{max} = 4.23$;
B2	0.33	1.00	0.50	1.00	0.22	$CI = 0.078$;
B3	0.20	2.00	1.00	0.50	0.23	$RI = 0.900$;
B4	0.25	1.00	2.00	1.00	0.28	$CR = 0.086$

Notes: The weight of four evaluation criteria: Stability of revegetated desert ecosystem (A); Soil moisture (B1); Soil (B2); Plant (B3); Topography (B4).

Table 3. Judgment matrix of the criterion B1 and its related factors.

B1	C1	C2	C3	Priorities	AHP Criteria
C1	1.00	5.00	4.00	0.95	$\lambda_{max} = 3.09$; $CI = 0.047$;
C2	0.20	1.00	2.00	0.26	$RI = 0.58$; $CR = 0.081$.
C3	0.25	0.50	1.00	0.18	

Notes: The weight of three evaluation criteria: Soil moisture (B1); Precipitation (C1); Surface soil moisture (0–40cm) (C2); Deep soil moisture (40-300cm) (C3).

Table 4. Judgment matrix of the criterion B2 and its related factors.

B2	C4	C5	C6	C7	C8	C9	Priorities	AHP Criteria
C4	1.00	0.33	2.00	1.00	5.00	0.50	0.27	
C5	3.00	1.00	5.00	3.00	9.00	2.00	0.78	$\lambda_{max} = 6.04$;
C6	0.50	0.20	1.00	0.50	3.00	0.33	0.15	$CI = 0.008$;
C7	1.00	0.33	2.00	1.00	5.00	0.50	0.27	$RI = 1.24$;
C8	0.20	0.11	0.33	0.20	1.00	0.14	0.06	$CR = 0.006$.
C9	2.00	0.50	3.00	2.00	7.00	1.00	0.47	

Notes: The weight of nine evaluation criteria: Soil (B2); Clay (C4); Bulk density (C5); Organic C (C6); Total N (C7); C/N (C8); EC (C9).

Table 5. Judgment matrix of the criterion B3 and its related factors.

B3	C10	C11	Priorities	AHP Criteria
C10	1.00	5.00	0.98	$\lambda_{max} = 2$; $CI = 0$;
C11	0.20	1.00	0.20	$RI = 0.00$; $CR = 0$.

Notes: The weight of two evaluation criteria: Plant (B3); Plant coverage (C10); Plant diversity (C11).

Table 6. Judgment matrix of the criterion B4 and its related factors.

B4	C12	C13	C14	Priorities	AHP Criteria
C12	1.00	1.00	2	0.63254	$\lambda_{max} = 3.02$; $CI = 0.008$;
C13	1.00	1.00	3	0.72389	$RI = 0.58$; $CR = 0.013$.
C14	0.5	0.33	1	0.27546	

Notes: The weight of three evaluation criteria: Topography (B4); Latitude (C12); Slope (C13); Altitude (C14).

Table 7. Overall weight of the ecosystem stability evaluation index.

Indices	B1 0.91	B2 0.22	B3 0.23	B4 0.28	Overall Priorities
C1	0.95	–	–	–	0.86
C2	0.26	–	–	–	0.23
C3	0.18	–	–	–	0.16
C4	–	0.27	–	–	0.06
C5	–	0.78	–	–	0.18
C6	–	0.15	–	–	0.03
C7	–	0.27	–	–	0.06
C8	–	0.06	–	–	0.01
C9	–	0.47	–	–	0.11
C10	–	–	0.98	–	0.25
C11	–	–	0.20	–	0.05
C12	–	–	–	0.63	0.18
C13	–	–	–	0.72	0.20
C14	–	–	–	0.28	0.08

$CI = 0.052; RI = 0.955; CR = 0.055 < 0.1$

3.2. Stability Index Definition, Measurement and Prediction

The above AHP analysis verified that precipitation (*C1*), surface soil moisture (*C2*) and plant coverage (*C10*) were the three main limiting factors that affected the stability of the ecosystem; the stability index (*SI*) was therefore defined as $SI = VAR (C1) \times VAR (C10) / VAR (C2)$. *VAR* denotes the variance of a random variable. Generally, we assumed that the natural vegetation has higher stability; thus, the greater the similarity of different successional stages of sand-binding vegetation, the higher its stability. As shown in Figure 3A–D, plant coverage and surface soil moisture depended on annual precipitation. In particular, the maximum annual rainfall in 2007 was 271.2 mm, which induced extensive plant growth; in contrast, in 2006, 2009 and 2010, the annual rainfall was only approximately 100 mm, and plant coverage and surface soil moisture were maintained at a relatively low level. However, the amplitude of plant coverage was less than that of surface soil moisture in different vegetative sites. The *SI* in the natural vegetation site was 0.005, and it was 0.011, 0.012, 0.013, 0.017 in the 1956a, 1964a, 1981a and 1987a vegetation sites, respectively. The above results suggest that with increasing years of sand-binding vegetation, the revegetated desert ecosystem would become more stable. However, when compared with the natural vegetation, ecological restoration in arid desert regions still occurred over a very long time scale.

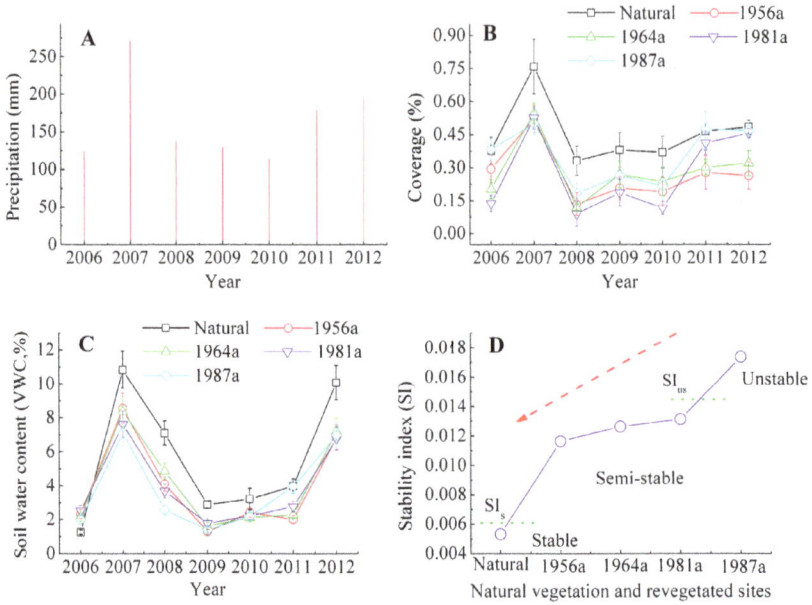

Figure 3. (**A**) Annual precipitation; (**B**) plant coverage; (**C**) soil moisture and (**D**) stability index of different vegetative sites and the control; The horizontal dashed lines represents the threshold of ecosystem stability, $SI_s = 0.004$, $SI_{us} = 0.015$.

At long time scales, we have simulated 219 years of rainfall, vegetation coverage and soil moisture changes from Equation (1), as shown in Figure 4A–C. Numerical results have shown that soil moisture and vegetation cover would increase in the future and that the soil moisture would be maintained at 2.5%, but the rate of increase was not very high. The vegetation coverage remained at 40%, with fluctuations, but it was still increasing, and the rate of increase of vegetation coverage was larger than that of soil moisture. The SI also fluctuated with ecosystem succession, most of which was closer to the natural vegetation, as shown in Figure 3D. Based on the above-measured data and its distribution patterns, the SI can be divided into several regimes, which we defined as follows: $SI < 0.006$ was stable, $0.006 \leqslant SI < 0.015$ was semi-stable and $0.015 \leqslant SI$ was unstable. Thus, the stable, semi-stable and unstable periods in our simulations were 191, 17 and 11 years within the total 219 years, respectively, which indicated that the revegetated desert ecosystem would be stable the majority of the time.

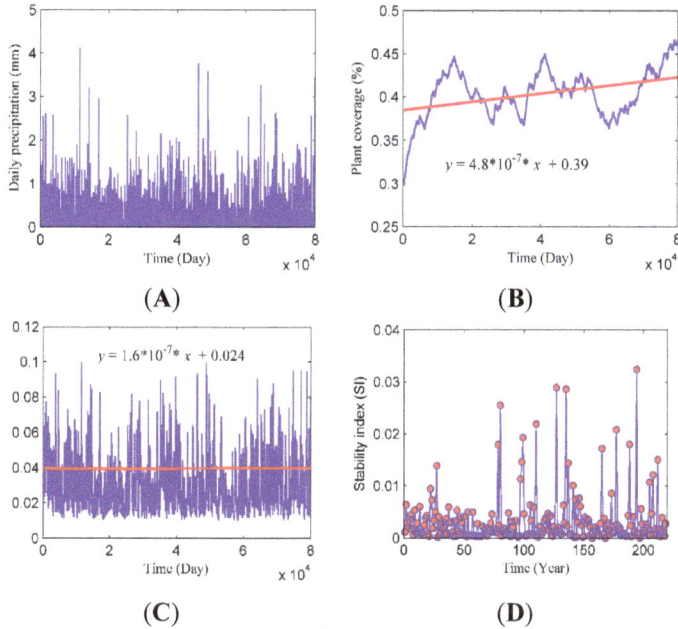

Figure 4. (A) Simulated results of daily precipitation; (B) plant coverage; (C) soil moisture and (D) stability index using Equation (1) over 219 years.

4. Discussion

Stability is an important indicator of ecosystem health and a necessary condition for ecosystem sustainability [4]. The assessment of ecosystem stability is helpful in revealing community dynamics and vegetation succession [2]. Debate is ongoing regarding the definition of stability [1], and each researcher has individual evaluation methods, which is because of regional differences or specific research objectives [14]. Therefore, in the present study, ecosystem stability was measured with the comparative method, and natural vegetation ecosystems were considered stable. In these ecosystems, natural vegetation indicators, such as plant coverage or soil moisture, were assumed to represent the standard, and ecosystems that are closer to this standard are more stable. This approach avoids theoretical controversy and provides a simple method for practical applications. In revegetated desert ecosystems, soil moisture, soil characteristics, plant cover and topography were selected as the integral influential factors for ecosystem stability. For the selection of useful variables, the AHP method was used because it effectively incorporates interdependent criteria and local problems involving both quantitative and qualitative issues. A key drawback in using the AHP method is the requirement of pair-wise comparison that must be completed by experts because such expert judgments may be affected by factors, such as fatigue and impatience during this process, especially when a

173

large number of criteria or alternatives are involved [45]. To avoid this drawback, a reasonable and manageable number of criteria were contained in the model. Another alternative is the Bayesian inference [46], which works very well if opinions among the experts are strongly divergent, and different prior parameters can be used to test for robustness [47,48]. In this study, based on results from the literature and information gleaned from discussions with ten experts, who are authorities on eco-hydrology in arid desert areas and were unified in their feedback, the AHP method was chosen in this study, and a total of fourteen criteria were determined. Using the AHP method, we verified that precipitation, surface soil moisture and plant coverage were the key limiting factors for artificial vegetation stability in arid desert regions. Precipitation was considered the sole source of water replenishment in this area, and soil water was the main driving force of the ecosystem's patterns and processes. Additionally, the changes in the vegetation patterns also affected the redistribution of precipitation and infiltration depth in different soil layers. Therefore, these ecohydrological processes and their feedback mechanisms were identified as the main problems affecting the restoration and reconstruction of certain ecological engineering projects in arid regions [6]. However, for other ecosystems (such as agro-ecosystems, grasslands or forests), ecosystem structure and function are more complicated than in revegetated desert ecosystems [36–38]; therefore, evaluating ecosystem stability is more difficult [3,4,6]. Thus, a Bayesian methodology that provides for semiautomatic searches of consensus building should be considered instead of AHP.

To quantify the stability indicators, we defined the *SI* as the integration of variance of three variables. Because precipitation and surface soil moisture were inextricably linked, they always varied at the same time. Therefore, we defined the ratio of the two as a coefficient, and then multiplied it by the variance of vegetation cover, which directly represents the ecosystem restoration or degradation within a certain period. Thus, the stability of the ecosystem is specific, quantitative and verifiable. Simulated results from Equation (1) have shown that the ecosystem may become unstable in years with high precipitation because years with high annual precipitation are often followed by several years of continuous drought [49,50], which dramatically changes soil moisture and vegetation cover. This greater fluctuation leads to instability of the revegetated ecosystem. Compared with other empirical stability indices, such as community numbers, biomass or plant diversity [51], our stability index is simple and practical. Furthermore, it reflects the intrinsic characteristics of ecosystem stability, such as stochastic dynamics and temporal dependence, and accurately reflects the entire ecosystem with real environmental fluctuations. The *SI* values of different vegetation sites were then compared with the natural vegetation community, and stable or unstable regimes were determined. Based on field observations, the results showed that revegetated ecosystems increase in stability over time after the establishment of sand-binding vegetation. This conclusion is also supported by other studies in which the establishment of planted sand-binding vegetation in the Shapotou region is suggested to promote the improvement and restoration of regional habitats and provide suitable conditions for an increase in biodiversity in the desert ecosystem [37].

In terms of the mechanism by which sand-fixing vegetation promotes stability in the Tengger Desert, studies present inconsistent results on the formation mechanisms of stable plant communities. Shen (1986) [52] considered that *A. ordosica* may form a relatively stable climax community or plagioclimax, with the originally planted shrubs, such as *C. korshinskii*, *H. scoparium* and *A. ordosica*, degrading nearly 20 years later and withdrawing from the community. Zhao *et al.* (1988) [53] predicted that the next stage of vegetation succession would be herb-dominated plant communities. Li (2005) [38] advanced a conceptual model and stated that the revegetated plants would lead toward herb-dominated vegetation, which is similar to the primary vegetation types of the adjacent steppified desert and desert steppe. These results suggest that vegetation adapts to the habitat and revegetated ecosystems form a new equilibrium with vegetation succession. However, these previous studies were conceptual models, and vegetation succession assessments cannot be quantified. In the present study, a simplified dynamical model originally developed by Baudena *et al.* (2007) [44] that has been widely used in vegetation pattern analyses in arid and semi-arid areas [54,55] was applied, and it was capable of inferring the vegetation pattern features and useful information on underlying processes, including the susceptibility of the system to abrupt shifts to a desert state (*i.e.*, unvegetated) as a result of climate change or anthropogenic disturbances [56]. Through our model simulations and the division of stability intervals, we predict that sand-fixing vegetation ecosystems will remain stable for a long period of time, although this stability will be interspersed with a number of semi-stable and unstable years. For certain theories or proposed mechanisms for the maintenance of ecosystem stability, such as the diversity theory or redundancy theory [57], self-organization of plant behaviors during a particular period of time is an essential component of vegetation succession over long time scales. Similar to banded or spot vegetation in North America, Africa and Australia [58–60], we could hypothesize that such self-organization reflects the normal vegetation successional pattern at different stages and suggest that vegetation may follow a "banded-spot-banded-spot (...)" pattern. Therefore, variations in vegetation composition, structure and responses to hydrological processes in the sand-binding areas are necessary stages of natural succession [61]. Thus, in artificial vegetation ecosystem management, the human should not be overly interfered with.

5. Conclusions

In this study, long-term monitoring and focused research was used to develop, measure and evaluate an index of ecosystem stability. Using the AHP method, we verified that rainfall (R), vegetation coverage (C) and surface soil moisture (S) are the three most influential factors in a revegetated desert area in the Tengger Desert, China. Over short time scales, the stability of the revegetated sandy ecosystem increased with years of succession. However, the stability of the artificial vegetation ecosystem may fluctuate with vegetation succession on a timescale of hundreds of years. The revegetated desert ecosystem was mostly stable based on our established theoretical system of ecosystem assessment and prediction, thus verifying the success of this method for desertification control and ecological restoration along a transport line in arid desert regions. Furthermore, our results provide new ideas for future artificial vegetation management and sustainable development in arid revegetated desert areas.

Acknowledgments: This work was supported by the National Key Basic Research program (2013CB429905) and Chinese National Natural Scientific Foundation (41201084; 41201086; 31170385).

Author Contributions: Conceived and designed the experiments: Lei Huang; Performed the experiments: Zhishan Zhang; Analyzed the data: Lei Huang; Contributed reagents/materials/analysis tools: Zhishan Zhang; Wrote the paper: Lei Huang. All authors read and approved the final manuscript.

Appendix

A.1. Analytic process hierarchy (AHP) methodology

AHP is a multi-criteria analysis method that is based on pair-wise comparisons of the components of a particular problem. This method enables a complex problem to be broken into a goal, criteria, solutions and other levels to provide a simple method for decision-making. Applying the AHP procedure involves three basic steps [62]: (1) Decomposition, or the hierarchy construction; (2) Comparative judgments, or defining and executing data collection to obtain pair-wise comparison data on elements of the hierarchical structure and (3) Synthesis of priorities, or constructing an overall priority rating.

In the first stage, a complex decision problem is structured as a hierarchy. This structure comprises a goal or focus at the topmost level, multiple criteria that define alternatives in the middle, and decision alternatives at the bottom. The second step is the comparison of the alternatives and the criteria. Once the problem has been decomposed and the hierarchy is constructed, prioritization procedure starts in order to determine the relative importance of the criteria within each level. The pairwise judgment starts from the second level and finishes in the lowest level. In each level, a nominal scale is used for the evaluation. The scale used in AHP for preparing the pairwise comparison matrix is a discrete scale from 1 (the two choice options are equally preferred) to 9 (one choice option is extremely preferred over the other), as presented in Table A1.

Table A1. Evaluation Scale in AHP.

Intensity of Importance	Definition	Explanation
1	Equal Importance	Two elements have equal importance regarding the element in higher level
3	Moderate Importance	Experience or judgement slightly favours one element
5	Strong Importance	Experience or judgement strongly favours one element
7	Very Strong Importance	Dominance of one element proved in practise
9	Extreme Importance	The highest order dominance of one element over another
2,4,6,8	Compromises between the Above	When compromise is needed
Adverse	Adverse Comparisions	The adverse evaluation of the same criteria, adverse of the same point under multiplication

Elements in each level are compared in pairs with respect to their importance to an element in the next higher level. Starting at the top of the hierarchy and working down, the pair wise comparisons at a given level can be reduced to a number of square matrices $A = (a_{ii})_{n \times n}$ as in the following:

$$A = \begin{bmatrix} a_{11} & a_{12} & ... & a_{1n} \\ a_{21} & a_{22} & ... & a_{2n} \\ ... & ... & ... & ... \\ a_{n1} & a_{n2} & ... & a_{nn} \end{bmatrix} \quad a_{ii} = 1, \; a_{ji} = 1/a_{ij}, \; a_{ij} \neq 0$$

After all pair wise comparison matrices are formed, the vector of weights, $w = (w_1, w_2, ..., w_n)$ is computed on the basis of Sattys eigenvector procedure. The computation of the weights involves two steps. First, the pair wise comparison matrix $A = (a_{ii})_{n \times n}$ is normalized by Equation (A1) and then the weights are computed by Equation (A2).

$$a_{ii}* = \frac{a_{ij}}{\sum\limits_{i=1}^{n} a_{ij}}, \text{ for all } j = 1, 2, ..., n \tag{A1}$$

$$w_i = \frac{\sum\limits_{j=1}^{n} a_{ij}*}{n}, \text{ for all } j = 1, 2, ..., n \tag{A2}$$

Satty (1980) showed that there is a relationship between the vector weights, w and the pair wise comparison matrix, A, as shown in Equation (A3).

$$Aw = \lambda_{max}w \tag{A3}$$

The λ_{max} value is an important validating parameter in AHP and is used as a reference index to screen information by calculating the Consistency Ratio (CR) of the estimated vector. To calculate the CR, the Consistency Index (CI) for each matrix of order n can be obtained from Equation (A4).

$$CI = \frac{\lambda_{max} - n}{n - 1} \tag{A4}$$

Then, CR can be calculated using Equation (A5):

$$CR = \frac{CI}{RI} \tag{A5}$$

Where RI is the random consistency index obtained from a randomly generated pair wise comparison matrix. Table A2 shows the value of the RI from matrices of order 1 to 10 as suggested by Satty [28]. If $CR < 0.1$, then the comparisons are acceptable. If, however, $CR \geqslant 0.1$, then the values of the ratio are indicative of inconsistent judgments. In such cases, one should reconsider and revise the original values in the pair wise comparison matrix A.

Table A2. Random Indicators.

n	1	2	3	4	5	6	7	8	9	10
RI	0	0	0.58	0.90	1.12	1.24	1.32	1.41	1.45	1.49

A.2. The Coupled Soil-Vegetation Model

The coupled soil-vegetation system (B1) was original presented in Baudena *et al.* (2007) [44], and we have improved some items in the equation as following, which would give a better description in our research area.

$$\begin{aligned} \frac{ds}{dt} &= I(s,t) - [bx_b(s) + (1-b)x_0(s)] \\ \frac{db}{dt} &= g(s)b(1-b) - u(s)b \end{aligned} \tag{A6}$$

where s is relative soil moisture averaged over the root zone ($0 \leqslant s \leqslant 1$), The fraction of vegetated sites is denoted by $b(0 \leqslant b \leqslant 1)$. The infiltration rate $I(s,t)$ is assumed to be equal to the rainfall rate, as long as the soil layer is not saturated; when rainfall exceeds the available water storage in the soil, the excess is converted into surface runoff. Thus,

$$I = \begin{cases} \frac{r}{nZr} & \frac{r\Delta t}{nZr} < 1 - s \\ \frac{1-s}{\Delta t} & \frac{r\Delta t}{nZr} \geqslant 1 - s \end{cases} \tag{A7}$$

where $\Delta t = 1$ day. n is soil porosity and Zr is the active soil depth in millimeters. r is daily rainfall, which is modeled as instantaneous events occurring according to a marked Poisson process of rate (mean frequency of rainfall events) λ, and with exponentially distributed depths with mean h. $x_b(s)$ was the water losses from vegetated soil, which include direct soil evaporation and plant transpiration, s^* is the critical soil moisture value below which plants start reducing transpiration by closing their stomata, and s_1 is the soil field capacity above which leakage occurs. The losses from evapotranspiration are assumed to increase linearly as a function of s until the moisture reaches a threshold s^*, above which the evapotranspiration takes place at a maximum value E_{max}, when the soil moisture exceeds the soil field capacity s_1, the leakage losses was start by an exponential growth and reaching the saturated hydraulic conductivity k_s at s =1.

$$x_b(s) = \begin{cases} \frac{E_{max}}{s^*} s & 0 \leqslant s < s^* \\ E_{max} & s^* \leqslant s < s_1 \\ E_{max} + k_s \frac{e^{\beta(s-s1)} - 1}{e^{\beta(1-s1)} - 1} & s_1 \leqslant s < 1 \end{cases} \tag{A8}$$

$x_0(s)$ was the water losses in bare soil; E_{soil} is pure soil evaporation before the leakage occurs. It increases linearly up to field capacity s_1, above which the leakage losses was start with the same expression as for the vegetated soil.

$$x_b(s) = \begin{cases} \frac{E_{soil}}{s_1} s & 0 \leqslant s < s_1 \\ E_{soil} + k_s \frac{e^{\beta(s-s1)} - 1}{e^{\beta(1-s1)} - 1} & s_1 \leqslant s < 1 \end{cases} \tag{A9}$$

$$x_b(s) = \begin{cases} \frac{E_{soil}}{s_1} s & 0 \leqslant s < s_1 \\ E_{soil} + k_s \frac{e^{\beta(s-s1)} - 1}{e^{\beta(1-s1)} - 1} & s_1 \leqslant s < 1 \end{cases} \tag{A10}$$

the colonization and extinction rates $g(s)$ and $u(s)$ depend on s, as seen in the Figure A1, it can be modeled as:

$$g(s) = \frac{0.05s^2}{1 + 12.36s^2}, \quad u(s) = 0.0006e^{-4.69s} \tag{A11}$$

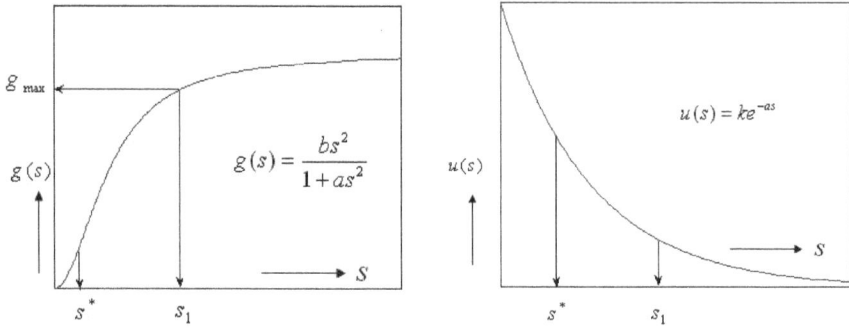

Figure A1. Sketch for vegetation colonization rate and extinction rate.

When the soil moisture was below the field capacity s_1, plant growth was slowly at first and then reached to its maximum g_{max} sharply with increased soil moisture, which have showed a typical "S" curve. The extinction rate is assumed to exponential decreased with soil moisture. The parameters for Matlab simulation were shown in Table A1.

Table A1. Parameters of the simplified ecohydrological box model.

Parameters	Symbol (Unit)	Value
Soil porosity	n	0.43
Active soil depth	Zr (cm)	40
Critical soil moisture below which plant undergoes water stress	s^*	0.11
Field capacity	$s1$	0.56
Pore size distribution parameter	β	12.7
Saturated hydraulic conductivity	Ks (cm/d)	800
Pure soil evaporation	E_{soil}(mm/d)	0.1
Maximum evapotranspiration rate	E_{max} (mm/d)	3.67
Average rainfall frequency	λ (/d)	0.15
Average precipitation depth	h (mm/d)	0.61

The parameters in the model were obtained from the directly measured data. The cutting ring method was used to determine soil porosity n and field capacity s_1 of root zone (3 repeats at the 0, 20 and 40 cm soil depths, respectively, and take the average). The saturated soil hydraulic conductivity K_s was measure by using of the tension infiltrometer model SW 080B (SDEC, Paris, France) in undisturbed field conditions; s* and another soil parameters β were determined according to related reference [63]. the historical precipitation information such as average rainfall frequency and depth were extracted from the receive data at a near weather station, and minor calibration was done by referring to the experiential relationship between precipitation and elevation. The depth of active soil or root zone depth, defined as the soil depth range in which 95% below-ground biomass were distributed, was determined by filed investigation. Pure soil evaporation E_{soil} were measured with micro-lysimeters, which were made using PVC pipes that were 30 cm high and 10 cm in internal diameter [64]. And maximum evapotranspiration rate E_{max} were determined with the combination of the stem heat balance technique (Dynamax Inc., Houston, TX, USA), the observations was measured continuous during the 2008-2012 growing season [65].

Conflicts of Interest: The authors declare no conflicts of interest.

References

1. McCann, K.S. The diversity-stability debate. *Nature* **2000**, *405*, 228–233.
2. Tilman, D.; Reich, P.B.; Knops, J.M.H. Biodiversity and ecosystem stability in a decade long grassland experiment. *Nature* **2006**, *441*, 629–632.
3. Allesina, S.; Tang, S. Stability criteria for complex ecosystems. *Nature* **2012**, *483*, 205–208.
4. Donohue, I.; Petchey, O.L.; Montoya, J.M.; Jackson, A.L.; McNally, L.; Viana, M.; Healy, K.; Lurgi, M.; O'Connor, N.E.; Emmerson, M.C. On the dimensionality of ecological stability. *Ecol. Lett.* **2013**, *16*, 421–429.
5. Hill, A.R. Ecosystem stability, some recent perspectives. *Prog. Phys. Geogr.* **1987**, *11*, 315–333.
6. Li, X.R.; Zhang, Z.S.; Huang, L.; Wang, X.P. Review of the ecohydrological processes and feedback mechanisms controlling sand-binding vegetation systems in sandy desert regions of China. *Chin. Sci. Bull.* **2013**, *58*, 1483–1496.
7. Grimm, V.; Wissel, C. Babel, or the ecological stability discussions, an inventory and analysis of terminology and guide for avoiding confusion. *Oecologia* **1997**, *109*, 323–334.
8. Cottingham, K.L.; Brown, B.L.; Lennon, J.T. Biodiversity may regulate the temporal variability of ecological systems. *Ecol. Lett.* **2001**, *4*, 72–85.
9. Deimeke, E.; Cohena, M.J.; Reissb, K.C. Temporal stability of vegetation indicators of wetland condition. *Ecol. Indic.* **2013**, *34*, 69–75.
10. Ives, A.R.; Carpenter, S.R. Stability and diversity of ecosystems. *Science* **2007**, *317*, 58–62.
11. Pimm, S.L. The complexity and stability of ecosystems. *Nature* **1984**, *307*, 321–326.
12. Haydon, D.T. Pivotal assumptions determining the relationship between stability and complexity: An analytical synthesis of the stability complexity debate. *Am. Nat.* **1994**, *144*, 14–29.
13. Townsend, S.E.; Haydon, D.T.; Matthews, L. On the generality of stability complexity relationships in Lotka-Volterra ecosystems. *J. Theor. Biol.* **2010**, *267*, 243–251.
14. Mougi, A.; Kondoh, M. Diversity of interaction types and ecological community stability. *Science* **2012**, *337*, 349–351.
15. Cleland, E.E. Biodiversity and ecosystem stability. *Nat. Education Knowl.* **2012**, *3*, 14.
16. Sterk, M.; Gort, G.; Klimkowska, A.; van Ruijven, J.; van Teeffelen, A.J.A.; Wamelink, G.W.W. Assess ecosystem resilience: Linking response and effect traits to environmental variability. *Ecol. Indic.* **2013**, *30*, 21–27.

17. Mitchell, R.J.; Auld, M.H.D.; Le Duc, M.G.; Marrs, R.H. Ecosystem stability and resilience, a review of their relevance for the conservation management of lowland heaths. *Perspect. Plant Ecol.* **2000**, *3*, 142–160.

18. Ives, A.R.; Dennis, B.; Cottingham, K.L.; Carpenter, S.R. Estimating community stability and ecological interactions from time series data. *Ecol. Monogr.* **2003**, *73*, 301–330.

19. Loreau, M.; de Mazancourt, C. Biodiversity and ecosystem stability, a synthesis of underlying mechanisms. *Ecol. Lett.* **2013**, *16*, 106–115.

20. May, R.M. Qualitative stability in model ecosystems. *Ecology* **1973**, *54*, 638–641.

21. Grimm, V.; Schmidt, E.; Wissel, C. On the application of stability concepts in ecology. *Ecol. Model.* **1992**, *63*, 143–161.

22. Wardle, D.A.; Bonner, K.I.; Barker, G.M. Stability of ecosystem properties in response to above-ground functional group richness and composition. *Oikos* **2000**, *89*, 11–23.

23. Hooper, D.U.; Chapin, F.S.I.; Ewel, J.J.; Hector, A.; Inchausti, P.; Lavorel, S.; Lawton, J.H.; Lodge, D.M.; Loreau, M.; Naeem, S.; *et al.* Effects of biodiversity on ecosystem functioning, a consensus of current knowledge. *Ecol. Monogr.* **2005**, *75*, 3–35.

24. Dale, V.H.; Beyeler, S.C. Challenges in the development and use of ecological indicators. *Ecol. Indic.* **2001**, *1*, 3–10.

25. Saaty, T.L. A scaling method for priorities in hierarchical structures. *J. Math. Psychol.* **1977**, *15*, 234–281.

26. Locantore, N.W.; Tran, L.T.; O'Neill, R.V.; McKinnis, P.W.; Smith, E.R.; O'Connell, M. An overview of data integration methods for regional assessment. *Environ. Monit. Assess.* **2004**, *94*, 249–261.

27. Zhang, R.Q.; Zhang, X.D.; Yang, J.Y.; Yuan, H. Wetland ecosystem stability evaluation by using analytical hierarchy process AHP, approach in Yinchuan Plain: China. *Math. Comput. Model.* **2013**, *57*, 366–374.

28. Saaty, T.L. *The Analytical Hierarchy Process: Planning, Priority Setting, Resource Allocation*; McGraw-Hill: New York, NY, USA, 1980.

29. Vaidya, O.S.; Kumar, S. Analytic hierarchy process, An overview of applications. *Eur. J. Oper. Res.* **2006**, *169*, 1–29.

30. Banai, R. Anthropocentric problem solving in planning and design, with analytic hierarchy process. *J. Archit. Plann. Res.* **2005**, *22*, 107–120.

31. Cheng, E.W.L.; Li, H.; Yu, L. The analytic network process (ANP), approach to location selection: A shopping mall illustration. *Onstr. Innov.* **2005**, *5*, 83–97.

32. Ho, W. Integrated analytic hierarchy process and its applications—A literature review. *Eur. J. Oper. Res.* **2008**, *186*, 211–228.

33. Barzekar, G.; Aziz, A.; Mariapan, M.; Ismail, M.H.; Hosseni, S.M. Using analytical hierarchy Pprocess (AHP), for prioritizing and ranking of ecological indicators for monitoring sustainability of ecotourism in northern forest: Iran. *Ecologia Balkanica* **2011**, *3*, 59–67.

34. Convertino, M.; Baker, K.M.; Vogel, J.T.; Lu, C.; Suedel, B.; Linkov, I. Multi-criteria decision analysis to select metrics for design and monitoring of sustainable ecosystem restorations. *Ecol. Indic.* **2013**, *26*, 76–86.

35. Le Houerou, H.N. Restoration and rehabilitation of arid and semiarid Mediterranean ecosystems in North Africa and West Asia, a review. *Arid Soil Res. Rehabil.* **2000**, *14*, 3–14.

36. Li, X.R.; He, M.Z.; Duan, Z.H.; Xiao, H.L.; Jia, X.H. Recovery of topsoil physiochemical properties in revegetated sites in the sand-burial ecosystems of the Tengger Desert: Northern China. *Geomorph* **2007**, *88*, 254–265.

37. Li, X.R.; Xiao, H.L.; Zhang, J.G.; Wang, X.P. Long-term ecosystem effects of sand-binding vegetation in the Tengger Desert: Northern China. *Restor. Ecol.* **2004**, *12*, 376–390.

38. Li, X.R. Influence of variation of soil spatial heterogeneity on vegetation restoration. *Sci. China Ser. D* **2005**, *35*, 361–370.

39. Eldridge, D.J.; Koen, T.B.; Harrison, L. Plant composition of three woodland communities of variable condition in the western Riverina, New South Wales, Australia. *Cunninghamia* **2007**, *10*, 189–198.

40. Loveland, P.J.; Walley, W.R. Particle size analysis. In *Soil and Environmental Analysis, Physical Methods*, 2nd ed.; Simth, K.A., Mullins, C.E., Eds.; Marcel Dekker: New York, NY, USA, 2001.

41. Nanjing Institute of Soil Research. *Analysis of Soil Physic-Chemical Features*; Shanghai Science and Technology Press: Shanghai, China, 1980.

42. Nelson, D.W.; Sommers, L.E. *Total Carbon, Organic Carbon and Organic Matter*. In *Methods of Soil Analysis Part 2: Chemical and Microbiological Properties*, 2nd ed.; Page, A.L., Miller, R.H., Keeney, D.R., Eds.; American Society of Agronomy, Soil Science Society of America: Madison, WI, USA, 1982; pp. 539–577.

43. Cinelli, M.; Coles, S.R.; Kirwan, K. Analysis of the potentials of multi criteria decision analysis methods to conduct sustainability assessment. *Ecol. Indic.* **2014**, *46*, 138–148.

44. Baudena, M.; Boni, G.; Ferraris, L.; von Hardenberg, J.; Provenzale, A. Vegetation response to rainfall intermittency in drylands, results from a simple ecohydrological box model. *Adv. Water Res.* **2007**, *30*, 1320–1328.

45. Lee, J.K.L.; Chan, E.H.W. The analytic hierarchy process AHP, approach for assessment of Urban renewal proposals. *Soc. Indicat. Res.* **2008**, *89*, 155–168.

46. Altuzarra, A.; Moreno-Jiménez, J.M.; Salvador, M. Consensus building in AHP-group decision making: A Bayesian approach. *Oper. Res.* **2010**, *58*, 1755–1773.

47. Cagno, E.; Caron, F.; Mancini, M.; Ruggeri, F. Using AHP in determining the prior distributions on gas pipeline failures in a robust Bayesian approach. *Reliab. Eng. Syst. Safe.* **2000**, *67*, 275–284.

48. Hahn, E.D. Decision making with uncertain judgments: A stochastic formulation of the analytic hierarchy process. *Decis. Sci.* **2003**, *34*, 443–466.

49. Mirza, M.Q.; Warrick, R.A.; Ericksen, N.J.; Kenny, G.J. Trends and persistence in precipitation in Ganges, Brahmaputra and Meghna river basins. *Hydrol. Sci. J.* **1998**, *43*, 845–858.

50. Wang, X.P.; Zhang, J.G.; Li, X.R.; Li, J.G. Distribution, trends and variability of precipitation in Shapotou region. *J. Desert Res.* **2001**, *4*, 260–264.

51. Lehman, C.L.; Tilman, D. Biodiversity, stability and productivity in competitive communities. *Am. Nat.* **2000**, *156*, 534–552.

52. Shen, W.S. The status of *Artemisia ordosica* in vegetation succession at Shapotou area. *J. Desert Res.* **1986**, *6*, 13–22.

53. Zhao, X.L. *Research on Problem of Controlling Sand of Vegetable*; Ningxia People's Publishing House: Yinchuan, China, 1988.

54. Baudena, M.; Provenzale, A. Rainfall intermittency and vegetation feedbacks in drylands. *Hydrol. Earth Syst. Sci.* **2008**, *12*, 679–689.

55. Baudena, M.; D'Andrea, F.; Provenzale, A. An idealized model for tree-grass coexistence in savannas. *J. Ecol.* **2010**, *98*, 74–80.

56. Borgogno, F.; D'Odorico, P.; Laio, F.; Ridolfi, L. Mathematical models of vegetation pattern formation in ecohydrology. *Rev. Geophys.* **2009**, *47*, RG1005.

57. Walker, B.H. Biodiversity and ecological redundancy. *Conserv. Biol.* **1992**, *6*, 18–23.

58. Klausmeier, C.A. Regular and irregular patterns in semiarid vegetation. *Science* **1999**, *284*, 1826–1828.

59. Rietkerk, M.; Dekker, S.C.; de Ruiter, P.C.; van de Koppel, J. Self-organized patchiness and catastrophic shifts in ecosystems. *Science* **2004**, *305*, 1926–1929.

60. Kefi, S.; RieterkI, M.; Alados, C.L.; Pueyo, Y.; Papanastasis, V.P.; ElAich, A.; de Ruiter, C. Spatial vegetation patterns and imminent desertification in Mediterranean arid ecosystems. *Nature* **2007**, *449*, 213–218.

61. Li, X.R. *Eco-hydrology of biological soil crusts in desert regions of China*; Higher Education Press: Beijing, China, 2012.

62. Lotfi, S.; Habibi, K.; Koohsari, M.J. An analysis of urban land development using multi-criteria decision model and geographical information system (A case study of Babolsar city). *Am. J. Environ. Sci.* **2009**, *5*, 87–93.

63. Laio, F.; Porporato, A.; Ridolfi, L.; Rodriguez-Iturbe, I. Plants in water-controlled ecosystems: active role in hydrologic processes and response to water stress—II Probabilistic soil moisture dynamics. *Adv. Water Resour.* **2001**, *24*, 707–723.

64. Zhang, Z.S.; Liu, L.C.; Li, X.R.; Zheng, J.G.; He, M.Z.; Tan, H.J. Evaporation properties of a revegetated area of the Tengger Desert North China. *J. Arid Environ.* **2008**, *72*, 964–973.

65. Huang, L.; Zhang, Z.S.; Li, X.R. The extrapolation of the leaf area-based transpiration of two xerophytic shrubs in a revegetated desert area in the Tengger Desert, China. *Hydrol. Res.* **2014**.

Catchment-Scale Modeling of Nitrogen Dynamics in a Temperate Forested Watershed, Oregon. An Interdisciplinary Communication Strategy

Kellie Vaché, Lutz Breuer, Julia Jones and Phil Sollins

Abstract: We present a systems modeling approach to the development of a place-based ecohydrological model. The conceptual model is calibrated to a variety of existing observations, taken in watershed 10 (WS10) at the HJ Andrews Experimental Forest (HJA) in Oregon, USA, a long term ecological research (LTER) site with a long history of catchment-scale data collection. The modeling framework was designed to help document and evaluate an evolving understanding of catchment processing of water, nitrogen, and carbon that has developed over the many years of on-going research at the site. We use the dynamic model to capture the temporal variation in the N and C budgets and to evaluate how different components of the complex system may control the retention and release of N in this pristine forested landscape. Results indicate that the relative roles of multiple competing controls on N change seasonally, between periods of wet/dry and growth/senescence. The model represents a communication strategy to facilitate dialog between disciplinary experimentalists and modelers, to produce a more complete picture of nitrogen cycling in the region. We view this explicit development of complete, yet conceptually simplified models as a useful and important way to evaluate complex environmental dynamics.

Reprinted from *Water*. Cite as: Vaché, K.; Breuer, L.; Jones, J.; Sollins, P. Catchment-Scale Modeling of Nitrogen Dynamics in a Temperate Forested Watershed, Oregon. An Interdisciplinary Communication Strategy. *Water* **2015**, *7*, 5345–5377.

1. Introduction

Human-induced changes to global biogeochemical cycles are large, having for example effectively doubled the global rate of annual N fixation [1] yet the ultimate response of ecosystems to these changes remains unclear. This lack of clarity is due in part to the inter-dependencies and feedbacks between system components as well as to an incomplete understanding of the complex processes governing biogeochemical cycles. These relationships result in the inherent complexity and non-linearity of environmental systems. Components comprising environmental systems include functionally discreet entities like the atmosphere, vegetation, and soils; spatial assemblages of these entities, including hillslopes, riparian transitions, and instream environments; and also the different elements that cycle through them, including nitrogen, carbon, water, and a host of others. The detailed study of each of these individual units tends to be well-covered by traditional disciplines, such as hillslope hydrology, aquatic ecology, soil biogeochemistry, forest ecology, *etc.*, but with less attention given to the relationships between components. That these dependencies, in some sense by definition, fall at interfaces between traditional disciplinary boundaries is an ongoing challenge for ecosystem science, though the emergence of cross-cutting disciplines, such as ecohydrology and hydrobiogeochemistry, is an indication that this challenge is being addressed. The use of numerical models, which are designed to capture relationships between system components, represents one method that can be used to contribute to our understanding of how ecosystems function, and how they may respond to future change in both climate and landcover.

The retention and release of N from temperate forested ecosystems provides a clear example of this need for understanding, at an integrated catchment scale. Its relevance is, of course, well established, with tremendous progress having been made throughout the last three decades [2–4]. Much of this progress has been driven by the need to better understand the response of forested catchments, which are commonly limited by N, to environmental changes, including climate change, disturbance, and increased N deposition [5]. Increasing N deposition provides a useful example, where the potential effects have been succinctly synthesized as the "nitrogen saturation" hypothesis [5,6]. Under this hypothesis, the chronically increased rates of N deposition invoke a variety of ecosystem responses, culminating in increased nitrate mobility in soils, nutrient imbalances in vegetation, forest decline, the acidification of soils, as well as increased surface water N loading leading to the potential for both acidification and eutrophication [5]. Numerous catchment scale experiments have been developed to test the saturation hypothesis [7,8], and along with them, models have been developed to predict the response of catchments to different deposition patterns [9]. Because nitrate in solution is the dominant form of N lost from these systems, the models that have been developed tend to emphasize inorganic N (e.g., [10]).

Despite the strides made in understanding catchment nitrogen cycling under the saturation hypothesis, in many temperate forested catchments, notably those on the Western side of both North and South America, current N deposition rates remain similar to those found historically [11]. In these areas that lie outside of urban or other point sources of atmospheric nitrogen, N continues to be a major limiting nutrient, losses of plant-available N remain very low, and nitrogen saturation is not currently part of the developmental trajectory [12]. The models that have been developed in regions of excess N deposition are not necessarily applicable in these places, a fact that underscores the suitability of low deposition regions as a useful counterpoint in the evaluation of ecosystem N cycling.

The HJ Andrews Experimental forest (HJA) in the Pacific Northwest of the USA is one such region. A variety of studies, based in small watersheds at HJA, and focused on different components of the nitrogen cycle have been developed since its inception in the 1960s. The synthesis of these various disciplinary studies is an ongoing effort. The application and development of a conceptual numerical model, which attempts to incorporate key components of the evolving understanding of N dynamics, provides an opportunity to inject some temporal dynamism into the ecosystem budget approach, and in this sense, contribute to the overall direction of field-based research and interpretation.

The overall objective of this paper is to outline a model formulated to describe the processes controlling N cycling in low deposition, small and primarily coniferous forest watersheds at HJA. The basis for this objective is twofold. First, some potential factors associated with the low deposition nature of the region are not explicitly captured by the current suite of standard ecosystem models. These include the relative importance of losses of organic nitrogen, as well as the potential importance of instream processing of nitrogen, and epiphytic and asymbiotic nitrogen fixation. Second, a range of disciplinary experimentalists, ranging from forest ecologists to soil biogeochemists to hillslope hydrologists and aquatic biologists collect data and develop expertise at the site. Much of this expertise has a direct bearing on nitrogen cycling, yet because it emerges from different disciplines it has been difficult to integrate it and develop a more complete understanding of the ecosystem as a whole. The model has been developed to explicitly capture a temporally dynamic N budget, by directly capturing observed rates and states that operate at HJA and forests like it (Figure 1).

The model represents a communication strategy to facilitate dialog between disciplinary experimentalists, to produce a more complete picture of nitrogen cycling in the region. This kind of modeling has seen recent growth in the environmental sciences [13,14] and we view this explicit development of complete, yet conceptually simplified models as a mechanism to more fully evaluate complex environmental dynamics. In particular, our work contributes to the idea that conceptual systems modeling can contribute to interdisciplinary science as a means of providing the capacity for individual researchers to contribute to evolving models of complex systems [13]. The models that result from such a process are not necessarily designed as predictive tools, but instead as a means to document key system details and how components may interact. Such a model may provide a useful point to begin the development of predictive modeling tools, either through detailed sensitivity and uncertainty analysis, or through the development of process-based algorithms. Nonetheless, roughly calibrated conceptual models, like we present here, present a useful framework for discussion.

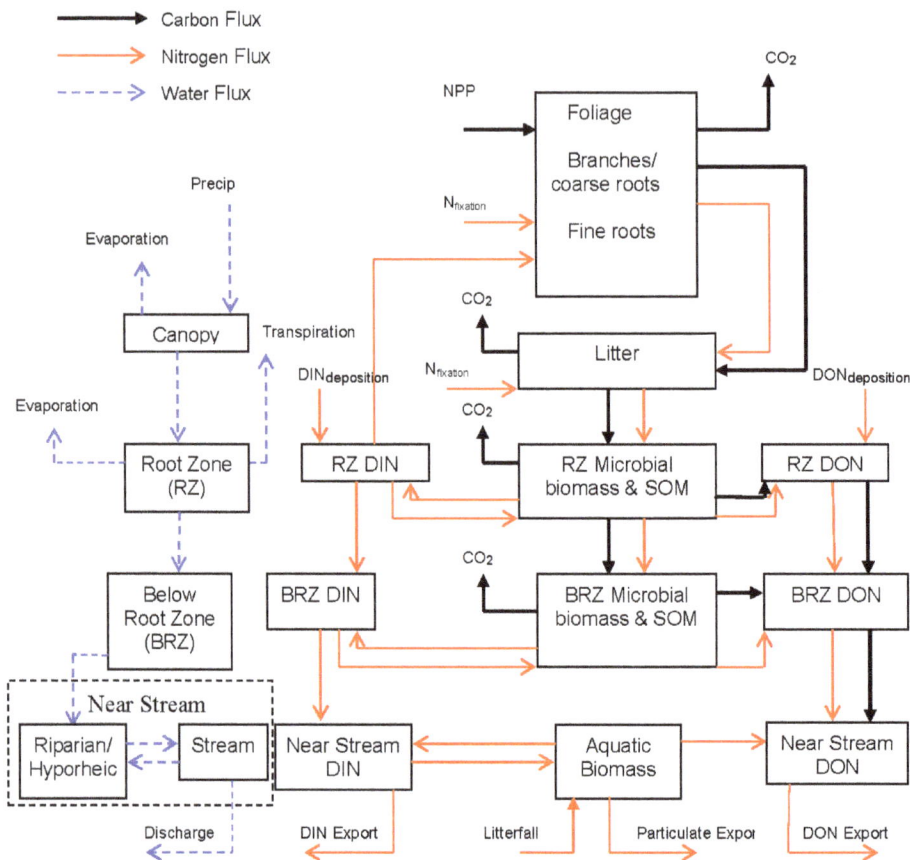

Figure 1. A flow chart outlining key model stocks and fluxes. Some minor flux terms have been left out of the figure to improve clarity. Note also that the litter input into aquatic biomass is derived directly from the stocks of foliage and branches/coarse wood.

2. Materials and Methods

2.1. Ecosystem Modeling

A wide variety of models have been developed to evaluate questions related to nitrogen dynamics, with the variation primarily manifested as different levels of complexity, different scales of application, and different questions of interest. These range from 3-dimensional physically-based research models, which are applicable across a broad range of environments and time scales (ecosys; [14]) to lumped data-based techniques (UNERF; [15]), which are reliant upon long term input output measurements, and maintain no physical basis. In this paper we are primarily interested in forested watershed nitrogen dynamics, which represents only a small subset of this broad range of models. For these ecosystems, modeling strategies generally incorporate associated carbon and water cycling, and while there is considerable overlap, the models tend to focus on one of three basic lines of inquiry. The first of these relates to forest productivity and succession. Productivity models are developed primarily to predict the successional evolution of above ground biomass (e.g., 3PG [16]) or focus on carbon dynamics and primary productivity (e.g., the PnET models [17]). Models in this group tend to include a more robust depiction of primary production, canopy processes, and carbon allocation, with somewhat less detail in terms of soil organic matter processing. The second general class of ecosystem models focuses more heavily on soil nutrient cycling and soil organic matter (SOM) dynamics. This group includes many of agricultural models, some of which have been adapted to represent forested landscapes (CENTURY [18]). A third approach focuses not on vegetation or SOM properties, but rather on hydrology and nitrogen export in an effort to provide a predictive tool to quantify nitrogen leaching potential (MERLIN [10]).

Regional to global scale biogeochemical models, a somewhat different class of simulation tools make use of many of the concepts outlined in the models above, coupling water, carbon, and nitrogen cycles to represent complete regional scale ecosystems. These models display less disciplinary focus, and are used primarily to evaluate the function of whole ecosystems under changing climate or deposition patterns. Representative models in this group include GEM [19] and BIOM-BGC [20] and INCA [21].

Most of these ecosystem scale models maintain a spatially lumped approach, though a number of spatially distributed simulation tools—those that include lateral interaction terms—have been developed (e.g., RHYSSYS [22]). With the exception of MEL [23] none of these models treat the production or mobility of dissolved organic nitrogen, an oversight representing a potentially significant structural error in terms of potential maximum amounts of sequestered carbon [23]. Additionally, while a variety of aquatic simulation models have been developed [2], these concepts have not been explicitly included in any of the ecosystem models.

The modeling we developed here relies fundamentally upon a variety of ideas and algorithms proposed within the range of existing ecosystem models. Our intention isn't to replace any of these models, but rather to select elements from them, and to put those elements together to represent a particular location with a particular set of processes.

2.2. Study Site

The HJ Andrews Experimental Forest (HJA) Watershed 10 (WS10) is located in the western Cascades of Oregon. Soils are predominantly composed of weakly developed Inceptisols with local areas of Alfisols and Spodsols made up of thick organic horizons over weathered parent materials [24]. The geology of WS10 is characterized by Miocene volcanics, primarily breccia and massive tuff, predominate [25]. Glacial, alluvial, and mass movement processes have resulted in a deeply dissected, locally steep drainage with highly variable regolith depths [25]. Vegetation is primarily comprised by Douglas-fir (*Pseudotsuga menziesii*), western hemlock (*Tsuga heterophylla*), and western red cedar (*Thuja plicata*). Average annual precipitation ranges from 2300 mm at lower elevations to over 3550 mm at the highest elevations and the climate is Mediterranean, with wet mild winters characterized by long duration low to moderate intensity frontal storms. WS10 is rather unique in that a significant debris flow in 1996 effectively removed the entire riparian area and most of the stream channel currently flows on bedrock. We limit our model explorations in WS10 to the pre-1996 period, when the riparian area was intact. A more complete description of WS10 and other small watersheds at HJA is provided in [26].

Over the 55 years of small watershed studies HJA, researchers have investigated recurring themes including material and elemental budgets, forest hillslope-stream interactions, biogeochemical and hydrologic responses to disturbances, and forest ecology. Framing this long-term research is a unique long term dataset documenting the seasonal input/output response, including organic and inorganic nitrogen and water, of six small catchments for over 30 years [27]. In this paper, we focus on one of the small catchments, WS10, where terrestrial [28,29] and aquatic [30] elemental budgets have previously been developed. These studies provide a key set of measurements which we revisit in this paper as calibration and evaluation terms for the numerical model.

2.3. The HJA-N Model

The HJA-N model is cast as a set of mass balance equations, and uses various strategies to represent rate terms. The formulation is designed to explicitly track fluxes and masses that pass through a set of roughly defined environmental storages, differentiated in both vertical and lateral terms. It is a dynamic model that incorporates transient input data at monthly time steps, and as such is useful for evaluating potential effects of seasonality on N cycling and long term trends in N cycling in response to environmental disturbances including climate change, increasing CO_2 concentration, and large scale vegetation manipulation. It is, however, not currently designed for the short time scales necessary to consider the time scale of storm events. Table 1 outlines the naming convention used in tables outlining the model. The full list of mass balance equations is included in Table 2, with the rate terms outlined in Tables 3–6. Model parameters are listed in Tables 7 and 8. The model was implemented using the Stella$^{©}$ systems modeling framework [31] to more fully communicate the developing model structure with the disciplinary experts providing the perceptual model of catchment process.

Table 1. Definition of naming convention used in the mathematical description of the model. Terms used in rate equations follow the form: $Component_{Mass,Location}$.

System Component		Mass		Additional Information	
Name	Short Name	Name	Short Name	Short Name	Name
Biomass	B	Carbon	C	Root Zone	R
Detritus	D	Particulate Nitrogen	N	Below Root Zone	B
Soil Organic Matter (SOM)	S	Available Nitrogen (DIN)	Na	Hyporheic Zone	HY
		Unavailable Nitrogen (DON)	Nu	In stream/aquatic	IS
Aquatic Biomass	A	Water	Wa	Canopy	C
				Wood	w
				Fine root	r
				Foliage	f

Table 2. The mass balance equations (differential equations) used to define the HJA-N model.

Equation	Defines
$\frac{dB_{Ci}}{dt} = P_{Ci} - T_{Ci} \rightarrow where\ i = foliage, wood, fine \cdot roots$	Carbon in biomass
$\frac{dB_{Ni}}{dt} = U_{if} + FS_{Ni} - T_{Ni} \rightarrow where\ i = foliage, wood, fine \cdot roots$	Nitrogen in biomass
$\frac{dD_{Ci}}{dt} = T_{Ci} - D_{Ci} - R_{Ci,D} \rightarrow where\ i = foliage, wood, fine \cdot roots$	Carbon in detritus
$\frac{dD_{if}}{dt} = FA_{Ni} + T_{Ni} - D_{Ni} \rightarrow where\ i = foliage, wood, fine \cdot roots$	Nitrogen in detritus
$\frac{dS_{C,R}}{dt} = D_{Cr} + D_{Cw} + D_{Cf} - R_{C,R} - T_C$	Carbon in the root zone SOM
$\frac{dS_{N,R}}{dt} = D_{Nr} + D_{Nw} + D_{Nf} - M_{Nu,R} - M_{Na,R} - T_N + I_{Na,R}$	Nitrogen in the root zone SOM
$\frac{dS_{C,B}}{dt} = T_C - R_{C,B}$	Carbon below the root zone SOM
$\frac{dS_{N,B}}{dt} = T_N - M_{Na,B} - M_{Nu,B} + I_{Na,B}$	Nitrogen below the root zone SOM
$\frac{dW_{Na,R}}{dt} = DP_{Na} + D_{Nu,R} + M_{Na,R} - U_{Na} - DN_{Na,R} - I_{Na,R} - Q_{Na,R}$	Dissolved available nitrogen in the root zone
$\frac{dW_{Nu,R}}{dt} = DP_{Nu} + M_{Nu,R} - D_{Nu,R} - Q_{Nu,R}$	Dissolved unavailable nitrogen in the root zone
$\frac{dW_{Na,B}}{dt} = Q_{Na,R} + D_{Nu,B} + M_{Na,B} - Q_{Na,B} - DN_{Na,B} - I_{Na,B}$	Dissolved available nitrogen below the root zone
$\frac{dW_{Nu,B}}{dt} = Q_{Nu,R} + M_{Na,B} - D_{Nu,B} - Q_{Nu,B}$	Dissolved unavailable nitrogen below the root zone
$\frac{dW_{Na,IS}}{dt} = Q_{Na,B} + M_{Na,IS} - I_{Na,IS} - DN_{Na,IS} - Q_{Na,IS}$	Dissolved available nitrogen in the aquatic environment
$\frac{dW_{Nu,IS}}{dt} = Q_{Nu,B} + M_{Nu,IS} - Q_{Nu,IS}$	Dissolved unavailable nitrogen in the aquatic environment
$\frac{dA_N}{dt} = BI_N + I_{Na,IS} - M_{Na,IS} - M_{Nu,IS} - P_N$	Aquatic Biomass (particulate, algae, benthic)
$\frac{dW_{W,C}}{dt} = DP_W - TF_w - E_W$	Water in the canopy
$\frac{dW_{W,R}}{dt} = TF_W - ET_w - VF_W$	Water in the root zone (we assume it moves vertically)
$\frac{dW_{W,B}}{dt} = IN_{W,R} - LF_w - SF_w$	Water below the root zone (we assume it moves horizontally)
$\frac{dW_{W,HY}}{dt} = LF_w + SF_w + HE_{W,IS} - HE_{W,HY}$	Water in the riparian/hyporheic zone
$\frac{dW_{W,IS}}{dt} = HE_{W,HY} - HE_{W,IS} - Q_{W,IS}$	Water in the channel

Table 3. Rate terms comprising the hydrological model. See Table 7 for parameter definitions.

Rate	Definition	Units
$DP_{W,C}$	Deposition of water (precipitation)	m/day
$TF_{W,C} = W_{W,C}k_{W,c}$	Throughfall from canopy	m/day
$E_{W,C} = W_{W,C}k_{W,e}$	Evaporation from canopy	m/day
$ET_{W,R}$	Evapotranspiration from root zone	m/day
$VF_{W,R} = W_{W,R}k_{W,v}$	Vertical movement of water within the root zone	m/day
$LF_{W,B} = W_{W,B}k_{W,l}$	Long (time) flowpaths of water below the root zone	m/day
$SF_{W,B} = W_{W,B}k_{W,s}$	Short (time) flowpaths of water below the root zone	m/day
$HE_{W,IS} = W_{W,IS}k_{W,IS}$	Hyporheic exchange of water from the stream	m/day
$HE_{W,HY} = W_{W,HY}k_{W,HY}$	Hyporheic exchange of water from the Hyporheic zone	m/day
$Q_{W,IS} = W_{W,IS}k_{W,q}$	Discharge of water from the stream	m/day

Table 4. Rate terms comprising vegetation mass balance equations. See Table 7 for definitions of parameter values.

Rate	Definition	Units
$P_{Ci} = P_n\eta_{Ci}$	P_n allocated to foliage($i = f$), wood ($i = w$) and fine roots ($i = r$)	kg C/m^2
$T_{Ci} = B_{Ci}\tau_{fi}$	Turnover of C from foliage ($i = f$), wood ($i = w$) and fine roots ($i = r$)	kg C/m^2
$U_n = u_{max}\left(\dfrac{W_{Na,R}}{u_{1/2}+W_{Na,R}}\right)$	Uptake of N	kg N/m^2
$U_{Ni} = U_N\eta_{Ni}$	Uptake allocated to foliage ($i = f$), wood ($i = w$) and fine roots ($i = r$)	kg N/m^2
$T_{Ni} = B_{Ni}\tau_{fi}$	Turnover of N from foliage ($i = f$), wood ($i = w$) and fine roots ($i = r$)	kg N/m^2
$FS_{Ni} = FS_{max}\eta_i\lambda$	Symbiotic fixation allocated to foliage ($i = f$), wood ($i = w$) and fine roots ($i = r$)	kg N/m^2
$D_{Ci} = B_{Ci}(1 - f_{co_2f})d_i\lambda$	Decomposition to SOM from foliage ($i = f$), wood ($i = w$) and fine roots ($i = r$)	kg C/m^2
$D_{Ni} = B_{Ni}e^{-(CN_{Df}/CN_{crf})^{k3}}d_i\lambda$	Decomposition of N from foliage ($i = f$), wood ($i = w$) and fine roots ($i = r$)	kg N/m^2
$FA_{Ni} = FA_{max}\lambda$	Asymbiotic fixation allocated to foliage ($i = f$), wood ($i = w$) and fine roots ($i = r$)	kg N/m^2
$R_{Ci,D} = B_{Ci}f_{co_2i}d_i\lambda$	Decomposition as resp from foliage ($i = f$), wood ($i = w$) and fine roots ($i = r$)	kg C/m^2

Table 5. Rate terms comprising soil-processing mass balance equations. See Table 7 for parameter definitions.

Rate	Definition	Units
$R_{Ci} = S_{Ci}r_i\lambda$	Respiration of C from root zone ($i = R$), below ($i = B$)	kg C/m^2
$T_{C,R} = S_{C,R}\tau_{t,R}$	Transfer of C from root zone to below	kg C/m^2
$M_{Nu,i} = \dfrac{R_{Ci}}{CN_{S,i}}f_{NuNa}\lambda$	Mobilization of unavailable N from RZ ($i = R$), below ($i = B$)	kg N/m^2
$M_{Na,i} = \dfrac{R_{Ci}}{CN_{S,i}}(1 - f_{NuNa})\lambda$	Mobilization of available N from RZ ($i = R$), below ($i = B$)	kg N/m^2
$T_{N,R} = S_{N,R}\tau_{t,R}$	Transfer of N out of root zone	kg N/m^2
$I_{Na,j} = i_{max}\left(\dfrac{W_{Na,j}}{i_{1/2j}+W_{Na,j}}\right)\lambda$	Immobilization of available N ($j = R$), below ($j = B$)	kg N/m^2
DP_{Na}	Deposition of DIN	kg N/m^2
$D_{Nu,i} = W_{Nu,i}d_{Nu,i}\lambda$	Breakdown of HMW DON from the root zone ($i = R$) or below ($i = B$)	kg N/m^2
$DN_{Na,i} = W_{Na,i}dn_i\lambda$	Denitrification from RZ ($i = R$), below ($i = B$), and instream ($i = IS$)	kg N/m^2
$Q_{i,R} = \dfrac{W_{i,R}}{W_{W,R}}W_{W,R}k_{W,v}s_{i,R}$	Hydrologic transfer of N from RZ. ($i = Na$—unavailable. $i = Nu$—available)	kg N/m^2
DP_u	Deposition of DIN	kg N/m^2
$Q_{i,B} = \dfrac{W_{i,B}}{W_{W,B}}W_{W,B}(k_{W,s} + k_{W,l})s_{i,B}$	Hydrologic transfer of N from brz. ($i = Na$ represents unavailable, $i = Nu$ represents available)	kg N/m^2

Table 6. Rate terms comprising instream-processing of N. See Table 7 for parameter definitions.

Rate	Definition	Units
$M_{i,IS} = W_{i,IS} m_{i,IS}$	Mobilization of available N from instream ($i = Na$ represents unavailable, $i = Nu$ represents available)	kg N/m²
$I_{Na,IS} = W_{Na,IS} i_{Na,IS} \lambda$	Immobilization of available N from instream	kg N/m²
$Q_{i,IS} = Q_{W,IS} \left(\frac{W_{i,IS}}{(W_{W,HY} + W_{W,IS})} \right)$	Discharge of available N from stream. ($i = Na$ represents unavailable. $i = Nu$ represents available)	kg N/m²
BI_N	Biological input of N to the stream	kg N/m²
$P_{N,IS} = A_n q_f^{-1} p_p$	Particulate export from the stream	kg N/m²

$$\text{when } Q_{W,IS}/q_{max} > 1, q_f = \left(\left(1 + \frac{(1 - Q_{W,IS}/q_{max})}{k_3} \right)^{k_2} \right), \text{Else } q_f = 1$$

Table 7. Parameters and auxiliary equations completing the production and allocation portions of the model.

Term	Definition	Units
ϕ_s	Incoming shortwave radiation	KJ/m²-day
ϕ_p	Photosynthetically active radiation (PAR)	KJ/m²-day
$\phi_{PA} = (1 - e^{-kLAI})\phi_p$	Absorbed PAR (Beer's Law)	KJ/m²-day
k	Light extinction coefficient	–
$LAI = B_{C,f}/SLA$	Leaf area	m²/m²
SLA	Specific leaf area	kg/m²
$\phi_{PAU} = \phi_{PA} \xi$	Utilized absorbed PAR	KJ/m²-day
$\xi = f_t f_a f_v f_{sw} f_n$	Growth modifier	–
$P_G = \alpha_c \phi_{PAU}$	Gross Primary Productivity (GPP)	kg/ha
α_c	Canopy quantum efficiency	Mol C/mol photon
$P_n = P_G Y$	Net canopy production (from 3PG)	kg C/m²
Y	Respiration fraction of GPP	–
$f_t = $ $\left(\frac{T_a - T_{min}}{T_{opt} - T_{min}} \right) \left(\frac{T_{max} - T_a}{T_{max} - T_{opt}} \right)^{\frac{T_{max} - T_{opt}}{T_{opt} - T_{min}}}$	Temperature modifier	–
$f_{sw} = 1/1 + [(1 - \theta/\theta_x)/c_\theta]^{n_\theta}$	Soil water modifier	–
$f_v = e^{-k_v V}$	Vapor pressure deficit modifier	–
$f_a = 1/1 + (a/r_a a_{max})^{n_a}$	Age modifier	–
If $CNtr/CN_{min} < 1$ $f_n = 1/1 + [(1 - CNtr/CN_{min})/n_1]^{n_2}$ otherwise $f_n = 1$	Nutrient modifier	-
$\eta_{if} = (1 - wf_i) * (1 - \eta_{ir})$	Allocation fraction of foliage where $i = C$ for Carbon and $i = N$ for Nitrogen	–
$\eta_{iw} = wf_i * (1 - \eta_{ir})$	Allocation fraction of wood where $i = C$ for Carbon and $i = N$ for Nitrogen	–
$\eta_{Cr} = a_1 \left[a_2 + \left(\frac{PAR_\mu}{PAR} \right) a_3 \right]^{-1}$	Allocation fraction root C	–
$\eta_{Nr} = a_4 \left[a_5 + \left(\frac{CN_r}{CN_{tr}} \right) a_6 \right]^{-1}$	Allocation fraction root N	–
wf_i	C ($i = C$) or N ($i = N$) remaining allocated to wood	–
$\lambda = q_{10}^{T/10}$	Temperature modifier	–
$a_1, a_2, a_3, a_4, a_5, a_6$	Root allocation parameters (hyperbola)	–

Table 8. Parameters and auxiliary equations completing the soil, aquatic, and hydrologic portions of the model.

Term	Definition	Units
u_{max}	Max uptake rate	kg N/m^2
$u_{1/2}$	Uptake $\frac{1}{2}$ saturation	kg N/m^2
$\tau_{ft}, \tau_{wt}, \tau_{rt}$	Turnover rate constants	year^{-1}
d_{fd}, d_{wd}, d_{rd}	Decomposition rate constants	day^{-1}
q_{10}	Coef by which D increases each 10 C	kg/m^2
T	temperature	C
$f_{CO_2 f}, f_{CO_2 w}, f_{CO_2 r}$	Fraction of D as CO_2	–
$CN_{Df}, CN_{Dw}, CN_{Dr}$	CN ratios of detrital pools	–
$CN_{crf}, CN_{crw}, CN_{crr}$	Critical CN, above which N is retained	–
r_R, r_B	Soil respiration coefficient	year^{-1}
$\tau_{t,R}$	Transfer rate of particulates	year^{-1}
$CN_{S,R}, CN_{S,B}$	CN ratios of SOM pools	–
i_{max}	Max immobilization rate	year^{-1}
$i_{1/2}$	Immobilization $\frac{1}{2}$ saturation	kg N/ha
dn_R, dn_B, dn_{IS}	Denitrification rate constant	year^{-1}
$s_{Na,R}, s_{Na,B}, s_{Nu,R}, s_{Nu,B}$	Sorption rate constant	year^{-1}
$d_{Nu,R}, d_{Nu,B}$	Decomposition rate of dissolved unavailaible N	day^{-1}
$m_{Na,IS}, m_{Na,IS}$	Instream mobilization rate constant	day^{-1}
$i_{Na,IS}, i_{Na,IS}$	Instream immobilization rate constant	day^{-1}
q_f	Hydro factor for particulate export	–
p_p	Max particulate export rate constant	day^{-1}
q_{max}	Hydro threshold	m/m
k_2	Hydro factor	–
k_3	Hydro factor	–
$k_{W,c}, k_{W,e}$	Canopy interception and evaporation rate terms	day^{-1}
$k_{W,v}, k_{W,l}, k_{W,s}$	Vertical, long, and short flowpath	day^{-1}
$k_{W,HY}, k_{W,IS}$	Hyporheic and instream exchange	day^{-1}
$k_{W,q}$	Instream water retention rate constant	day^{-1}

2.3.1. Measured Data

Atmospheric inputs included precipitation, air temperature, radiation, and atmospheric deposition from observational records at the primary meteorological station at the Andrews Forest, composited to 3-weekly sampling intervals. Stream outputs, used for model evaluation, include discharge and fluxes of dissolved organic and inorganic N from stream chemistry sampling at Watershed 10 (WS10) in the Andrews Forest, also at a 3-weekly sampling resolution [27].

2.3.2. Hydrologic Model

The hydrologic model is conceptually similar to models such as HBV [32] in that process descriptions involving filling and drainage of storages in the model are based on first order assumptions. While a variety of more sophisticated techniques are readily available, our simplifications are consistent with the monthly timescales of both input and output data. A complete description of the rate terms is included in Table 3. Five pools are used to define the hydrology (Table 2). These pools represent the canopy, root zone, below the root zone, instream aquatic environment, and as well as a separate storage representing the riparian/hyporheic zone.

Interception is treated as a linearly decreasing function of canopy storage. Canopy evaporation is calculated independent of evapotranspiration and, along with canopy throughfall, it is treated as a first order loss term from canopy water storage. Evapotranspiration from the rooting zone is calculated using a simple air temperature index approach, which is limited by water content, following from [33]. The runoff generation model is comprised of two vertically-oriented storages. The storages conceptually correspond to the rooting zone and to the region below the rooting zone and above bedrock, which is considered to be impermeable. Surface soils in the region are highly permeable and surface ponding or infiltration excess overland flow has not been observed [34]. Following from these observations, modeled infiltration capacity assumed to be larger than the rainfall rate, and all throughfall enters the upper soil layer. The upper water storage feeds water vertically into the lower storage unit. Downslope flows are assumed to occur within the lower storage as saturated subsurface stormflow, which has been demonstrated in the catchment during high input events [26]. Water exiting the lower soil zone enters the near-stream zone where exchanges between the riparian zone, hyporheic zone and surface water are depicted again using a series of first order storage terms.

2.3.3. Vegetation Model

Carbon and nitrogen in pools representing wood, foliage, and fine roots are included in the model. The woody pool includes coarse roots, logs, as well as branches (Table 9). The three pools were utilized primarily because they are consistent with a variety of measurements that have been made at HJA and because they are functionally useful in that CN ratios and decomposition rates from these three pools are distinct.

Biomass production follows closely from the 3-PG model [16]. Gross primary production (GPP) is estimated based upon measured net shortwave radiation and using a simple empirical relationship between shortwave radiation and the photosynthetically active fraction (PAR). Beer's law is utilized to approximate light attenuation through the single layer canopy and the fraction of incident PAR absorbed by the vegetation. The leaf area index (LAI) is calculated based upon the simulated foliar biomass, where the specific leaf area is assumed to be a species dependent constant.

A collection of five functional modifiers are utilized to reflect role of environmental conditions in limiting the quantity of absorbed radiation utilized by the vegetation. These modifiers relate to soil moisture, vapor pressure deficit, stand age, air temperature, and in an extension of the original 3-PG concept [16], we have also included the availability of plant-available nitrogen in the root zone. The resulting estimate of utilized radiation governs, in combination with an estimate of canopy quantum efficiency, the estimated GPP. Net primary production (NPP) is then estimated as a constant fraction of GPP. Live allocation of NPP (as carbon) to the three major biomass stocks follows from [16]. Nitrogen storages follow from the production of carbon, and are based on targeted CN ratios for each of the pools. NPP is initially allocated to fine roots, as a function of absorbed and utilized radiation. More limiting growth conditions (captured as the five modifiers defined above and resulting in utilized radiation) result in a larger allocation to roots, following directly from [19]. After fine root NPP is calculated, the remaining fraction NPP is allocated to woody material and foliage using set fractions developed to maintain targeted CN ratios of wood and foliage.

Uptake of N is a rate term which transfers mass from the dissolved inorganic nitrogen (DIN) pool into the living biomass N pools. This rate term is similar to that employed within MERLIN [10]. Michaelis-Menten kinetics are used to develop a non-linear rate which depends upon DIN availability and also plant nutritional requirements (see Table 4), as inferred from NPP and the targeted CN ratio of the three vegetation pools. The dependence of uptake on production is incorporated by allowing K_{Nup} to vary linearly with NPP. The rate of change, or the $\frac{1}{2}$ saturation constant likely varies in time, dependent upon the plant CN ratio [8], however the additional complexity is not incorporated into the current model. N_{up} is allocated to each of the three live biomass storages based upon deviations of the current CN ratios from target live CN ratios (defined as CN/CN_t) for the fine root components. As the CN ratio for the fine roots deviates further from the targeted value, a larger percentage of N_{up} is allocated to the fine roots. The portion of N_{up} which is not allocated to the fine roots is portioned between the wood and foliar components based upon a constant allocation fraction.

Stoichiometric N requirements of living biomass are also satisfied through N fixation, which occurs primarily in the canopy, given the presence of lichen. The rate of symbiotic fixation is calculated using a maximum fixation rate modified using the air temperature modifier [26]. Nitrogen fixed within the canopy is distributed based upon the nitrogen allocation fractions. Nitrogen is also introduced into the system through asymbiotic fixation, calculated analogously to symbiotic fixation. The moisture status of substrate may plays a role in the rate of fixation, but given the overall degree of model complexity, we did not attempt to include this factor directly. The important point is that we have tried to include key features, and to parameterize them based upon available observations and/or acceptable estimates. A symbiotic fixation rate of 2.8 kg/ha-year has been estimated at WS10 [29] and for our modeling we used a maximum value of 4 kg/ha-year, modified by temperature to result in somewhat more dynamic value that is approximately similar to the older estimate. Asymbiotic fixation was not estimated by [29], but in the intervening years it has become clear that it is a potential N source; one which we did include in our modeling. Without additional information, we assumed the maximum rate was

1 kg/ha-year as an addition into each of the Dead Biomass pools. Turnover of each of the vegetation pools is assumed to proceed as a first order loss rate.

Transfers of C and N from plant residue into SOM are based upon fixed turnover rates, with fluxes dependent upon C and N concentrations and air temperature. The dependency of turnover rates on other physical factors, such as, moisture status, ET, CO_2 concentration, fire patterns or surface to volume ratios are not incorporated. A more mechanistic model could provide better estimates, but our goal was to balance model simplicity with an interest in capturing key stocks and flows. Here we felt that simplicity in the SOM turnover was justified. The plant availability of N is largely determined by decomposition; however the microbial populations present within decomposing material typically immobilize any available N prior to its release to SOM. This results in a typical pattern comprised of an initial decrease in the CN ratios of fresh plant residues, with release of N and stability of CN ratios after only some period time [35].

This observation is incorporated into the model through the specification of the stable CN ratio below which N is transferred to SOM. As substrate CN ratios fall below these critical CN values, N is lost to SOM at the same rate as C. Initial CN ratios of the different residue pools exert a strong influence on N losses through decomposition in this lumped model. Refer to Table 4 for a complete description of the rate terms defining the vegetation sub-model.

2.3.4. Soils Model

The soil organic matter (SOM) sub model is defined similarly to that utilized in the PNET-CN model of [17] in that the number of SOM pools is very small, particularly when compared with standard SOM models. The current version of HJA-N includes two SOM pools, the first representing root zone SOM and the second representing below root zone SOM, with carbon and nitrogen explicitly represented in both (Table 2). Although the inclusion of additional pools could be used to more precisely describe the wide distribution of temporal SOM stability, an evaluation of the simpler definition against the long term measured data represents a useful first step, and is consistent with the soil nitrogen budget developed in WS10 in the early 1980s [29].

Four additional below-ground nitrogen pools are included to represent DIN and DON in both the rooting and below rooting zones. A kinetic sorption isotherm is used to separate soil bound nitrogen from dissolved forms, assuming that the proportion of each stays constant through time. Hydrologic losses are defined based upon the flux rates calculated by the hydrologic components of the model and the concentrations of freely available DON and DIN. Landscape scale denitrification rates are not well understood, but the model does maintain a first order denitrification loss pathway from the DIN storage.

The soil respiration model is defined similarly to that for respiration from plant residues as a first order rate, which includes temperature dependence based upon the q10-based temperature modifier. The respiration rate is assumed to represent the production of both CO_2 and DOC. The mobilization of both DIN and DON is calculated as a proportion of the soil respiration rate.

The incorporation of DON production and loss is a key feature of the model. Very few ecosystem models include DON as a component of the nitrogen cycle, however [19] proposed and evaluated four potential definitions of DOC mobilization, and then used soil CN ratios to proportionally estimate the production of DON. These definitions included a constant loss model, a first order model, a model based upon soil CN ratios, and lastly a model where the rate of mobilization was proportional to the microbial respiration rate. The last of these rate definitions is consistent with our definition of SOM production, and as such was incorporated into the model.

At WS10 we have a long term record of streamwater DON and DIN export, but production rates in soils have not been studies. Our model reflects this in its simplicity. We assume that the production of dissolved N, in total, is proportional to the soil respiration rate, depending upon the soil CN ratios. This overall production rate of dissolved nitrogen is then split into DIN and DON assuming a fixed portioning constant. The DON pool includes an additional respiration term that is used to simulate the continuous decomposition processes of DON, which we assume result in the further production of DIN.

A more compelling definition of these dissolved N pools would separate plant available N from unavailable N, rather than organic from inorganic [36,37]. Such a distinction would recognize the fact that organic nitrogen is a term that represents a wide variety of compounds, with a significant range in molecular weights [38]. This would then allow for the lower molecular weight fraction of that distribution to interact more directly with the vegetation and microorganisms. However, in this case we are limited by available long term records of aquatic DON and DIN, which do not support such distinction. To be consistent with these data, we make the simplifying assumption that the DON pools, throughout the model domain, are unavailable forms. A complete description of the rate terms defining the SOM sub-model is included in Table 5.

2.3.5. Aquatic Environment Model

Most watershed-level studies of nutrient retention and release focus primarily upon terrestrial processes, and most watershed-level ecosystem models maintain this focus, and do not explicitly include nutrient dynamics within near stream areas. Yet it is known that the aquatic and hyporheic environment in small streams can exert significant influence on both the quantity and the forms of exported nutrients [32,33]. The residence time of water and solutes can also be extended based upon hyporheic exchange flows [39]. A 15N addition experiment [40] demonstrated that 32.5% of N added over a 6-week period during the growing season was retained by a second order stream in HJA. HJA-N explicitly includes a set of stocks (Table 2) and flux terms (Table 6) designed to capture the potential role of the aquatic system in regulating the export of terrestrial N fluxes.

The model makes use of three pools to represent nitrogen in the aquatic environment, and in this version of the model carbon is not accounted for within the aquatic environment. The pools that are included correspond to DON, DIN, and the aquatic biomass. These pools are assumed to represent the combination of the channel and hyporheic zones. The aquatic biomass pool contributes, through a first order respiration model, to both the DIN and DON pools. In addition, gross immobilization of DIN, as an addition input into the aquatic biomass, is also included as a first order term. Nitrogen is lost from the stream system as DON and DIN export, and also through a first order denitrification term. The aquatic biomass also includes a loss rate associated with particulate export, conceptually associated only with the near stream area environment. Inputs of particulate matter from the upslope region are defined based upon the turnover terms (litterfall and mortality) of foliage and woody material.

The loss of aquatic biomass is treated as a first order rate that is activated only above a discharge-based threshold. At high flows, accumulated biomass is quickly lost from the system, with periods of accumulation during lower flow conditions.

2.3.6. Control Capacity

To facilitate a discussion of the seasonal variation in the features controlling nitrogen cycling, we propose the nitrogen control capacity as a set of normalized rate terms which can be derived from the temporally varying model results to provide insights into these features. Hydrology controls nitrogen dynamics through export of dissolved nitrogen compounds. To capture this flushing behavior, we define the term transport control as:

$$\text{Hydrologic Flushing} = Q_{DIN,IS} + Q_{DON,IS} + Q_{DIN,RZ} + Q_{DIN,BRZ} + Q_{DON,RZ} + Q_{DON,BRZ} \quad (1)$$

where $Q_{DIN,IS}$ and $Q_{DON,IS}$ are the simulated export rates of DON and DIN from the stream and $Q_{DIN,RZ}$, $Q_{DIN,BRZ}$, $Q_{DON,RZ}$ and $Q_{DON,BRZ}$ are the rates of movement of dissolved N through the watershed. Hydrologists tend to view nitrogen dynamics through the lens of the flushing hypothesis and this term is designed to capture the contribution of flushing to the movement of N in the system. The contributions of simulated DIN and DON to the flushing index were normalized by the area of the stream, rather than the area of the watershed. We used an area of 767 m^2, following from [30].

The vegetation controls N cycling through nitrogen mobilization, which we define as the difference between litter decomposition rates and uptake. We have elected to group litter and vegetation together, though clearly they could also be treated independently. Under this definition, the above ground control term is defined as:

$$\text{Vegetation Processes} = D_{N,f} + D_{N,w} + D_{N,r} - U_{DIN,RZ} \tag{2}$$

where $U_{DIN,RZ}$ is the uptake rate into vegetation and $D_{N,f}$, $D_{N,w}$, $D_{N,r}$, are the decomposition rates contributing nitrogen from foliar, woody, and root litter respectively.

Soil control is defined analogously as the of the net mobilization rate, in this case:

$$\text{Soil Processes} = M_{DIN,RZ} + M_{DON,RZ} + M_{DIN,BRZ} + M_{DON,BRZ} - (I_{DIN,RZ} + I_{DIN,BRZ}) \tag{3}$$

where $M_{DIN,RZ}$ and $M_{DON,RZ}$ are the mobilization rates of DIN and DON within the root zone, respectively, $M_{DIN,BRZ}$ and $M_{DON,BRZ}$ are the mobilization rates of DIN and DON below the root zone, respectively, and $I_{DIN,RZ}$ and $I_{DIN,BRZ}$ are the immobilization rate of DIN in the root zone and below the root zone, respectively. Note that DON is assumed to be unavailable to plants or the microbial complex, and as such has no immobilization rate.

We then define the aquatic control in a somewhat different fashion, including the net mobilization of nitrogen, but also the simulated rate of particulate export.

$$\text{Instream Processes} = M_{DIN,IS} + M_{DON,IS} - I_{DIN,IS} + P_{N,IS} \tag{4}$$

where $M_{DIN,IS}$ and $M_{DON,IS}$ are the mobilization rates of DIN and DON from the aquatic biomass, $I_{DIN,IS}$ is the immobilization rate of DIN from the aquatic biomass, and $P_{N,IS}$ is the export rate of particulate N, which again originates from the aquatic biomass. Similar to the simulated values of stream DIN and DON, the instream process control was normalized by stream area, rather than the full watershed area.

For comparative purposes, the four values are then normalized by the overall sum of the included rate terms to produce a ratio of control for each term, which varies throughout the model timeframe, given the system dynamics.

3. Results and Discussion

3.1. Results

The model includes a variety of rate terms, many of which have not been independently measured. In order to accommodate the resulting uncertainty we approached model evaluation using a parameter adjustment strategy based primarily upon expert judgment. During this phase of application, the model was evaluated against both measurements and, for those terms where measurements in WS10 were unavailable, more qualitative estimates of reasonability. The model was run for a total of 80 years, using a repeated 20-year input dataset, which was based on the 3-week compositing of inputs and outputs from 1968 to 1988. It is important to note that the watershed was clearcut in 1975, and that the effects of the harvest were evident in the observed N export. Reported here are only the last 20 of those years, with the first 60 years acting as a period to allow the differential equations which make up the model to come to a relatively steady state with respect to the initial values of all of the state variables.

3.1.1. Evaluation of Budget Estimates of N and C Stocks

The average modeled results of the key nitrogen and carbon stocks are consistent with budget-based measurements that are available from WS10 [29], or have been taken from similar forested regions [41]. Key features of the results include a dominance of carbon storage in woody material (65% of total carbon storage) and nitrogen storage in soils (75% of total nitrogen storage) (Figure 2). Differences between the modeled values and the measurements were anticipated, particularly because the measured stocks were not necessarily binned in the same manner as the model description, and because we are comparing average model results representing 20 years of simulation to measurements that were developed to represent a full year, and in the case of the Wind River data [41], at different location. Direct comparisons between the available budget-based measurements and the model-based estimates (Figures 3 and 4; Table 1) indicate that the model is able to capture the general magnitude of carbon and nitrogen storage, and also the differences between the key environmental compartments.

3.1.2. Model Evaluation against Observations

The long-term record, which includes stream water discharge, and DON and DIN export is rare, and provides an opportunity to constrain model operation. A comparison of the time series records to the modeled result is included in Figure 5 and demonstrates that model effectively captures the seasonal pattern that is outlined by the measured discharge and DON. For these variables, the modeled Nash Sutcliffe efficiency [42] values are 0.71 and 0.53 respectively. The efficiency for DIN is −0.12, which clearly indicates that the model does not capture the measured response. This apparent failure of the model is likely because we did not attempt to simulate impacts of the clear-cut harvest that occurred in 1975. The removal of the trees, and pre-treatment activity, led to elevated DIN export after the harvest. The effects on DIN of this disturbance have been explored by [43] and are clearly evident in the long-term record and has been attributed to reduced uptake of N by vegetation. It is worth noting that the calculated Nash-Sutcliffe for years prior to the management activity (1968 to 1973) is higher, 0.33, lending additional support to the suggestion.

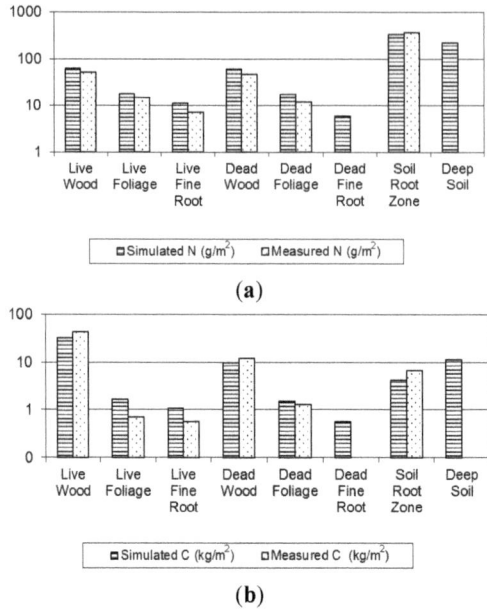

(a)

(b)

Figure 2. A comparison of simulated stocks of nitrogen (a) and carbon (b) against budget estimates from [29]. Note that estimates of observed C stocks (except for SOM, labelled above were originally derived by [28] and that we assumed the carbon content of dry mass was 50%. Additionally, we assumed that 50% of the category "Fallen foliage and fine woody litter" from [29] were dead foliage and that 50% were dead wood.

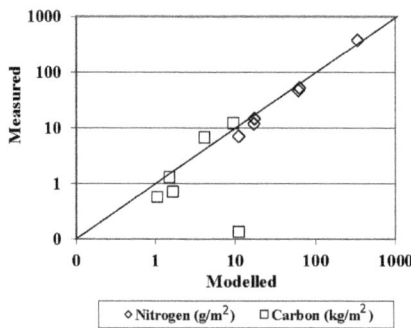

Figure 3. Measured *versus* modeled annual yearly storage of nitrogen and carbon. These data are also plotted in Figure 2. Each data point represents a different storage term, as outlined in Figure 2. A one to one line is included for reference.

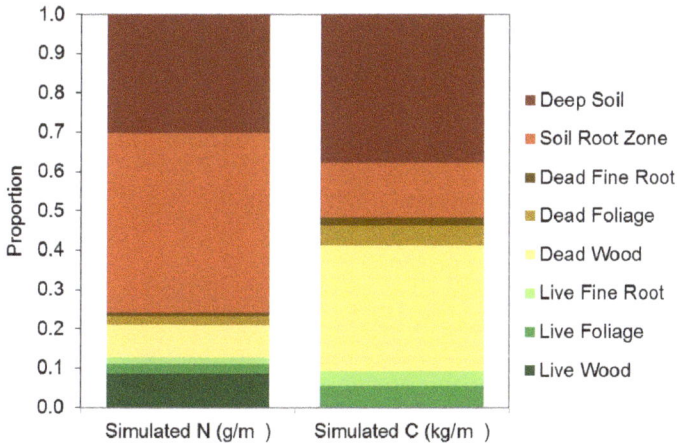

Figure 4. Model results for average annual storage of nitrogen and carbon. Results are normalized by the total sum of nitrogen and carbon.

Evaluation of against Budget Estimates of N and C Fluxes

Given that the model is capable of representing key long term output measurements of water, DIN and DON, and also the overall trends in storage, the next step is to evaluate the degree to which the model is working for the right reasons—this we accomplish through an evaluation of the internal rate terms. These rates include respiration, mobilization, immobilization, internal solute fluxes and the aquatic processes that follow from them. Here we return to the existing budget studies which provide a set of annual flux estimates which we utilize in calibration, and to better understand model function. Rate terms are broadly grouped into four categories representing carbon fluxes, the sources of nitrogen, SOM dynamics, and processes occurring in the aquatic environment. This division is not to suggest that these categories are independent of one another, but only to facilitate presentation of results. In all cases, we present continuous model results (Figures 6–9) and in addition, time-integrated average yearly values (included in Figures 6–9 and Tables 9 and 10), which can be directly compared against yearly values from the budget studies.

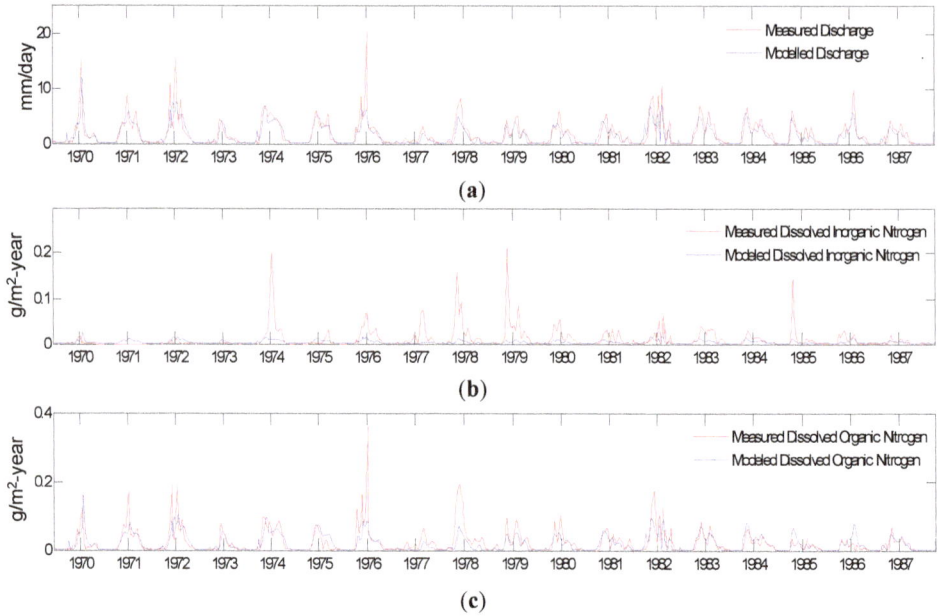

Figure 5. Measured *versus* model time series for stream water discharge (**a**), dissolved inorganic nitrogen (**b**) and dissolved organic nitrogen (**c**). Nash-Sutcliffe efficiencies for discharge is 0.71, for DON is 0.53. For DIN the efficiency is -0.13, as discussed in the text, this is likely due to the clearcut that occurred in the watershed in the early 1970s. The model did not attempt to incorporate the effects of the management.

Table 9. Comparison of modeled pool sizes (averaged over the 20 year simulation period) against measured values as reported in (*a*) [29] from WS10 and from (*b*) [41] from Wind River, WA. We assumed a carbon content of 50% to estimate C from the dry mass reported in [29]. The standard deviation of the modeled values is included in parenthesis.

Ecosystem Component		C (kg/m²)			N (g/m²)		CN	
		a	*b*	Model	*a*	Model	*a*	Model
Vegetation	Wood	42.29	37.87	31.77 (0.04)	52.17	63.3 (0.11)	810.63	501.57
	Foliage	0.69	0.94	2.17 (0.04)	14.44	18.2 (0.38)	48.13	119.20
	Fine Roots	0.06	0.36	1.12 (0.04)	0.69	11.7 (0.16)	80.65	96.34
Detritus	Wood	13.31	9.87	9.54 (0.05)	47.1	61.1 (0.58)	282.59	156.05
	Foliage	1.28	1.78	2.01 (0.02)	11.76	18.1 (0.15)	108.84	110.73
	Fine Roots		0.48	0.593 (0.01)		6.4 (0.11)		93.06
Soil	Root Zone	6.65	9.3	4.28 (0.06)	372.4	338.5 (4.23)	17.88	12.65
	Deep Soil			12.80 (0.33)		220.5 (5.39)		58.07

(a)

(b)

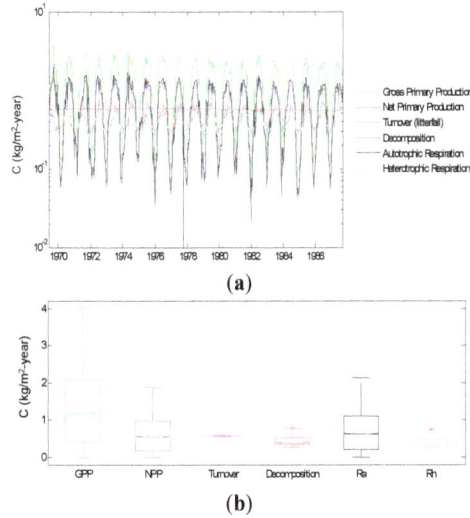

Figure 6. Key modeled rate terms associated with carbon cycling. (a) The plot represents the modeled time series, as indicated in the legend, across the 20 year simulation time frame. These data are summarized and presented as a box and whisker plot (b). Note that the y axis is log transformed in the upper plot, but is linear in the lower plot. The combination is provided to more clearly outline the range in variation.

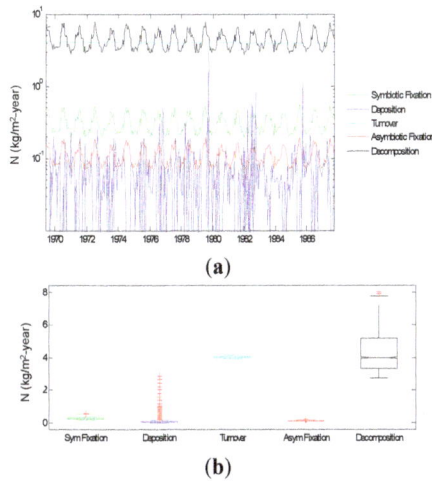

(a)

(b)

Figure 7. Key modeled rate terms associated with the input of nitrogen into the system. (a) The plot represents the modeled time series, as indicated in the legend, across the 20 year simulation time frame. These data are summarized and presented as a box and whisker plot (b). Note that the y axis is log transformed in the upper plot, but is linear in the lower plot.

(a)

(b)

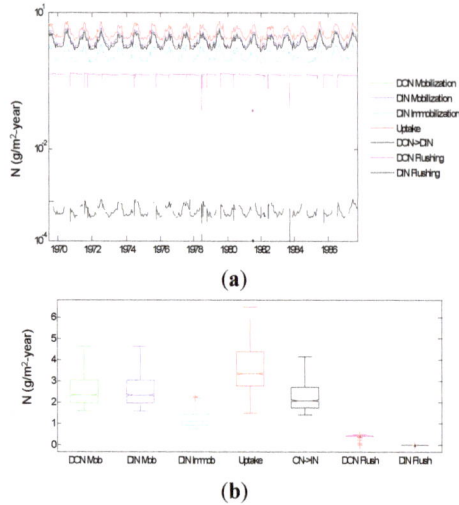

Figure 8. Key modeled rate terms associated with root zone SOM. (**a**) The plot represents the modeled time series, as indicated in the legend, across the 20 year simulation time frame. These data are summarized and presented as a box and whisker plot (**b**). Note that the y axis is log transformed in the upper plot, but is linear in the lower plot.

Table 10. Simulated average annual rates defining the carbon model. The rates in bold are summaries across each of the individual compartments.

Rate	g/m^2-year
Gross primary production	935.28
Growth wood	44.34
Growth foliage	177.34
Growth fine roots	217.96
Total growth (NPP)	**439.63**
Mortality wood	32.56
Mortality foliage	176.62
Mortality fine roots	217.35
Total mortality	**426.53**
Decomposition from wood	4.37
Decomposition from foliage	127.54
Decomposition from fine roots	185.66
Total decomposition	**672.16**
Respiration from wood	39.35
Respiration from foliage	0.01
Respiration from fine roots	32.76
Respiration from SOM RZ	173.98
Respiration from SOM BRZ	44.16
Respiration heterotrophic	332.78
Respiration Autotrophic	495.70
Total respiration	**828.48**
DOC Production RZ	17.4
DOC Production BRZ	4.42
Total DOC production	**21.82**

206

(a)

(b)

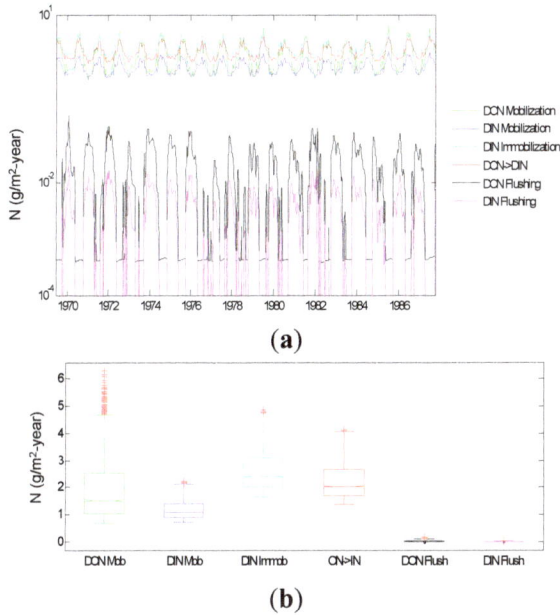

Figure 9. Key modeled rate terms associated with SOM processing occurring below the root zone. (**a**) The plot represents the modeled time series, as indicated in the legend, across the 20 year simulation time frame. These data are summarized and presented as a box and whisker plot (**b**). Note that the y axis is log transformed in the upper plot, but is linear in the lower plot.

Table 11. Simulated average annual rates defining the nitrogen model. The rates in bold are summaries across each of the individual compartments.

Rate	g/m^2-year
DON_Deposition	0.04
DIN_Deposition	0.03
Total deposition	**0.08**
Symbiotic fixation allocated to wood	0
Symbiotic fixation allocated to foliage	0
Symbiotic fixation allocated to roots	0.43
Total symbiotic fixation	**0.43**
Uptake allocated to wood	0.04
Uptake allocated to foliage	1.79
Uptake allocated to roots	1.87
Total uptake	**3.70**
Mortality wood	0.06
Mortality foliage (litterfall)	1.81
Mortality roots	2.33
Total turnover	**4.21**
Asymbiotic fixation allocated to wood	0.08
Asymbiotic fixation allocated to foliage	0.08
Asymbiotic fixation allocated to roots	0.076
Total asymbiotic fixation	**0.22**

Table 11. Simulated average annual rates defining the nitrogen model. The rates in bold are summaries across each of the individual compartments.

Rate	g/m²-year
Decomposition of wood	0.27
Decomposition of foliage	1.93
Decomposition of roots	2.42
Total decomposition	**4.64**
DON mobilization (RZ)	2.58
DIN mobilization (RZ)	2.58
DIN Immobilization (RZ)	1.19
DON breakdown to DIN (RZ)	2.28
DON transport in RZ	0.34
DIN transport in RZ	0.00
Denitrification RZ	0.00
DON mobilization (BRZ)	1.92
DIN mobilization (BRZ)	1.17
DIN Immobilization (BRZ)	2.67
DON breakdown to DIN (BRZ)	2.23
DON transport in BRZ	0.62
DIN transport in BRZ	0.020
Denitrification BRZ	0.01
Particulate input to aquatic	1.88
DON mobilization	0.00
DIN immobilization	0.00
DIN mobilization	0.00
Denitrification from aquatic	0.00
DIN export	0.00
DON export	0.02
Particulate export	0.02

3.1.3. Carbon Fluxes

The key rate terms capturing carbon dynamics are gross and net primary production, turnover (including mortality and litterfall), and both autotrophic and heterotrophic respiration. Model results for each rate term were integrated over the 20 year simulation period, and the yearly average over that time period is presented in Table 9. The average yearly gross primary productivity is estimated as 1.4 kg/m²-year, with the daily rates ranging from nearly 0 kg/m²-year (periodically during the winter period) to over 3 kg/m²-year (Figure 6). The average yearly rate is somewhat lower than the range of 1.4 to 3.3 kg/m²-year estimated by [41] for a similar old growth forest, and within the range of 1.08–1.92 kg/m²-year estimated using remote sensing techniques across a Douglas-fir western hemlock forest on Vancouver Island, Canada by [44]. The simulated values are considerably lower than the 11.1–21.7 kg/m²-year estimated by [28] at HJA in 1970, and different respiration estimates explain the discrepancy. We assumed a respiration value of 47% of GPP, a heuristic model generated as part of the collaborative model development process. The older budgets of [28] resulted in respiration values of 92%–94% of GPP. Given the assumption that Ra is 47% of GPP, the yearly average NPP from the model is 0.60 kg/m²-year. This value is similar to other estimates of NPP reported at other similar sites, for example [41] estimated a range of NPP from 0.45–0.741 kg/m^2-year at similar location in Washington State, [28] estimated a range from 0.41 to 0.66 kg/m^2-year at HJA and [45] used a model to estimate a stable value of 0.50 kg/m^2-year at HJA. Our model estimates are within the wide range of variation possible for NPP. The constant fraction model to estimate R_a explains the similarity in the time series representing NPP and R_a in Figure 6. Turnover is treated as a first order model which does not include any outside dependence (for example air temperature, soil water content, *etc.*). This is reflected in the low temporal variation displayed for the turnover rate. The modeled value of average yearly turnover (0.572 kg/m^2-year) is similar to that measured by [41] which ranged from 0.370 to 0.690 kg/m^2-year. Modeled heterotrophic respiration (R_h) is defined to include the production of CO_2 and DOC from litter and soils. The average yearly value of 0.397 kg/m^2-year is of within the range of 0.341 to 0.509 kg/m^2-year estimated by [43].

3.1.4. Input of N to SOM

External inputs of N include deposition, and both symbiotic and asymbiotic fixation. Internal inputs of N to SOM include both mortality/litterfall and decomposition. Decomposition and turnover rates—internal recycling—are simulated to be at least an order of magnitude larger than the external rates (Figure 7). The external rates of deposition (which include both DIN and DON) are simply measurements, and the fixation rate terms have been calibrated to mimic the few estimates that are available for these sites. Symbiotic respiration in WS10 have been estimated to be 0.280 g/m^2-year [29], which is of a similar order to our average yearly modeled estimate of 0.305 g/m^2-year. The modeled estimate of asymbiotic fixation is 0.107 g/m^2-year. Actual asymbiotic fixation rates of, on average, 0.45 umol/g/day have been estimated for the tree species which dominated W10 prior to the clearcut in 1975 (*Psuedotsuga menziesii*) [46]. Given the model's average annual litter estimate of 43.55 kg/m^2, this measured fixation rate is equivalent to 0.100 g/m^2-year, in close agreement with the model rate. A complete listing of time integrated average nitrogen fluxes is included in Table 10.

3.1.5. Root Zone N Dynamics

The dynamics of nitrogen in the root zone are defined primarily by a series of seven rate terms, including the mobilization and flushing of DON and DIN, the immobilization and uptake by vegetation of DIN, and the breakdown of DON to produce DIN, which occurs through respiration processes. There is also potential for denitrification as a pathway of loss, however given the well aerated nature of soils in WS10, we assume that denitrification does not occur in the root zone. These terms are outlined in Figure 8, as both time series and box plots. These results indicate a net mobilization (mineralization) of inorganic nitrogen of 2.51 g/m^2-year, and that mobilization of DON is 2.51 g/m^2-year. The values are equivalent because, without additional data, we simply assume that total nitrogen mobilization was proportional to respiration and that the product was half organic and half inorganic nitrogen. Mobilized DON is continuously decomposed to further add to the DIN pool, most of which is utilized through plant uptake. The average yearly uptake rate is 3.62 g/m^2-year, similar to the 2.29 g/m^2-year estimated by [13]. The rate of flushing for DON (0.337 g/m^2-year) is considerably larger than for DIN, which was effectively 0 for our simulations. This result is consistent with the well-established high N retention capacity of these watersheds. The rates of flushing periodically fall to zero, indicated as breaks in the time series in Figure 8. This occurs when amounts of DIN or DON are not sufficient to support all of the simulated loss pathways.

3.1.6. N Dynamics below the Root Zone

Below root zone dynamics are similar to those simulated in the root zone, with flushing rates of both DON and DIN significantly lower than the rates of internal recycling (Figure 9). However, flushing rates of DON are larger than from within the rooting zone (467.7 mg/m^2-year root zone compared to 283.6 mg/m^2-year below the root zone), while flushing of DIN is somewhat higher from below the rooting zone (0.36 mg/m^2-year root zone compared to 23.61 mg/m^2-year below the root zone) primarily because vegetation is no longer able to mediate the flux of DIN from below the rooting zone. The model results indicate that the region below the root zone is a moderate nitrogen sink, with a net mobilization value of -99.40 mg/m^2-year. No measurements exist within this region of the watershed, and Figure 9 therefore represents only one possible result which is consistent with both simulated inputs from the root zone, and more importantly, measured outputs of water and dissolved nitrogen from the catchment.

The discussion of the aquatic environment focuses on mobilization and immobilization of N, as well as flushing. Particulate export of N dominates the model results in the aquatic region, with simulated values of 1.73 g/m^2-year. A value of 2.53 g/m^2-year was estimated by [30] for a particular year. The next largest rate term in the aquatic environment, DON export, is 0.024 g/m^2-year, and because of the approximately two orders of magnitude difference, particulate export has not been included in Figure 10. Results indicate net immobilization of DIN (3.42 mg/m^2-year) occurs in the aquatic region, consistent with [47] for a somewhat larger stream in HJA. The dynamic model results also indicate, however, that during the growing season, the instream biomass can function as a source of N, primarily because DIN limitation caps the growth rate, yet DIN mobilization is treated with first order dependence on the aquatic biomass, and so proceeds at a relatively higher rate during periods of low DIN availability (Figure 10). Nevertheless, even excluding simulated particulate export, model results indicate that more nitrogen is lost from the stream environment on a yearly basis in dissolved forms (24.8 mg/m^2-year) that is retained within it by the aquatic biomass (3.42 mg/m^2-year).

3.1.7. Aquatic Environment

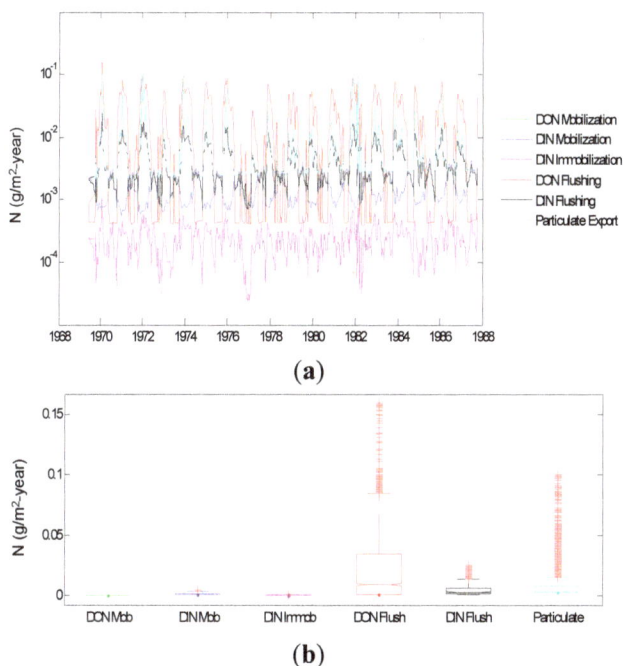

Figure 10. Key modeled rate terms associated with the aquatic environment. (**a**) The plot represents the modeled time series, as indicated in the legend, across the 20 year simulation time frame. These data are summarized and presented as a box and whisker plot (**b**). Note that the y axis in log transformed in the upper plot, but is linear in the lower plot.

3.2. Discussion

HJA-N was constructed in an effort to explore relationships between biotic and abiotic processes in the retention and release of nitrogen from small watersheds. An elementary, yet key, finding is that the model can be parameterized so as to produce results that are consistent with a wide range of measurements from WS10 or from similar sites. This result is a prerequisite for any further analysis. Having established consistency with available measurements, the model results can be further evaluated to provide a number of intriguing insights into how watershed components interact over seasonal timescales to recycle nitrogen. The important contribution of the model is that it allows us to quantify the seasonal variability of rate terms, greatly extending

211

the budget based estimates of storage and fluxes which comprise a significant amount of available measurements [28–30,41].

3.2.1. Relative Importance of Various System Components

A key theme of this research is the development of quantitative, internally consistent estimates of the relative roles of watershed components in the retention and release of nitrogen over seasonal timescales. The overall goal is the exploration of the temporally varying relationship between these components—vegetation, soils, hydrology, and the aquatic environment—in regulating the release of nitrogen from the system. In the hydrologic literature, much work in this direction has focused on the concept of hydrologic flushing [48]. At the same time, it is often assumed in both the soils and forest ecology literature (e.g., [49]) that the temporal variation in net mobilization is ultimately responsible for the availability of any nitrogen that might be flushed out of the system by the hydrology. This mobilization potential is particularly relevant in regions, like the PNW, where atmospheric deposition remains low. All the while, the role of the stream channel, as well as the riparian vegetation [47] in immobilizing significant amounts available nitrogen from the aquatic system, and hence modifying cross-weir export measurements, frequently remains unnoticed in catchment studies. And perhaps even more importantly in systems where nitrogen is strongly retained, the particulate export of dissolved organic nitrogen, over which the aquatic system exerts significant control, is often of a similar magnitude, if not considerably larger, than the export of dissolved nitrogen [30].

Our modeling work indicates that hydrologically-mediated fluxes are much smaller in magnitude (DON + DIN export of 22.1 mg/m^2-year) than the mobilization fluxes that occur within the vegetation (747.7 mg/m^2-year N mobilization, decomposition-uptake) or within the root zone SOM (3838.3 mg/m^2-year) N mobilization. This finding is consistent with the wide variety of observational work at HJA [29] and has also been demonstrated at other forested watersheds [43]. However, the hydrologically-mediated fluxes are larger in magnitude to the immobilization potential of the aquatic environment (3.42 mg/m^2-year) (Figures 6–10). These observations provide a useful means of understanding the system and ranking system components as to their role in the regulation of nitrogen cycling. In addition, the model results provide data that can be evaluated at finer seasonal time scales.

Control Capacity

We interpret the N control capacity ratio as the degree of control that each system components exerts on the release dynamics of nitrogen cycling within the model domain (Figure 11). There is, of course, a degree of subjectivity in the definitions, nonetheless evaluating the results in this fashion provides a unique mechanism to evaluate the relative importance of components, and how this degree of control varies with time. These kinds of analyses are available only through the use of the continuously varying results, which are not typically measured over long periods, providing significant further utility in the application and development of conceptual numerical simulation.

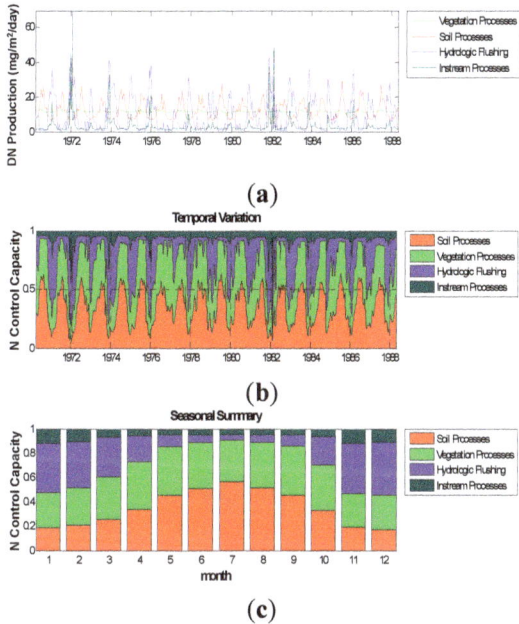

Figure 11. The nitrogen control capacity. (a) The plot outlines the summed rate terms which are used to estimate the control capacity of the watershed components as indicated in the legend. These four time series correspond directly to Equations (1)–(4); (b) The plot represents the control capacity, where the time series from the upper plot have been normalized by the total rate of nitrogen movement. (c) The plot represents the monthly averages of the control capacity, again including comparisons of the four different watershed components.

As expected given the well-established capacity of these types of watersheds to retain N, vegetation and soils exert a primary control on N dynamics. Hydrologic flushing of nitrogen is well-represented during the winter period, though even during periods of elevated N export and flushing, vegetation and soils still represent important controls (Figure 11). The explanation behind the result is clear—the Mediterranean climate in the Pacific Northwest results in significant winter moisture, which produces increased soil moisture, higher soil water flux, and larger stream discharge than seen during the dry summer period. In addition, the lower temperatures that dominate during the winter period suppress primary production and SOM dynamics. During the growing season, however, hydrologic flushing moves into the background, while control capacity of other components increases. This change in control is most evident in June–August, when mobilization rates tend to increase and flushing rates decrease. SOM contributes a larger portion of available nitrogen than the vegetation during this period of time, and this is primarily because uptake increases and the vegetation acts as a stronger sink of N that SOM during the summer period. This difference is in part a reflection of our lumping of litter and live biomass in our definition of control capacity. Particulate flushing of nitrogen mimics hydrologic flushing because of increased mobilization of aquatic biomass during the wet season, but the signal is muted when compared to hydrologic flushing. The lower seasonal variability develops at least in part because particulate inputs are derived directly from the litterfall/mortality model which did not include seasonal effects.

These results lend credence to the idea that understanding the dynamics of nitrogen, carbon, and water in ecosystems requires a multidisciplinary approach [39]. This approach certainly includes attention to flushing behavior, but the level of attention given to flushing must be on par with that given to production of available nitrogen, which may be dissolved or in particulate forms. Furthermore, the seasonality of the system imposes a series of constraints that result in predictable temporal variation in the activity of different system components. This variation is difficult to approach through standard field-based budget techniques; however numerical modeling can be used to extend budget results to provide a clearer picture of the seasonality.

3.2.2. Limitations of the Modeling Framework

The model framework and analyses presented here represent a step in our evolving understanding of how small catchments at HJA function. There are, however, a number of limitations to this work. The modeling focuses on an approximately monthly time series of input output data, which provides insight into seasonal dynamics. However, this time step is too coarse to understand the finer time dynamics that are often the focus of experimentalists working within the wide variety of contributing disciplines. At the same time, it may be too coarse to understand the longer term evolution of catchments, both in terms of nitrogen availability and changing climate. We envision significant potential to redevelop these modeling ideas to correspond to these different timescales—but the model as presented herein is not suited to either. Similarly, while we included a simple model of instream particulate retention and release, extreme events are not included. These events include for example large storms, mass movement, and fire, and in terms of nitrogen control, it is clear that over long time scales, these are at least as significant as the biotic controls and flushing that we explored here. These limitations simply mean that interpreting the results outside of the timeframe over which the model was run is not possible.

The model outlined here includes a variety of parameters that cannot be measured directly. For example to separate net nitrogen mineralization into gross mineralization and gross immobilization requires a set of measurements that are very difficult to perform at point scales [50]. And further, the relationship between these point scale observations for catchment level simulations is not well-established. Yet competition between vegetation and microbes is a key feature of nitrogen cycling [51], and microbial immobilization and mobilization, as well as plant uptake must be included in a model designed to explore the effects of this competition. The parameters involved in this portion of the model (and others) are developed based solely upon educated guesses, and calibration procedures relying upon evaluation of the resulting rate terms. In the case of models developed for predictive purposes, simpler tends to be more effective, and incorporating net mineralization, which can be more readily measured, would make more sense. Nonetheless, while some of our decisions clearly reduce the predictive capability of the model through increased parameter uncertainty, the more complete structural definition provides a framework to outline both what is known, and what hypothesized about how these catchments function with respect to N cycling.

Some of the most interesting aspects of watershed nitrogen work to emerge in the last two decades involve the role of spatially disaggregated watershed components in the processing of key stocks. This spatial dependence is evident throughout the literature, including terrestrial biogeochemistry, aquatic processes, catchment hydrologic processes, hyporheic zone interactions, climate science, and forest ecology. As constructed, HJA-N does include some quasi spatial distribution, but this distribution simply separates the upslope processes from the near stream processes. From both a biotic and abiotic standpoint, this is a large simplification. An important next step is the incorporation of similar mass balance equations within a spatially-distributed model that would better allow for the incorporation of key differences in spatial distribution of ecosystem processes.

Lastly, a more complete evaluation of the model is needed to provide further guidance in terms of parameter sensitivity and associated model uncertainty. Sensitivity analyses also have the potential to provide insight into the degree of understanding we have regarding different model components, and as such assist in the prioritization of experimental studies designed to improve the both the general understanding of the system, and the predictive capability of the model.

4. Conclusions

The primary goal behind the development of HJA-N was to distill knowledge from a variety of disciplinary scientists, including key observational datasets, to succinctly describe N dynamics in WS10 at HJA. In doing so, we have produced a temporally dynamic simulation, which produces results that are consistent with existing water, N, and C budgets. One of the motivations was to construct a tool that could be used to quantify the relative roles of vegetation, hydrology, soils and SOM, and the near stream zone in controlling the release of N in retentive regions like HJA. The key finding is that each of the different catchment elements plays a significant role in the retention of N, and that those roles vary seasonally. While in and of itself, this is not entirely surprising, the ability to quantify those contributions provides a useful means to more fully understand how the catchment functions with respect to N dynamics.

Acknowledgments: This paper represents a summary of discussions with a wide variety of researchers working at the HJ Andrews experimental forest. Discussions with Mark Harmon, Kate Lajtha, Roy Haggerty, Steve Perakis, Barbara Bond, Dick Waring, Sherri Johnson, and Jeff McDonnell were instrumental, and comments from two anonymous reviewers significantly improved the paper. Data were provided by the HJ Andrews Experimental Forest research program, funded by the National Science Foundation's Long-Term Ecological Research Program (DEB 1440409), US Forest Service Pacific Northwest Research Station, and Oregon State University.

Author Contributions: Kellie Vaché developed and ran the model, and wrote the manuscript; Lutz Breuer contributed to the modeling and the development of the manuscript; Julia Jones and Phil Sollins contributed to the modeling and text.

Conflicts of Interest: The authors declare no conflict of interest.

References

1. Galloway, J.N.; Dentener, F.J.; Capone, D.G.; Boyer, E.W.; Howarth, R.W.; Seitzinger, S.P.; Asner, G.P.; Cleveland, C.C.; Green, P.A.; Holland, E.A.; *et al.* Nitrogen cycles: Past, present, and future. *Biogeochemistry* **2004**, *70*, 153–226.

2. Matson, P.; Lohse, K.A.; Hall, S.J. The globalization of nitrogen deposition: Consequences for terrestrial ecosystems. *AMBIO J. Hum. Environ.* **2002**, *31*, 113–119.

3. Van Breemen, N.; Boyer, E.W.; Goodale, C.L.; Jaworski, N.A.; Paustian, K.; Seitzinger, S.P.; Lajtha, K.; Mayer, B.; van Dam, D.; Howarth, R.W.; *et al.* Where did all the nitrogen go? Fate of nitrogen inputs to large watersheds in the northeastern USA. *Biogeochemistry* **2002**, *57*, 267–293.

4. Perakis, S.S.; Sinkhorn, E.R. Biogeochemistry of a temperate forest nitrogen gradient. *Ecology* **2011**, *92*, 1481–1491.

5. Aber, J.D.; Nadelhoffer, K.J.; Steudler, P.; Melillo, J.M. Nitrogen saturation in northern forest ecosystems. *BioScience* **1989**, *39*, 378–286.

6. Stoddard, J.L. Long-term changes in watershed retention of nitrogen. Its causes and aquatic consequences. In *Environmental Chemistry of Lakes and Reservoirs, Edition: Advances in Chemistry Series, Edition Advances in Chemistry Series*; American Chemical Society: Washington, DC, USA, 1994; Volume 237, pp. 223–284.

7. Meixner, T.; Fenn, M. Biogeochemical budgets in a Mediterranean catchment with high rates of atmospheric N deposition—Importance of scale and temporal asynchrony. *Biogeochemistry* **2004**, *70*, 331–356.

8. Curtis, C.J.; Evans, C.D.; Goodale, C.L.; Heaton, T.H. What have stable isotope studies revealed about the nature and mechanisms of N saturation and nitrate leaching from semi-natural catchments? *Ecosystems* **2011**, *14*, 1021–1037.

9. Whitehead, P.G.; Lapworth, D.J.; Skeffington, R.A.; Wade, A. Excess nitrogen leaching and C/N decline in the Tillingbourne catchment, southern England: INCA process modelling for current and historic time series. *Hydrol. Earth Syst. Sci.* **2002**, *6*, 455–466.

10. Cosby, B.J.; Ferrier, R.C.; Jenkins, A.; Emmett, B.A.; Wright, R.F.; Tietema, A. Modelling the ecosystem effects of nitrogen deposition: Model of Ecosystem Retention and Loss of Inorganic Nitrogen (MERLIN). *Hydrol. Earth Syst. Sci. Discuss.* **1997**, *1*, 137–158.

11. Fenn, M.E.; Haeuber, R.; Tonnesen, G.S.; Baron, J.S.; Grossman-Clarke, S.; Hope, D.; Jaffe, D.A.; Copeland, S.; Geiser, L.; Rueth, H.M.; *et al.* Nitrogen emissions, deposition, and monitoring in the Western United States. *BioScience* **2003**, *53*, 391–403.

12. Fenn, M.E.; Baron, J.S.; Allen, E.B.; Rueth, H.M.; Nydick, K.R.; Geiser, L.; Bowman, W.D.; Sickman, J.O.; Meixner, T.; Johnson, D.W.; *et al.* Ecological effects of nitrogen deposition in the Western United States. *BioScience* **2003**, *53*, 404–420.

13. Heemskerk, M.; Wilson, K.; Pavao-Zuckerman, M. Conceptual models as tools for communication across disciplines. *Conserv. Ecol.* **2003**, *7*, 8.

14. Grant, R.F.; Roulet, N.T. Methane efflux from boreal wetlands: Theory and testing of the ecosystem model Ecosys with chamber and tower flux measurements. *Glob. Biogeochem. Cycles* **2002**, *16*.

15. Eshleman, K.N. A linear model of the effects of disturbance on dissolved nitrogen leakage from forested watersheds. *Water Resour. Res.* **2000**, *36*, 3325–3335.

16. Landsberg, J.J.; Waring, R.H. A generalised model of forest productivity using simplified concepts of radiation-use efficiency, carbon balance and partitioning. *For. Ecol. Manag.* **1997**, *95*, 209–228.

17. Aber, J.D.; Ollinger, S.V.; Driscoll, C.T. Modeling nitrogen saturation in forest ecosystems in response to land use and atmospheric deposition. *Ecol. Model.* **1997**, *101*, 61–78.

18. Parton, W.J. The CENTURY model. In *Evaluation of Soil Organic Matter Models*; Springer: Berlin, Germany; Heidelberg, Germany, 1996; pp. 283–291.

19. Rastetter, E.B.; Ryan, M.G.; Shaver, G.R.; Melillo, J.M.; Nadelhoffer, K.J.; Hobbie, J.E.; Aber, J.D. A general biogeochemical model describing the responses of the C and N cycles in terrestrial ecosystems to changes in CO_2, climate, and N deposition. *Tree Physiol.* **1991**, *9*, 101–126.

20. Running, S.W.; Coughlan, J.C. A general model of forest ecosystem processes for regional applications I. Hydrologic balance, canopy gas exchange and primary production processes. *Ecol. Model.* **1988**, *42*, 125–154.

21. Whitehead, P.G.; Wilson, E.J.; Butterfield, D. A semi-distributed Integrated Nitrogen model for multiple source assessment in Catchments (INCA): Part I—Model structure and process equations. *Sci. Total Environ.* **1998**, *210*, 547–558.

22. Band, L.E.; Tague, C.L.; Groffman, P.; Belt, K. Forest ecosystem processes at the watershed scale: Hydrological and ecological controls of nitrogen export. *Hydrol. Process.* **2001**, *15*, 2013–2028.

23. Rastetter, E.B.; Shaver, G.R. A model of multiple-element limitation for acclimating vegetation. *Ecology* **1992**, *73*, 1157–1174.

24. Ranken, D.W. Hydrologic Properties of Soil and Subsoil on a Steep, Forested Slope. Master's Thesis, Oregon State University, Corvallis, OR, USA, 1974.

25. Swanson, F.J.; James, M.E. Geology and Geomorphology of the HJ Andrews Experimental Forest, Western Cascades, Oregon. 1975. Available online: http://catalog.hathitrust.org/Record/007405299 (accessed on 20 March 2015).

26. McGuire, K.J.; McDonnell, J.J.; Weiler, M.; Kendall, C.; McGlynn, B.L.; Welker, J.M.; Seibert, J. The role of topography on catchment-scale water residence time. *Water Resour. Res.* **2005**, *41*.

27. Vanderbilt, K.L.; Lajtha, K.; Swanson, F.J. Biogeochemistry of unpolluted forested watersheds in the Oregon Cascades: Temporal patterns of precipitation and stream nitrogen fluxes. *Biogeochemistry* **2003**, *62*, 87–117.

28. Grier, C.C.; Logan, R.S. Old-growth *Pseudotsuga menziesii* communities of a Western Oregon watershed: Biomass distribution and production budgets. *Ecol. Monogr.* **1977**, *47*, 373–400.

29. Sollins, P.; Grier, C.C.; McCorison, F.M.; Cromack, K., Jr.; Fogel, R.; Fredriksen, R.L. The internal element cycles of an old-growth Douglas-fir ecosystem in Western Oregon. *Ecol. Monogr.* **1980**, *50*, 261–285.

30. Triska, F.J.; Sedell, J.R.; Cromack, K., Jr.; Gregory, S.V.; McCorison, F.M. Nitrogen budget for a small coniferous forest stream. *Ecol. Monogr.* **1984**, *54*, 119–140.

31. Richmond, B.; Peterson, S. *An Introduction to Systems Thinking*; High Performance Systems, Incorporated: Honolulu, HI, USA, 2001.

32. Bergström, S.; Singh, V.P. The HBV model. In *Computer Models of Watershed Hydrology*; Singh, V.P., Ed.; Water Resources Publications: Highlands Ranch, CO, USA, 1995; pp. 443–476.

33. Shuttleworth, W.J. *Evaporation. Handbook of Hydrology*; Maidment, D.R., Ed.; McGraw-Hill: New York, NY, USA, 1993.

34. Harr, R.D. Water flux in soil and subsoil on a steep forested slope. *J. Hydrol.* **1977**, *33*, 37–58.

35. Chen, H.; Harmon, M.E.; Griffiths, R.P. Decomposition and nitrogen release from decomposing woody roots in coniferous forests of the Pacific Northwest: A chronosequence approach. *Can. J. For. Res.* **2001**, *31*, 246–260.

36. Hedin, L.O.; Armesto, J.J.; Johnson, A.H. Patterns of nutrient loss from unpolluted, old-growth temperate forests: Evaluation of biogeochemical theory. *Ecology* **1995**, *76*, 493–509.

37. Perakis, S.S. Nutrient limitation, hydrology and watershed nitrogen loss. *Hydrol. Process.* **2002**, *16*, 3507–3511.

38. Neff, J.C.; Chapin, F.S., III; Vitousek, P.M. Breaks in the cycle: Dissolved organic nitrogen in terrestrial ecosystems. *Front. Ecol. Environ.* **2003**, *1*, 205–211.

39. Bond, B.J.; Jones, J.A.; Moore, G.; Phillips, N.; Post, D.; McDonnell, J.J. The zone of vegetation influence on baseflow revealed by diel patterns of streamflow and vegetation water use in a headwater basin. *Hydrol. Process.* **2002**, *16*, 1671–1677.

40. Haggerty, R.; Wondzell, S.M.; Johnson, M.A. Power-law residence time distribution in the hyporheic zone of a 2nd-order mountain stream. *Geophys. Res. Lett.* **2002**, *29*, 18.

41. Harmon, M.E.; Bible, K.; Ryan, M.G.; Shaw, D.C.; Chen, H.; Klopatek, J.; Li, X. Production, respiration, and overall carbon balance in an old-growth Pseudotsuga-Tsuga forest ecosystem. *Ecosystems* **2004**, *7*, 498–512.

42. McCuen, R.H.; Knight, Z.; Cutter, A.G. Evaluation of the Nash–Sutcliffe efficiency index. *J. Hydrol. Eng.* **2006**, *11*, 597–602.

43. Sollins, P.; McCorison, F.M. Nitrogen and carbon solution chemistry of an old growth coniferous forest watershed before and after cutting. *Water Resour. Res.* **1981**, *17*, 1409–1418.

44. Coops, N.C.; Black, T.A.; Jassal, R.P.S.; Trofymow, J.T.; Morgenstern, K. Comparison of MODIS, eddy covariance determined and physiologically modelled gross primary production (GPP) in a Douglas-fir forest stand. *Remote Sens. Environ.* **2007**, *107*, 385–401.

45. Abdelnour, A.; McKane, R.B.; Stieglitz, M.; Pan, F.; Cheng, Y. Effects of harvest on carbon and nitrogen dynamics in a Pacific Northwest forest catchment. *Water Resour. Res.* **2013**, *49*, 1292–1313.

46. Hicks, W.T.; Harmon, M.E.; Griffiths, R.P. Abiotic controls on nitrogen fixation and respiration in selected woody debris from the Pacific Northwest, USA. *Ecoscience* **2003**, *10*, 66–73.

47. Ashkenas, L.R.; Johnson, S.L.; Gregory, S.V.; Tank, J.L.; Wollheim, W.M. A stable isotope tracer study of nitrogen uptake and transformation in an old-growth forest stream. *Ecology* **2004**, *85*, 1725–1739.

48. Hornberger, G.M.; Bencala, K.E.; McKnight, D.M. Hydrological controls on dissolved organic carbon during snowmelt in the Snake River near Montezuma, Colorado. *Biogeochemistry* **1994**, *25*, 147–165.

49. Lajtha, K.; Crow, S.E.; Yano, Y.; Kaushal, S.S.; Sulzman, E.; Sollins, P.; Spears, J.D.H. Detrital controls on soil solution N and dissolved organic matter in soils: A field experiment. *Biogeochemistry* **2005**, *76*, 261–281.

50. Bosatta, E.; Ågren, G.I. Theoretical analyses of interactions between inorganic nitrogen and soil organic matter. *Eur. J. Soil Sci.* **1995**, *46*, 109–114.

51. Kaye, J.P.; Hart, S.C. Competition for nitrogen between plants and soil microorganisms. *Trends Ecol. Evol.* **1997**, *12*, 139–143.

Spatial Quantification of Non-Point Source Pollution in a Meso-Scale Catchment for an Assessment of Buffer Zones Efficiency

Mikołaj Piniewski, Paweł Marcinkowski , Ignacy Kardel , Marek Giełczewski , Katarzyna Izydorczyk and Wojciech Frątczak

Abstract: The objective of this paper was to spatially quantify diffuse pollution sources and estimate the potential efficiency of applying riparian buffer zones as a conservation practice for mitigating chemical pollutant losses. This study was conducted using a semi-distributed Soil and Water Assessment Tool (SWAT) model that underwent extensive calibration and validation in the Sulejów Reservoir catchment (SRC), which occupies 4900 km^2 in central Poland. The model was calibrated and validated against daily discharges (10 gauges), NO$_3$-N and TP loads (7 gauges). Overall, the model generally performed well during the calibration period but not during the validation period for simulating discharge and loading of NO$_3$-N and TP. Diffuse agricultural sources appeared to be the main contributors to the elevated NO$_3$-N and TP loads in the streams. The existing, default representation of buffer zones in SWAT uses a VFS sub-model that only affects the contaminants present in surface runoff. The results of an extensive monitoring program carried out in 2011–2013 in the SRC suggest that buffer zones are highly efficient for reducing NO$_3$-N and TP concentrations in shallow groundwater. On average, reductions of 56% and 76% were observed, respectively. An improved simulation of buffer zones in SWAT was achieved through empirical upscaling of the measurement results. The mean values of the sub-basin level reductions are 0.16 kg NO$_3$/ha (5.9%) and 0.03 kg TP/ha (19.4%). The buffer zones simulated using this approach contributed 24% for NO$_3$-N and 54% for TP to the total achieved mean reduction at the sub-basin level. This result suggests that additional measures are needed to achieve acceptable water quality status in all water bodies of the SRC, despite the fact that the buffer zones have a high potential for reducing contaminant emissions.

Reprinted from *Water*. Cite as: Piniewski, M.; Marcinkowski, P.; Kardel, I.; Giełczewski, M.; Izydorczyk, K.; Frątczak, W. Spatial Quantification of Non-Point Source Pollution in a Meso-Scale Catchment for an Assessment of Buffer Zones Efficiency. *Water* **2015**, *7*, 1889–1920.

1. Introduction

1.1. Water Management Context

Fulfilling the requirements of the European Water Framework Directive [1] and the Nitrates Directive [2] by reducing pollution emissions to water ecosystems remains a major challenge faced in water management. Particularly, the main issue is the reduction of non-point pollution that originates from agricultural land. The contributions of agriculture to the pool of nitrogen and phosphorus compounds in water ecosystems are high.

In Poland, the large share of farmland consisting of highly fragmented arable land and strongly dispersed developments has resulted in major pressure from pollution emission sources, including (1) pressure from agriculture related to the use of inappropriate farming practices (transport of organic and mineral nitrogen and phosphorus compounds from fertilizers to the environment) and (2) pressure from scattered households that are not connected to sewage systems.

Thus, the development of N and P reduction strategies is a major task for water authorities throughout Europe. One example of activities that are undertaken to achieve sustainable water management goals in agricultural catchments is the EU-funded EKOROB project (Ecotones for reducing diffuse pollution). The main objective of this project is to develop an Action Plan for reductions of diffuse pollution in the Pilica River catchment (Poland) and will help achieve a good ecological status for the water in the Sulejów Reservoir, particularly by reducing eutrophication and decreasing the frequency and intensity of cyanobacterial blooms.

The Action Plan is based on the ecohydrology concept [3–5], which assumes that the basis for integrated river basin management is the quantification of catchment-scale processes that are part of the hydrological cycle. The concept of ecohydrology involves quantifying hydrological processes at the basin scale and the entire hydrological cycle to quantify ecological processes. This quantification includes the patterns of hydrological pulses along the river continuum and the identification of various human impacts on point and non-point sources of pollution [6]. Thus, this quantification should be the first step when developing regulatory processes for sustainable water use and ecosystem protection. Although many mathematical tools are available for this task, the Soil and Water Assessment Tool (SWAT) [7] is one of the most widely used and comprehensive tools.

1.2. The Use of SWAT for Quantifying Emission Sources

SWAT is a comprehensive hydrological/water quality model that is increasingly being used to address a wide variety of water resource problems across the globe [8]. Several studies have investigated the spatial variability and distribution of various pollutant emissions/losses in catchments of different sizes [9–12]. Niraula *et al.* [9] calibrated SWAT (and a less complex GWLF model) for a small catchment in Alabama and used it to identify Critical Source Areas (CSAs) for sediment, TN and TP based on the loadings per unit area (yield or emission or losses) at the sub-basin level. Another application of SWAT in a medium-sized Greek catchment resulted in similar findings, but with a finer level of Hydrological Response Units (HRUs) [10]. Wu and Chen [11] investigated the influences of point source and diffuse pollution on the water quality of a relatively large catchment in south China by using SWAT. These authors concluded that diffuse pollution overwhelmingly surpasses point source pollution for all constituents except TP. In addition, these authors identified CSAs at the HRU and sub-basin levels. Finally Wu and Liu [12] calibrated SWAT for a large catchment in Iowa and showed a relationship between the shares of agricultural areas with sediment and NO_3-N emissions by using the calibrated model.

1.3. Riparian Buffer Zones and Their Modeling in SWAT

Riparian buffer zones (ecotones, vegetative filter strips) are an effective Best Management Practice (BMP) for buffering aquatic ecosystems against nutrient losses from the agricultural landscapes. Buffer zones are strips of permanent vegetation (including herbs, grasses, shrubs or trees) that are adjacent to aquatic ecosystems and used to maintain or improve water quality by trapping and removing various non-point source pollutants from overland and shallow sub-surface flow [13–16].

For pollutants transported in surface runoff, the process of sediment and nutrient trapping by buffer zones is reasonably well understood, particularly for grass filter strips (*cf.* review [17]). Reductions in the surface flow velocities due to the increased hydraulic roughness of the vegetation in the buffer enhanced particle deposition. Vought *et al.* [18] reported that a buffer strip with a width of 10 m can reduce phosphorus loads, which are typically bound to sediments, by as much as 95%. Buffer zone are effective for removing sediments and other suspended solids contained in surface runoff; however, soluble forms of nitrogen and phosphorus are not removed as effectively as sediments [19]. Results collected from 44 fields (row crops with slopes range from 1%–14%) showed that a 10-m buffer zone reduced the total suspended solids, soluble phosphorus and nitrate-nitrogen contents by 64%, 34% and 38%, respectively [20]. The efficiencies of narrow buffer zones (5–10 m) in Norway varied from 81%–91%, 60%–89% and 37%–81% for particles, total phosphorus and total nitrogen, respectively [21].

Buffer strips are normally less efficient for removing nitrate than orthophosphates from surface runoff. In contrast with orthophosphate, nitrogen is very labile and is not largely adsorbed within the soil [18]. However, the impacts of the sub-surface flow efficiency of the buffer zones on reducing nitrogen are well described in the literature and often reach a concentration reduction of

90% [17]. A meta-analysis of nitrogen removal in riparian buffers based on data from 65 individual riparian buffers from published studies indicated a mean removal effectiveness of 76.7% [22]. However, the efficiency of buffer zones for reducing phosphorus in shallow groundwater is not well documented, and some studies suggest that riparian zones are ineffective for removing dissolved phosphorus or could even release phosphorus to the water [23,24].

Buffer zones along small streams that are more exposed to pressure from agriculture are more efficient than buffer zones along larger rivers. A key factor for determining the efficiency of buffer zones is their continuity. Continuous and narrow riparian buffers are more efficient than wider and intermittent buffers [25]. Hence, an important issue for effective water management is the selection of priority areas that have the highest emissions of diffuse pollution. Next, concentrating measures, such as buffer zones, should be applied in these areas. Catchment-scale water quality modeling is one possible solution for quickly identifying priority areas.

Numerous examples are available regarding the application of SWAT for simulating the affects of buffer zones on diffuse pollution [26–29]. Older versions of SWAT used a very simplistic equation that was only based on filter width for calculating the HRU-level reduction rate of buffer zones. This equation was based on empirical data from the US regarding buffer strip efficiency [27,28]. Since then, SWAT has undergone certain modifications to address variable source areas within watersheds and vegetated buffers adjacent to streams [26,30]. The new VFS sub-model currently used in SWAT reduces the sediment, nitrate and phosphorus loading in streams as a function of estimated reductions in runoff. Hence, the new VFS sub-model only affects contaminants present in surface runoff and neglects nutrient trapping in shallow groundwater. As mentioned previously for nitrogen in sub-surface flow, buffer zones are very efficient measures [17]. However, little consensus has been reached for phosphorus [23,24]. This result suggests that the buffer zone efficiency is case-specific and depends on local conditions. Hence it is equally as important to apply existing models as it is to measure the efficiency of existing buffer zones in the field to gain more confidence regarding their behavior.

1.4. Objective

Two objectives of this paper are:

1. Spatial quantification of NO_3-N and TP emissions from major pollution sources in a meso-scale catchment using SWAT.
2. Simulation of buffer zone effects on the mitigation of pollution losses when applied in Critical Source Areas through the combined use of the default SWAT VFS sub-model and local field monitoring data.

The term "meso-scale catchment" refers to catchments with an order of magnitude between 10 and 10^3 km^2 [31]. A part of the Pilica catchment selected as the case study in this paper, a demonstration catchment of the EKOROB project, satisfies this condition.

2. Materials and Methods

2.1. Study Area

The study was conducted in the Sulejów Reservoir catchment (hereafter referred to as the SRC). Sulejów is a shallow and eutrophic artificial reservoir that was built in 1974 and is situated in the middle course of the Pilica River in central Poland. Two main tributaries supply water to the Sulejów Reservoir: the Pilica and Luciąża Rivers. At its full capacity, this reservoir covers an area of 22 km^2, with a mean depth of 3.3 m and a volume of 75×10^6 m^3. The Sulejów Reservoir was used as a drinking water reservoir for Łódź agglomeration until 2004 and is currently an important recreational site that has been extensively studied (*cf.* review [32]). Microcystis aeruginosa is the dominant species of bloom-forming cyanobacteria in the reservoir and produces microcystin-LR, microcystin-YR, and microcystin-RR [33–35].

The SWAT model is used in this study for the entire SRC area upstream of the dam, which occupies 4933 km^2 (Figure 1). This area consists of the Pilica (contributing 79.8% of area) and the Luciąża (15.3%) River catchments and a direct reservoir sub-catchment with several smaller streams (4.9%). The elevation of the SRC varies from 154 m in the lowland areas in the north to 499 m in the highland areas in the south. The distribution of land cover in the SRC area is as follows: 44.4% arable land, 38.6% forest areas, 12.3% grasslands, and 4.7% urban areas (mainly low-density residential areas), with the remaining land occupied by other types of land cover (data according to Corine Land Cover 2006). The predominate soil types in this area are loamy sands and sands. The climate of this area is typical for central Poland, with a mean annual temperature of 7.5 °C and mean January and July temperatures of −4 °C and 18 °C, respectively. The mean annual precipitation is 600 mm. The highest amounts of precipitation occur in June/July, and the lowest amounts of precipitation occur in January. The flow regime is characterized by early spring snow-melt induced floods and summer low flows with occasional summer floods. The quantitative pressure on surface water resources is relatively low. The fish farming industry scattered around the catchment and the Cieszanowice Reservoir constructed in 1998 on the Luciąża River (volume of 7.3×10 m^3 at full capacity) are the only considerable sources of flow alteration.

In contrast, multiple point and non-point pollution sources in the area result in elevated N and P loads flowing into the Sulejów Reservoir, which eventually contribute to toxic cyanobacterial blooms in its waters. These different sources will be described systematically in terms of SWAT model inputs in Sub-Section 2.4.

2.2. SWAT Model

2.2.1. General Features

SWAT is a physically based, semi-distributed, continuous-time model that simulates the movement of water, sediment, and nutrients on a catchment scale with a daily time step. The basic calculation unit, referred to as a "hydrological response unit" (HRU) is a combination of land use, soil, and slope overlay. Both water balance components, which is a driving force behind affect all processes that occur in a watershed, and water quality, output parameters are computed separately for each HRU. Water, nutrients and sediment leaving HRUs are aggregated at the sub-basin level and routed through the stream network to the main outlet to obtain the total flows and loadings for the river basin.

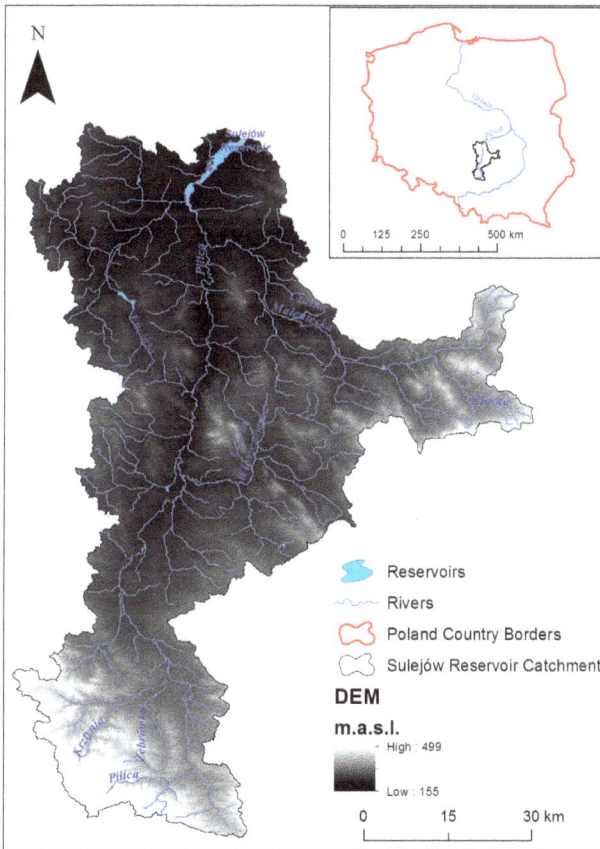

Figure 1. Study area.

225

In this study, channel routing was modeled using a variable storage coefficient approach. The modified USDA Soil Conservation Service (SCS) curve number method for calculating surface runoff and the Penman-Monteith method for estimating potential evapotranspiration (PET) were selected. In the model, snow-melt estimations are based on the degree-day method. The SWAT adapted plant growth model, which is used to assess the removal of water and nutrients from the root zone, transpiration, and biomass/yield production, is based on EPIC [36]. The in-stream water quality component allows us to control nutrient transformations in the stream. The in-stream kinetics used in SWAT for nutrient routing are adapted from QUAL2E [37].

SWAT simulates the movement and transformation of several forms of nitrogen and phosphorus in the watershed. In the nitrogen cycle, the main processes are denitrification, nitrification, mineralization, plant uptake, decay, fertilization, and volatilization. In the phosphorus cycle, the main processes are mineralization, fertilization, decay, and plant uptake. The nutrient transport pathways from upland areas to stream networks correspond to the following hydrological transport pathways: surface runoff, lateral subsurface flow and groundwater flow. Additionally point source discharges of water and contaminants can be defined that are directly input into the water routed through the stream network.

From the point of view of modeling buffer zones in SWAT, it is important to note that HRUs are lumped and non-contiguous geographic units within each sub-basin. A SWAT model setup may consist of thousands of such units, and each of them may represent one field, a portion of a field, or, more likely, portions of many fields [38].

2.2.2. Runoff-Reduction-Based Buffer Zone Sub-Model

A key characteristic of the buffer zone sub-model implemented in SWAT is that it works at the HRU level and reduces the loads of sediment, nitrate and phosphorus that enter the stream as a function of estimated reductions in runoff. Hence, the sub-model only affects contaminants that are present in surface runoff and neglects the potential affects of buffer zones on shallow groundwater. This sub-model was developed and evaluated using measured data derived from the literature and included data that were collected using differing experimental protocols and under conditions with different soils, slopes and rainfall intensities. When measured data were unavailable, predictions from the process-based Vegetative Filter Strip MODel (VFSMOD) [39] and its companion program, UH, were used. The UH (upland hydrology) utility uses the curve number approach (USDA-SCS, 1972), unit hydrograph and the Modified Universal Soil Loss Equation (MUSLE) [40] and allowed us to generate a database of sediment and runoff loads that enter the VFS. VFSMOD simulations were used to evaluate the sensitivities of various parameters and correlations between the model inputs and predictions. Consequently, an empirical model for runoff reduction by VFSs was developed, as described by the following equation:

$$RR = 75.8 - 10.8 \ln(R_L) + 25.9 \ln(K_{sat})$$ (1)

where RR is the runoff reduction (%); R_L is the runoff loading to the buffer zone (mm); and K_{sat} is the saturated hydraulic conductivity (mm $\cdot$ h^{-1}). An important consideration is that SWAT conceptually partitions VFS sections within an HRU into two parts: a short part that occupies 10% of its length and receives flow from the 0.25–0.75 of the field area and a part that occupies 90% of its length and receives the remaining amount of flow. A fraction of the flow through the most heavily loaded 10% that is fully channelized is not subject to the VFS sub-model. Although the buffer zone width is an essential and intuitive characteristic that influences its trapping efficiency, it is not implemented as a parameter of the VFS sub-model in SWAT. Instead, the drainage area to buffer zone area ratio ($DAFS_{ratio}$) that is negatively correlated with the buffer zone width is combined with the HRU-level predicted runoff to estimate R_L.

The sediment reduction model based on VFS data removes sediment by reducing runoff velocity and enhancing infiltration in the VFS area, which is described by the following equation:

$$S_R = 79.0 - 1.04 S_L + 0.213 R_R$$ (2)

where S_R is the predicted sediment reduction (%) and S_L is the sediment loading (kg $\cdot$ m^{-2}).

The nitrate reduction model was only based on runoff reduction, as described by the following equation:

$$NN_R = 39.4 + 0.584 R_R$$ (3)

where NN_R is the reduction of nitrate nitrogen (%).

The model for total phosphorus reduction was based on sediment reduction, which is described by the following equation:

$$TP_R = 0.9 S_R$$ (4)

where TP_R is the total phosphorus reduction (%); and S_R is the sediment reduction (%).

2.3. SWAT Setup of the SRC

Table 1 lists all major data items and sources used to create the SWAT model setup of the SRC. The specific applications of this data at different stages of model development are described below.

Table 1. Data items and sources used to create the SWAT model setup of the SRC.

Data	Source
Digital Elevation Model	CODGiK (Central Agency for Geodetic and Cartographic Documentation)
Water cadastre GIS layers	RZGW (Regional Water Management Authority in Warsaw)
Corine Land Cover 2006	GDOS (General Directorate of the Environmental Protection)
Orthophotomap	GUGiK (Head Office of Geodesy and Cartography: geoportal.gov.pl)
Agricultural statistics	The Local Data Bank of GUS (Central Statistical Office)
Agricultural soil map 1:100,000	IUNG-PIB (Institute of Soil Science and Plant Cultivation - National Research Institute)
Forest soil maps 1:25,000	RDLP (Regional Directorate of State Forests in Łódź, Radom and Katowice)
Atmospheric deposition of nitrogen	GIOS (Chief Inspectorate of Environmental Protection
Climate data	IMGW-PIB (Institute of Meteorology and Water Management - National Research Institute)

The automatic watershed delineation of the SRC was based on the Digital Elevation Model (DEM) and stream network GIS layer. A 5 m resolution DEM characterized by a mean elevation error of 0.8–2.0 m was created from the ESRI TIN DEM available from CODGiK (Polish Central Geodetic and Cartographic Agency). This DEM resulted in the division of the catchment into 272 sub-basins with average areas of 18.1 km^2 (Figure 2). The Corine Land Cover (CLC) 2006 layer was used as the primary data source for the land use/land cover map. However, this layer was enhanced by several supplementary datasets and analyses:

- The (open) drainage ditch layer was used to sub-divide the CLC grasslands class into those under (code: FES2) and beyond (code: FESC; *cf.* Figure 3) the influences of drainage. It was assumed that the influences of the drainage ditches occurred within a 100 m buffer around the ditches.
- The orthophotomap was used to identify which SWAT crop database classes should be assigned to the "Heterogeneous agricultural areas" CLC class (code 2.4). Based on a manual, case-by-case investigation, the following three classes were most frequently assigned: "Agricultural land generic" (AGRL), "Urban low density" (URLD) and "Mixed forests" (FRST).
- The commune-level (39 units) agricultural census statistical data from 2010 were used to sub-divide the "Non-irrigated arable land" CLC class (code 2.1.1) into classes that represented particular crops that were cultivated in the SRC. This subdivision was done using a set of GIS techniques, including the "Create Random Raster" tool in ArcGIS. Thus, a 100 m resolution raster dataset that represented the random (yet preserving the commune level of crop distribution) locations of 6 major crops was created and combined with

the final land cover map used as SWAT input. Although it may seem risky to generate random crop locations, we believe that this approach does not significantly impact the modeling result due to the lumped nature of the HRUs within each sub-basin.

Overall, the following five crops were distinguished as well as a fallow/abandoned land class (BERM): spring barley (BARL), rye (RYE), potato (POTA), corn silage (CSIL) and head lettuce (LETT). Figure 3 shows the final distributions of all land cover classes in the SWAT model setup.

The numerical soil map (scale 1:100,000) from the Institute of Soil Science and Plant Cultivation (IUNG) and numerical soil maps (scale 1:25,000) from the Regional Directorate of State Forests were used to create a user soil input map with 27 soil classes. By overlaying the land use and soil maps, 3401 HRUs were delineated in the catchment. The following area thresholds were used in the HRU delineation: 30 ha for land cover and 50 ha for soils. Thus, when using this method, all land cover types below the first threshold in each sub-basin were removed and aggregated into the remaining classes.

The meteorological data required by SWAT (precipitation, solar radiation, relative humidity, wind speed, and maximum and minimum temperatures) were acquired from the Institute of Meteorology and Water Management-National Research Institute (IMGW-PIB) for 1982–2011. Precipitation data were obtained from 49 stations, whereas data for other variables from 17 stations. To improve the spatial representation of climate inputs, spatial interpolation of all variables (apart from solar radiation, which was only available for one station) was performed before reading the SWAT input files. For precipitation, the Ordinary Kriging (OK) method was applied. However, the Inverse Distance Weighted method was used for the other variables. Szcześniak and Piniewski [41] showed that the OK method outperforms the SWAT default method for precipitation. For other weather variables, the interpolation process did not significantly affect the modeling results.

2.4. Pollution Sources in the Model Setup

Parameterization of point and non-point source pollution plays a critical role in water quality modeling and has attracted considerable attention in this paper. The following anthropogenic pollution sources were identified: (1) diffuse pollution from agricultural areas; (2) sewage treatment plants and septic tanks and (3) fish ponds. Atmospheric deposition of nitrogen was also considered.

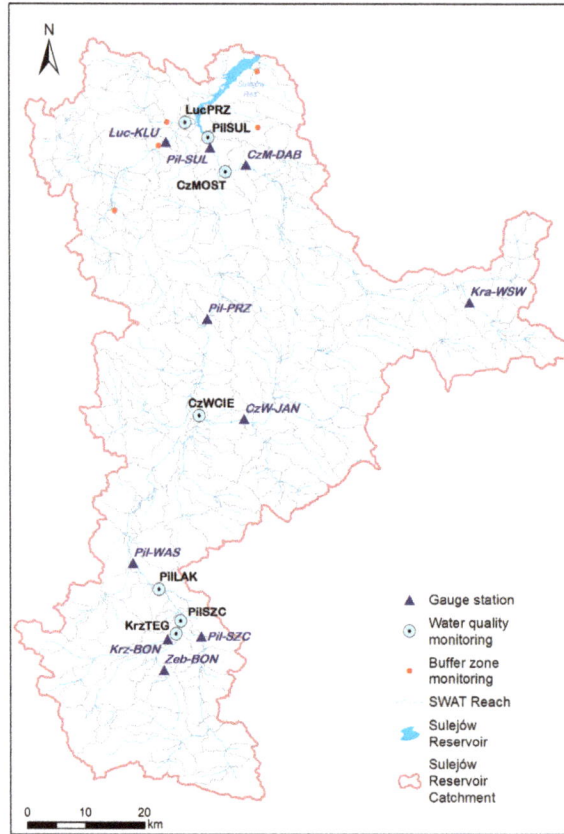

Figure 2. Delineation of the SRC into sub-basins and the gauging station locations used for calibration and validation.

2.4.1. Atmospheric Deposition

The mean concentration of nitrogen in precipitation measured in Sulejów near the inlet of the Sulejów Reservoir was obtained from the Chief Inspectorate of Environmental Protection (GIOS). The monthly results covered the time period of 1999–2010. Precipitation samples were collected daily, and an integrated sample was created and measured in the laboratory each month. Between 1999 and 2010, the mean annual concentrations varied significantly between 1.39 and 2.2 mg N $\cdot$ L^{-1}. The final value input into the model was subject to calibration and equaled 1.48 mg N $\cdot$ L^{-1}.

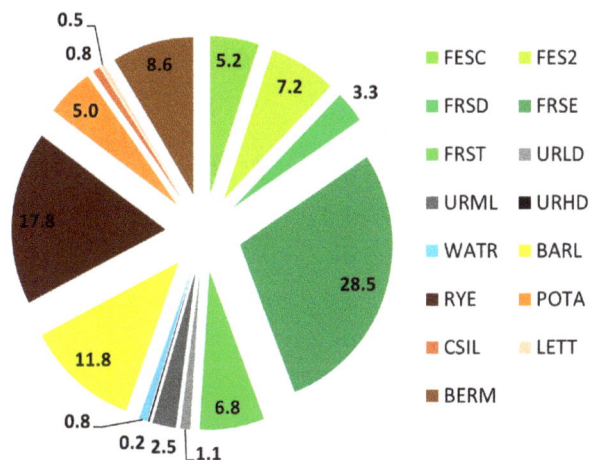

Figure 3. Land cover classes used as input for the SWAT model of the SRC. All codes as in the default SWAT plant database with one exception: FES2 has the same parameters as FESC, but is under the influence of open drainage ditches.

2.4.2. Fertilizers

Diffuse N and P pollution from agricultural fields mainly results from fertilizer use. SWAT enables us to define crop- and soil-specific management practices scheduled by date for each agricultural HRU. Typical management practice schedules, including the dates, types and amounts of fertilization, were obtained by consulting local extension service experts. First, derived management schedules were assigned to agricultural HRUs by using the ArcSWAT interface. However, this approach typically leads to bias in the total amounts of spatially-averaged fertilizer when compared with data from external sources. In this study, we used commune-level data from the Central Statistical Office (as for 2010) to determine mineral fertilizer use and livestock population. The livestock population was used to calculate the amount of available organic fertilizer (manure or slurry). In the final step conducted using GIS software, correction factors for fertilizer rates were defined for the sub-basins that overlapped with different communes. The commune layer intersected the sub-basin layer so that the total amounts of fertilizer used annually in different communes (expressed in tons of N and P) could be distributed over the SWAT sub-basins proportionally to the area of agricultural land in each sub-basin. Simultaneously, we aggregated the total fertilizer use per sub-basin from the model output based on initially implemented management schedules. Next, correction factors were calculated for each sub-basin as the ratios of total fertilizer use at the sub-basin level from census data to the total fertilizer use obtained from the SWAT output files. In the final step, each HRU fertilizer rate in the operation schedules was adjusted using the calculated correction factors. After this adjustment, the bias in the spatially averaged

231

amounts of fertilizer largely decreased. Figure 4 shows the final, sub-basin-averaged rates of mineral and organic fertilizers applied in the SWAT model of the SRC.

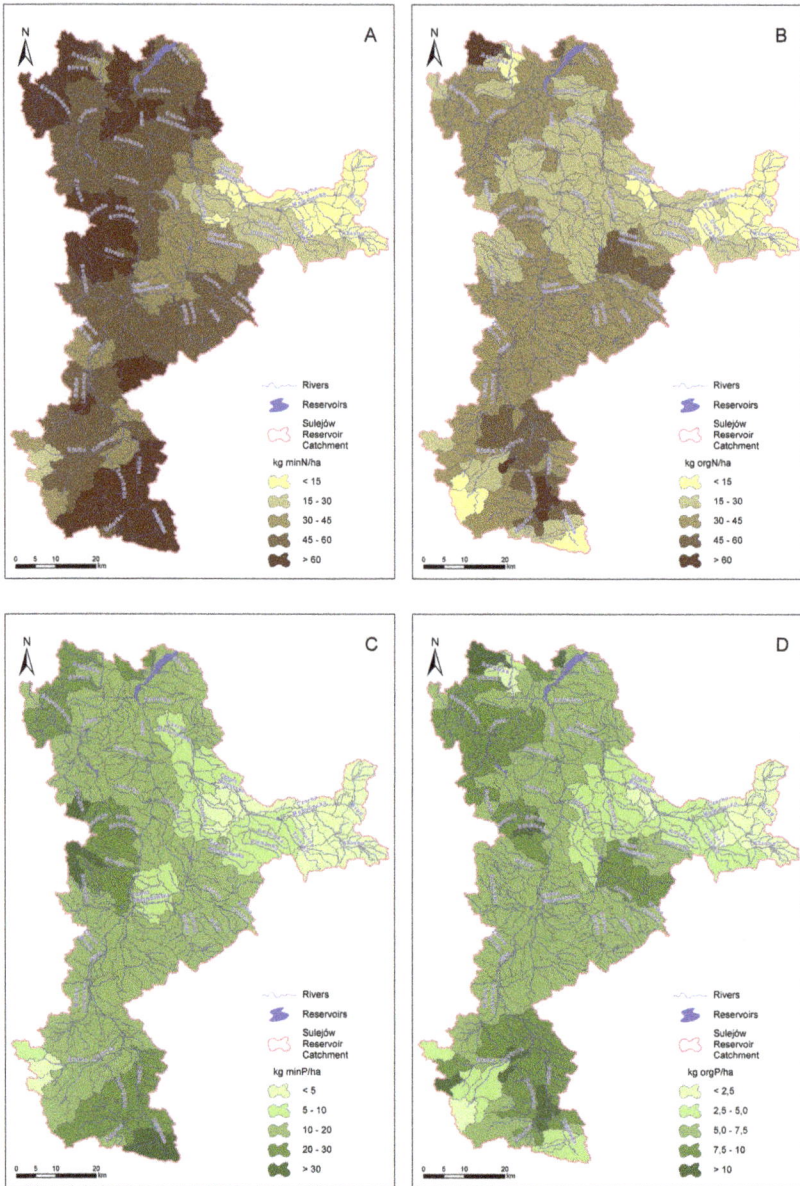

Figure 4. Sub-basin-averaged N and P fertilizer rates: (**A**) mineral nitrogen; (**B**) organic nitrogen; (**C**) mineral phosphorus; (**D**) organic phosphorus.

2.4.3. Sewage

Twenty three sewage treatment plants discharging an average of more than $2 \, L \cdot s^{-1}$ of treated sewage water annually were identified in the SRC and used in the model setup (Figure 5A). The largest plant of the SRC situated in Piotrków Trybunalski discharges its sewage water downstream of the Sulejów Reservoir and was therefore neglected during model setup. For each sewage treatment plant, discharge and nutrient loads were expressed as constant or mean monthly values depending on the available data. These values were obtained directly from plant operators in most cases by using a telephone/electronic survey.

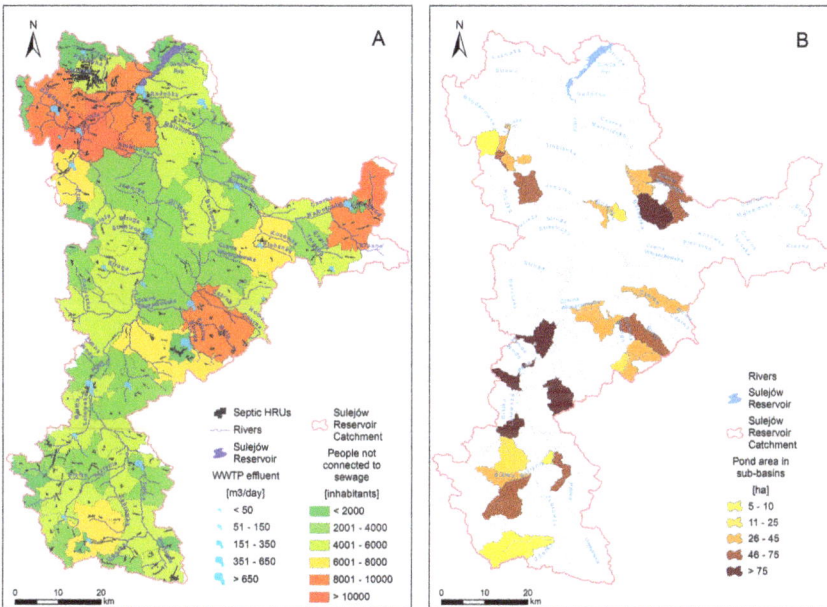

Figure 5. Sewage treatment plants (**A**) and fish ponds (**B**).

Even though water management in Poland is undergoing rapid modernization, which is manifested, for example, by investments in treatment plants, many rural and suburban areas remain disconnected from sewer systems. In such cases, domestic septic systems are usually used for sewage treatment. However, one common problem associated with domestic septic systems in Poland is leaking septic tanks [42]. The SWAT model uses a biozone algorithm [43] to simulate the effects of on-site wastewater systems. The type of septic tank widely used in Poland can be approximately represented by the so-called "failing systems" in SWAT. To identify approximate locations of septic systems in the SRC, commune-level data for the number of people disconnected from wastewater treatment plants were used, which were obtained from the Local Data Bank of the Central Statistical Office (GUS). This number was estimated as 229,000 people for the entire SRC. Spatial analysis of these data made it possible to identify 202 HRUs with the land cover classes "URLD" (urban low density) or "URML" (urban medium-low density), in which the septic function was initiated (Figure 5A). The water quality parameters of the sewage effluents were specified based on the available literature [44]. For example, the TN and TP concentrations were 60 mg $\cdot$ L^{-1} and 20 mg $\cdot$ L^{-1}, respectively.

2.4.4. Fish Ponds

An important feature of the SRC is carp breeding in earth ponds at traditional land-based farms adjacent to the river channels. The area of ponds identified in the 32 sub-basins was significant (Figure 5B). The ponds were represented in SWAT by defining monthly water use parameters (water withdrawn from the reaches of the river for filling the ponds in the spring and maintaining the desired water level until late summer) and point source discharges (representing water release to adjacent reaches of the river in October to empty the ponds before winter). The quantities of the abstracted and released water were calculated based on the estimated pond volume. The water quality characteristics of the discharged water remain largely unknown. Thus, a literature review of the effects of carp breeding on water quality in Central Europe [45,46] was used to define the mean concentrations of the different constituents in the released water: 2.96 mg TN $\cdot$ L^{-1} and 0.7 mg TP $\cdot$ L^{-1}.

2.4.5. Summary

Table 2 lists three main anthropogenic point pollution sources and quantifies the mean annual TN and TP loads that originated from these sources and entered the stream network of the SRC. These estimates are very uncertain for each pollution source. The quantity of released water and the TN and TP concentrations both vary temporally and spatially. Table 2 shows that the order of magnitude is the same for all variables. For TN, the loads from the sewage treatment plants are slightly greater than those from the septic tank effluent and fish pond releases. For TP, the fish pond releases are the major source and the treatment plants and septic tanks are the second and third sources, respectively. However, in our opinion, the TP load from septic tank effluents is underestimated because SWAT does not simulate the downward movement of P to the groundwater. In addition, while the loads from treatment plants (in SWAT) are usually constant with time, the loads from the fish ponds only occur in October. In contrast, the loads from septic tank effluents are variable with time because the travel time between the bottom of the tank to the nearest river depends on the soil physical properties and hydrological conditions.

Table 2. Mean annual TN and TP loads entering the stream network in the SRC that originated from different pollution sources.

Pollution Source	TN (kg/year)	TP (kg/year)
Sewage treatment plants	81,989	6,913
Septic tank effluent	65,572	1,462
Fish pond releases	53,313	9,273
Total	200,874	17,648

Table 2 does not include the loads from the two remaining sources of atmospheric deposition and agricultural production. Because these sources are land-based sources, the load that enters the soil profile is generally known (e.g., Figure 4) but the load that enters the stream network is not. The load that enters the stream network is definitely smaller than the load entering the soil profile due to soil retention, plant uptake etc., and its direct estimation would require modeling.

2.5. Spatial Calibration Approach

SWAT-CUP is a program that allows to use a number of different algorithms to optimize the SWAT model. In addition, SWAT-CUP can be used for sensitivity analysis, calibration, validation and uncertainty analysis [47]. In this paper, we applied SWAT-CUP version 2009 4.3 and selected the optimization algorithm SUFI-2 (Sequential Uncertainty Fitting Procedure Version 2), which is an inverse modeling program that contains elements of calibration and uncertainty analysis [48]. Although SUFI-2 is a stochastic procedure, it does not converge with any "best simulation" and quantifies standard goodness-of-fit measures, such as the Nash-Sutcliffe Efficiency (NSE) or R^2 for each model run. Hence, SUFI-2 indicates the "best simulation" among all of the performed runs, which corresponds to the run with the highest/lowest value of the earlier defined objective function. In this study, we used the widely used NSE as an objective function. The NSE can range from $-\infty$ to 1, where 1 is optimal. Moriasi *et al.* [49] recommended the value of 0.5 as the threshold for satisfactory model performance for a monthly time step, mentioning that under certain circumstances (e.g., daily time step, high uncertainty of observations) this requirement could be made less stringent. We also tracked other goodness-of-fit values, such as R^2 and percent bias (PBIAS). The PBIAS measures the average tendency of the modeled data to be larger or smaller than their observed counterparts. Positive values indicate model underestimation bias, and negative values indicate model overestimation bias.

Calibration was performed in three steps, beginning with continuous daily discharge, continuing with irregular (approximately one measurement per month) and daily NO_3-N loads and ending with TP loads. The calibration period was from 2006 to 2011, and the validation period was from 2000 to 2005. Figure 2 presents the locations of the 10 flow gauging stations (data acquired from IMGW-PIB) and 7 water quality monitoring stations (concentration data acquired from the General Inspectorate of Environmental Protection), from which the time series were used for calibration and validation. The average daily loads (kg · day^{-1}) on the sampling dates were calculated based on observed daily discharge data (m^3 · day^{-1}) at the closest flow gauging station. If the flow gauging station was situated at another location than the water quality station, discharge data were scaled using catchment area ratios. We evaluated the relationships between the NO$_3$-N and TP concentrations and discharge for all studied gauges and concluded that the correlations were too low (median R^2 equal to 0.2 for NO$_3$-N and 0.03 for TP) to use any regression-based methods for continuous load estimation.

Three parameter sets, one for discharge, one for NO$_3$-N, and one for TP, and their initial ranges applied in SUFI-2 (Electronic Supplement, Tables S1–S3) were chosen based on the previous applications of the SWAT model under Polish conditions [41,50,51], and on the sensitivity analysis performed in the SRC.

In most SWAT studies, calibration is restricted to the catchment outlet. In some cases, especially in small (*i.e.*, <100 km^2) catchments, this approach is justified and sometimes inevitable due to data scarcity. However, wide variations occur in the runoff that is produced in different sub-areas of large river basins due to variations in the physical catchment properties and the associated hydrological processes [52]. Variations in water quality may be even higher due to natural and anthropogenic factors. One of the most effective methods used to account for this type of variation is to perform spatially distributed calibration (*i.e.*, multi-site or multiple gauges, hereafter referred to as "spatial calibration"), as performed by [52–54].

Spatial calibration is a much more complex task than single-gauge calibration, and its complexity depends on the number of gauges used and the spatial dependencies between them. We used the widely applied approach (e.g., [54,55]) of the "regionalization" of parameter values sequentially from upstream to downstream nested catchments. This approach was applied in three steps: discharge, NO$_3$-N and TP.

After successful calibration and validation, the optimal parameter values were written into the SWAT project and the model was executed for the joint calibration and validation period from 2000 to 2011. Hereafter, this simulation is referred to as the "Baseline" scenario.

2.6. Buffer Zone Efficiency Monitoring in Shallow Groundwater

The monitoring program of the buffer zone efficiency for reducing nitrate and phosphate pollution in shallow groundwater was conducted in 12 transects located in 6 different areas within the SRC. All investigated buffer zones were located between arable fields and stream channels and had variable widths (ranging from 10 to 50 m) and hydrogeological structures (from high to low permeability). The predominant type of land cover of the buffer zones were cultivated meadows, with narrow tall herb fringes and common reed bed communities adjacent to the stream channels.

The groundwater well network was installed in January 2011. Two wells were installed for each transect, one at the edge of the arable land (inlet) and one at the edge of the buffer zone of the stream bank. The wells consisted of HDPE pipes (ϕ 50 mm; Eijkelkamp) that were installed in hand-drilled or machine-drilled holes. The bottom 1 m of each HDPE pipe was perforated. The lithology (granulometric estimation and thickness) was determined through visual inspection of the cores that were collected with the auger during installation.

Groundwater samples were collected monthly from February 2011 until February 2014. Once the water level was measured, the water filling the well bottom were pumped out. Next, the groundwater was sampled by using submersible pumps (Eijkelkamp). During each sampling, temperature, conductivity, and pH were measured in situ. The nitrate, nitrite, ammonium, and phosphate levels were measured using ion chromatography (Dionex ICS-1000, Sunnyvale, CA, USA).

The percent effectiveness of the riparian buffer zones (RR for reduction rate) was calculated by assessing the degree by which the NO_3-N and PO_4-P levels were reduced along the buffer zone.

$$RR_X = \frac{c_{in} - c_{out}}{c_{in}} \cdot 100\% \qquad (5)$$

where X denotes a measurement variable, NO_3-N or PO_4-P, c_{in} denotes the inlet concentration and c_{out} denotes the outlet concentration. The values of RR_X were calculated separately in the first step for each year and transect.

The goal was to derive one reduction rate value per variable based on the entire set of sampling results for application in the buffer zone scenario model in SWAT. The mean annual c_{in} across all investigated transects ranged from 0.08 to 31.4 mg NO_3-N $\cdot$ L^{-1} and from 0.05 to 1.49 mg PO_4-P $\cdot$ L^{-1}. Because it was observed that positive RR_X values mainly occur if the inlet concentration exceeds a certain threshold (which usually corresponds to high diffuse pollution in a neighboring field), all measurements with mean annual c_{in} values below 5.65 mg NO_3-N $\cdot$ L^{-1} and 0.166 mg PO_4-P $\cdot$ L^{-1} were removed before conducting further calculations. The thresholds were set according to Polish water legislation. Concentrations above these thresholds are in third or higher classes of groundwater quality (where first and second classes denote very good and good quality, respectively). Consequently, only nine transects located in five different areas (Figure 2) were retained for analysis. Thus, the following values of c_{in}, c_{out} and RR were obtained (mean values across all transects and years):

- For NO_3-N: $c_{in} = 17.6$, $c_{out} = 7.91$ and $RR_{NO_3-N} = 56\%$;

- For PO$_4$-P: $c_{in} = 0.76$, $c_{out} = 0.18$ and $RR_{PO_4-P} = 76\%$

2.7. Buffer Zone Scenario Assumptions

The VFS SWAT sub-model simulates reductions in sediment and nutrient contents in surface runoff and neglects the role of lateral and groundwater flow in nutrients that contribute to the stream. The field measurements described in Section 2.6 clearly indicated the efficiency of VFS in the reduction of nitrates and soluble phosphorus concentrations in shallow groundwater. Thus, the buffer zone scenario implemented in SWAT in this study consisted the following two items:

- The application of the default SWAT VFS function mimicking reduction of nutrients in surface runoff using the default values of all parameters describing the VFS action; and
- The adjustment of groundwater quality parameters related to nutrient concentrations mimicking reduction of nutrients in shallow groundwater.

At the model parameterization stage, the soluble phosphorus concentrations in the groundwater *GWSOLP* were specified at the HRU level based on the available field measurements in the wells situated on arable land fields. SWAT does not dynamically model the pool of P in the groundwater. Thus, the concentration remained constant throughout the simulation period. To reflect the role of the buffer zone, the *GWSOLP* values were multiplied by the estimated phosphorus reduction rate RR_{PO_4-P}.

Unlike phosphorus, the groundwater nitrate pool was modeled in SWAT, which allowed for fluctuations in nitrate loadings in the groundwater over time. To reflect the reduction of nitrate in the buffer zone, the values of the *HLIFE_NGW* parameter (the half-life of nitrate in the shallow aquifer) were adjusted. This parameter accounts for nitrate losses due to biological and chemical processes; thus, this parameter can be manipulated to approximate reductions of nitrate due to the acting buffer zone. HRU-specific values of *HLIFE_NGW* were decreased by empirical factors, and the nitrate concentrations in the groundwater were reduced by a value of RR_{NO_3-N} relative to the concentrations before the change.

The buffer zone scenario was only implemented in the HRUs that used arable land as a type of land cover and were characterized by high N and P emissions to surface waters. The arable land HRUs accounting for the top 20% of nitrogen and phosphorus emissions were selected. Hence, buffer zones were only tested in Critical Source Areas (CSAs) (*i.e.*, areas with disproportionately high pollutant losses). As proven by the field monitoring results described in Section 2.6, the buffer zone efficiency rapidly decreases when the input concentrations are low (*i.e.*, when the upland field is extensively cultivated). This finding suggests that applying buffer zones in low-emission areas is not efficient. Thus, the application of buffer zones was restricted to the CSAs.

3. Results and Discussion

3.1. Calibration and Validation

3.1.1. Discharge

Table 3 presents the model performance measures for the calibration and validation periods of the three modeled variables. Figure 6 shows simulated *vs.* observed hydrographs for the two main gauges (the last stations on the Pilica and Luciąża before the Sulejów Reservoir). The hydrographs for the remaining 8 gauges are shown in Figure S1 of the Electronic Supplement. The goodness-of-fit values and a visual inspection of the hydrographs both demonstrate good model performance for simulating daily flows in the SRC. However, a few deficiencies were noted.

- During the validation period, the model generally underestimates discharge across the entire range of flow variability. The median value of PBIAS is 0.21;
- The peaks of the largest floods are generally slightly underestimated by most gauges;
- The timing of the flood peaks is sometimes advanced by 1–3 days compared with the timing of the peaks identified in the observed data;
- For the three upstream gauges with relatively small catchment areas (less than 360 km^2) the values of NSE were smaller than 0.5 for either the calibration or validation period.

Table 3. Median values of selected goodness-of-fit measures for discharge, NO$_3$-N and TP for calibration and validation periods.

Variable	NSE cal.	NSE val.	R^2 cal.	R^2 val.	PBIAS cal.	PBIAS val.
Discharge	0.64	0.61	0.70	0.72	0.07	0.21
NO$_3$-N loads	0.56	−0.04	0.69	0.28	0.01	0.26
TP loads	0.48	0.08	0.71	0.25	0.05	0.47

As observed in previous SWAT applications in Poland [51,54], we observed a clear relationship between the model performance indicators and the area upstream of the calibration gauge, at least for NSE and R^2 (Figure 7). The larger catchment size, the higher values of NSE and R^2. No relationship of this type can be identified for the absolute value of PBIAS.

The hydrological conditions for the validation period were much wetter than those during the calibration period, which potentially resulted in the observed differences, particularly the high positive value of PBIAS. The mean discharge at the main outlet in Sulejów was higher. Snow melt floods were dominant during the validation period and storm floods were dominant during the calibration period.

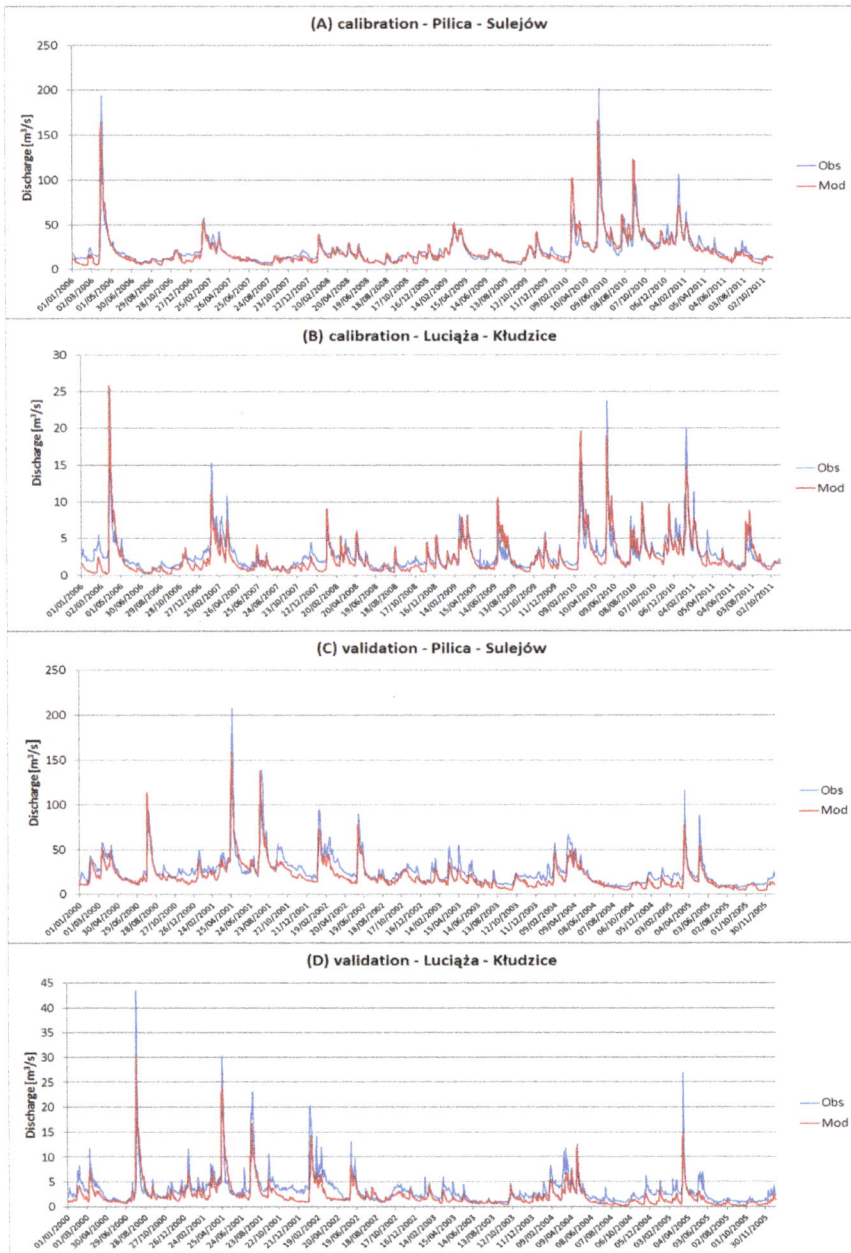

Figure 6. Calibration and validation plots for discharge at the Sulejów gauging station (the Pilica River) and the Kłudzice gauging station (the Luciąża River).

240

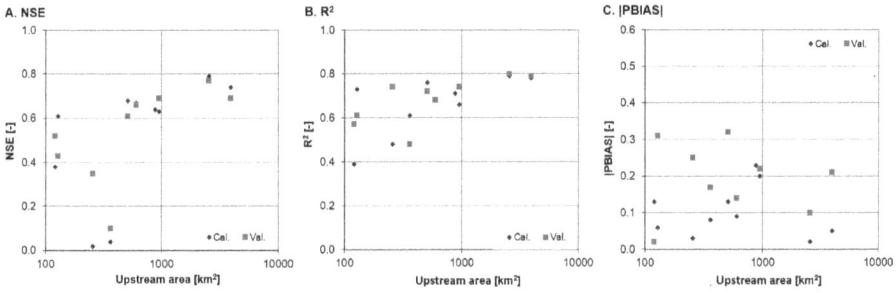

Figure 7. Relationship between the area upstream of a gauge and the model performance measures for discharge for the calibration and validation periods: (**A**) NSE; (**B**) R^2; (**C**) |PBIAS|.

3.1.2. NO$_3$-N Loads

The goodness-of-fit statistics for NO$_3$-N loads are not as good as those for discharge (Table 3). During the calibration period, the results are highly variable depending on the gauge. The three problematic gauges with low NSE and R^2 values are situated in the headwater highland part of the SRC. It is very likely that these values are affected by the low performance measure values for discharge simulations in this part of the studied catchment. By contrast, the results are very good for the Czarna Maleniecka and Czarna Włoszczowska Rivers (*cf.* Figure S2 and Table S4 of the Electronic Supplement). In addition, a reasonable fit was observed between the simulations and observations of the two main rivers entering the Sulejów Reservoir (Figure 8A,B).

The model performance during the validation period is slightly worse than during the calibration period. As shown in Figure 8F, the model failed to capture one very large peak. However, a more detailed analysis shows that the modeled peak lagged by 5 days. This lag resulted from the lag in the flood peak from snow melt. Another issue that is visible during the validation period is the considerable bias for most gauges (with a median of 0.26). This bias also reflects the bias in the modeled discharges. Overall, some of the problems identified during hydrology calibration and validation were transposed to the calibration of NO$_3$-N loads. However, it should be noted that the model preserves several important aspects of the NO$_3$-N loads and concentration dynamics (e.g., the highest values during the winter and spring and the lowest values during the summer and autumn).

3.1.3. TP Loads

As with NO$_3$-N, the goodness-of-fit statistics for the TP loads are not as good as those for discharge (Table 3; Figure S3 of the Electronic Supplement). For the calibration period, the comments mentioned in Section 3.1.2 are largely valid for

241

TP. Particularly, lower performance measure values were also noted in the small headwater sub-catchments of the SRC in the south. The very good fit between the simulations and observations is shown in Figure 8C,D.

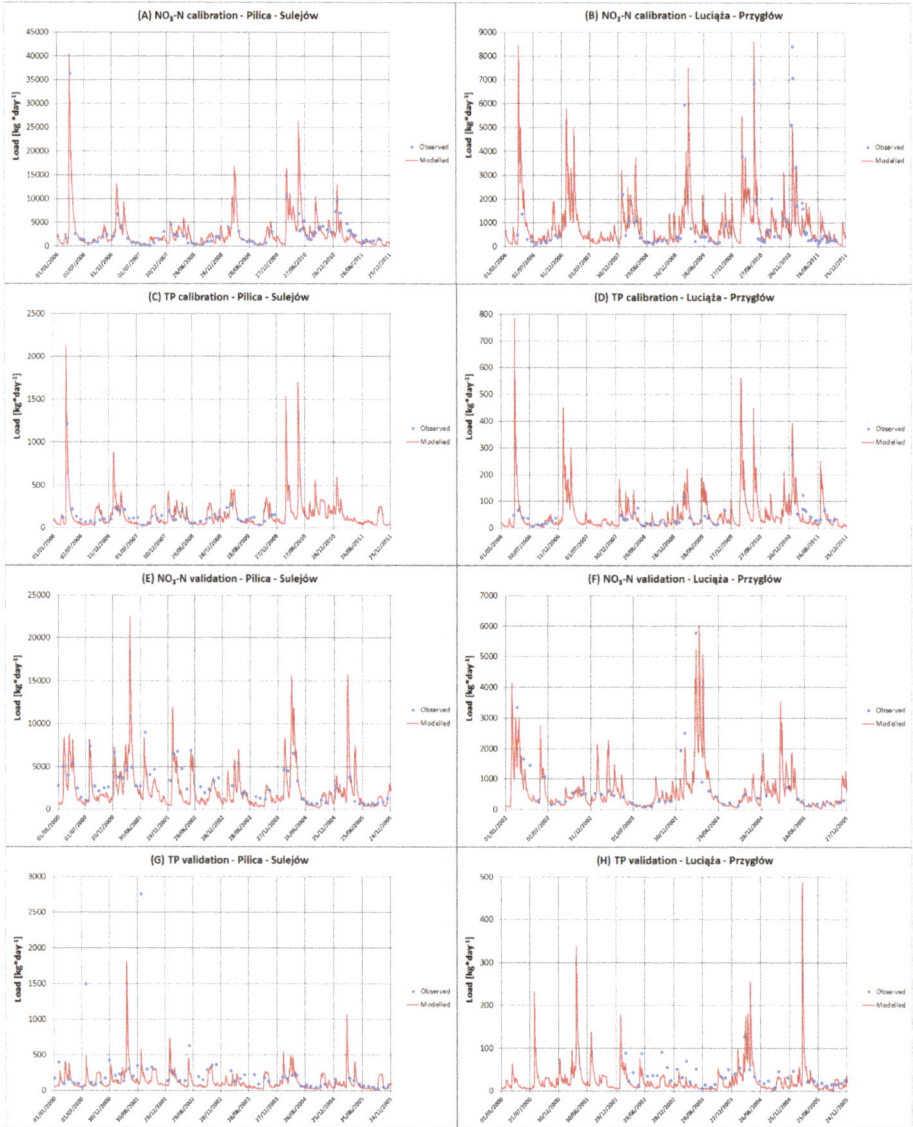

Figure 8. Calibration and validation plots for the NO₃-N and TP loads at the Sulejów (the Pilica River) and Przygłów gauging stations (the Luciąża River).

However, the model significantly underestimates the observed TP loads in most of the gauges, with a median PBIAS value of 0.47. This high bias cannot be explained by the underestimation of discharge alone. In some cases (e.g., for the large peak in TP loads shown in Figure 8G), the modeled flood peak occurred 5 days before the measured flood peak, which clearly affected the high underestimation of the TP load when water samples were measured.

3.2. Spatial Variability in NO_3-N and TP Emissions

In this section, we present an analysis of the calibrated model outputs for the Baseline scenario (2000–2011).

Figure 9 shows the mean emissions at the sub-basin level of NO_3-N (A) and TP (B) from land areas to the stream network. These emissions include all of the possible pathways of the studied constituents from the sub-basins to SWAT reaches via surface runoff, sub-surface runoff, tile drain outflow and base flow. The results are expressed per unit of catchment (not just agricultural) area. Therefore, the results indirectly incorporate the effects of different areas of agricultural land in different sub-basins. For nitrogen and phosphorus, the spatial variability of the calculated emissions is very high. The difference between the sub-basins with the highest and the lowest emissions is two orders of magnitude for NO_3-N and three orders of magnitude for TP.

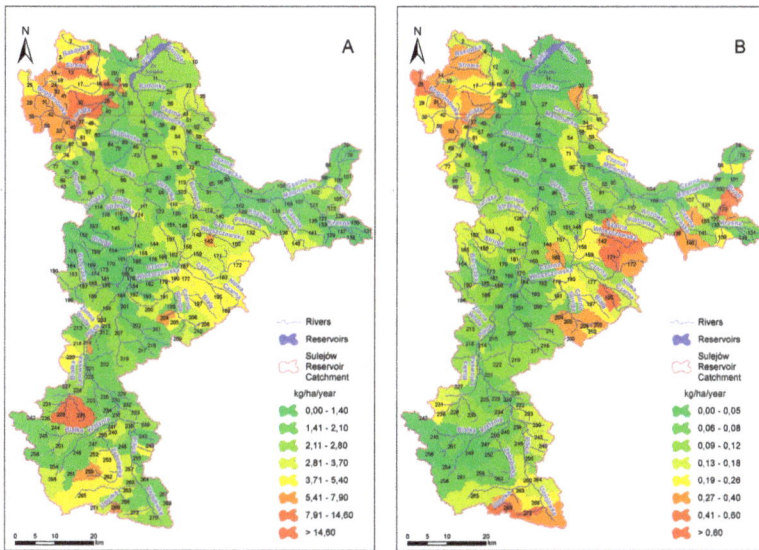

Figure 9. Mean emissions of NO_3-N (**A**) and TP (**B**) from land areas to stream networks for the baseline period 2000–2011 in the SRC. The units are in kilograms per hectare of sub-basin area per annum.

For NO_3-N, the highest emissions are concentrated in two regions of the SRC: (1) the Bogdanówka and Strawa sub-catchments in the northwest and (2) the Białka Lelowska catchment in the south. Both areas are characterized by relatively high proportions of agricultural land and high fertilizer rates (*cf.* Figure 4). The first area has the highest share of inhabitants not connected to sewage systems (*cf.* Figure 5A). The second area is covered by a large patch of loess soils that are less permeable than the neighboring sands and loamy sand. In addition, two large areas are present with moderately high emission rates: (1) the upper Pilica and Żebrówka sub-catchments in the south and (2) the Biała and Nowa Czarna sub-catchments in the central portion of the SRC. As previously observed, it is clear that both agriculture and septic tank effluents play a critical role in the emission levels in these two areas.

The regions of high TP emissions only partly overlap with the regions of high NO_3-N emissions. The new regions with high TP emissions (in which NO_3-N emissions are not too high) include the headwater parts of the Czarna Włoszczowska and Czarna Maleniecka sub-catchments in the east and the Udorka and Uniejówka sub-catchments in the south. The latter area is also known for intensive head lettuce farming and for using large amounts of fertilizers. Moderately high TP emissions can be found in the Czarna Struga sub-catchment in the central region and in a few smaller isolated sub-catchments that are scattered around the SRC. Most of the mentioned regions overlap with areas receiving relatively high P fertilizer rates. However, this result does not only occur for the headwater area of the Czarna Maleniecka sub-catchment in the east. In this case, the emissions can be explained by the high septic tank effluent emissions from the households that are not connected to sewage systems (*cf.* Figure 5).

3.3. Buffer Zone Scenario Results

Figure 10 illustrates the locations of the agricultural HRUs with the highest NO_3-N and TP emission rates that were identified as the CSAs. Overall, 20% of the HRUs with the highest NO_3-N emissions are responsible for 36% of the total load, and the same amount of HRUs with the highest TP emissions is responsible for 51% of the total load. This finding shows that the magnitude of TP losses is more diversified than the magnitude of NO_3-N losses. The areas with the highest density of selected 606 HRUs largely correspond with the high emission regions described in Section 3.2. In addition, Figure 10 shows the mean difference in NO_3-N (A) and TP (B) emissions between the "Buffer zone" scenario and the Baseline scenario (negative values should be interpreted as the estimated reduction levels that are reached by applying buffer zones). The values are expressed in kg per unit of sub-basin area; thus, they are affected by the HRU-level efficiency of the buffer zone and the percentage of the selected HRUs in the sub-basins. The mean HRU-level reductions reached 0.82 kg

NO$_3$-N $\cdot$ ha^{-1} and 0.18 kg TP $\cdot$ ha^{-1} (the values per hectare of HRU area), and the 90th percentiles reached 1.64 kg NO$_3$-N $\cdot$ ha^{-1} and 0.28 kg TP $\cdot$ha^{-1}, respectively. However, at the sub-basin level, the efficiency is significantly reduced, with mean values of 0.16 kg NO$_3$-N $\cdot$ ha^{-1} and 0.03 kg TP $\cdot$ ha^{-1} (the values per hectare of sub-basin area), and 90th percentiles of 0.31 kg NO$_3$-N $\cdot$ ha^{-1} and 0.09 kg TP $\cdot$ ha^{-1}. When expressed as a percentage, the average reduction across all of the sub-basins where buffer zones were "implemented" is considerably higher for TP than for NO$_3$-N (19.4% compared to 5.9%).

A spatial analysis of Figure 10 results in the observation that the highest reductions of NO$_3$-N or TP generally correspond with the areas with the highest emissions (*cf.* Figure 9). However, this result does not occur in sub-basins where at least one of the two following circumstances occur: (1) the percentage of HRUs selected for this measure is low and (2) high emissions result from septic tank effluents rather than from agriculture. In several cases, the baseline emission level from some sub-basins was low. Thus, although the percent reduction reached 10% or 15%, it was too low in terms of the absolute value to appear on the map.

As mentioned in Section 2.7, the "Buffer zone" scenario implemented in SWAT in this study consisted of two items: (1) the application of the default SWAT VFS function dealing only with surface runoff and (2) the incorporation of field monitoring-based reduction rates to the shallow groundwater component in SWAT. To verify how each item contributed to the final result, we created two additional scenarios, one that only incorporates feature No. 1 ("BZ-VFS"), and another that only incorporates feature No. 2 ("BZ-GW"). Next, we estimated the sub-basin level reduction rates for the "BZ-VFS" and "BZ-GW" scenarios and compared them with the results from the original "Buffer zone" scenario. Overall, the effect of effect of VFS (72% of the total load reduction) for NO$_3$-N was dominant over the effect of field monitoring-based parameters (28% of the total load reduction). By contrast, the contributions of each component to the total reduction of TP emissions in the SRC were similar: 46% and 54%, respectively. These modeling results showed that shallow groundwater reduction mechanisms are more effective for TP than for NO$_3$-N, which agrees with the calculated reduction rates from Section 2.6.

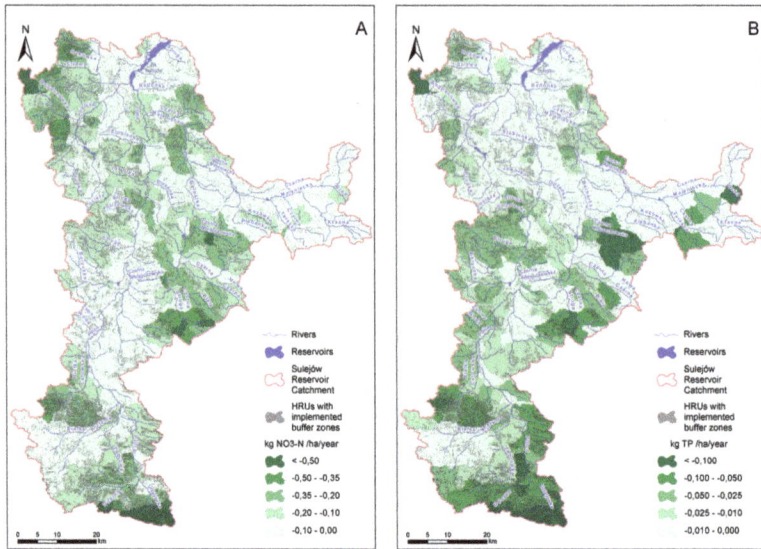

Figure 10. The modeled effects of buffer zones on agricultural pollution emissions from a land to stream network. The mean difference in NO_3-N (**A**) and TP (**B**) emission between the "Buffer zone" scenario and the Baseline scenario. Units are in kilograms per hectare of sub-basin area per annum.

3.4. Discussion

The performance of the SWAT model for simulating daily discharge in the SRC was spatially variable but generally good or satisfactory. The main downside was underestimation bias during the validation period, which occurred because of the significantly wetter hydrological conditions during this period compared with the calibration period. Because the calibrated parameter values are very sensitive to climatic conditions, the values calibrated for dry and short periods might not be suitable for simulating the opposite conditions [56,57], which results in lower performance statistics. Unfortunately, this bias in the validation period for discharge translates into an even greater bias for the NO_3-N and TP loads during this period. However, the reported values of PBIAS for most of the gauges are within the ranges of satisfactory performance for discharge (+/−25%) and the NO_3-N and TP loads (+/−70%) [49]. In summary, our results support the findings of Ekstrand *et al.* [58], who applied SWAT to model the TP losses in five catchments in central Sweden. Overall, Ekstrand *et al.* [58] observed that obtaining satisfactory results for a validation period often depends on whether the range of hydroclimatological conditions is similar (as in calibration).

An additional problem is that evaluating water quality simulations using a daily time step and only one measurement per month is not an optimal. Typically, model simulations are less accurate when shorter time steps are considered than when longer time steps are considered [59]. If sampling is performed during a flood event, which occurred several times (as shown in Figure 8), it is likely that (1) discharge estimations from SWAT for the sampling date are very far from the observations because of common underestimations and lag problems associated with flood peaks (*cf.* Section 3.1.1) We analyzed a few events with different magnitudes that occurred during different seasons and reaches and noted that SWAT was not capable of reproducing this kind of effect with reasonable accuracy. Regarding the problem of capturing peaks, we analyzed all NO_3-N and TP daily validation plots case by case. In six out of 11 plots (three per variable), we identified situations in which the observed peaks lagged behind or preceded the event by 2–15 days. Next, we matched the modeled and observed peaks (between two and three per plot) and recalculated the performance statistics. The model performance improved for each case and for each indicator (Electronic Supplement, Table S5). Increases in the NSE ranged between 0.15 and 0.57, increases in the R^2 value ranged between 0.1 and 0.54 and the positive values of PBIAS decreased by 2%–18%. This result demonstrates that the validation results were significantly impacted by a small number of missed peak events by the majority of gauges.

Furthermore, in five out of 11 cases, we identified another reason for poor validation results. We compared the mean observed discharges and loads between the calibration and validation periods (Electronic Supplement, Table S6). In all analyzed cases, (1) the PBIAS during the validation period was significantly higher than during the calibration period; (2) the PBIAS during the validation period was larger than or equal to 0.4; (3) the mean observed discharge during the validation period was significantly greater than that during the calibration period; and (4) the mean observed NO_3-N and TP loads were much greater during the validation period. These results clearly demonstrate that more than the hydrological conditions differed between these two periods. In addition, this analysis shows that the mean nutrient concentrations in some of the gauges were significantly greater in 2000–2005 than in 2006–2011. The first decade of the twentieth century in Poland has been marked by rapid development in the number of sewage treatment plants and by an increasing treatment level [60]. Because the majority of the input data used to build the model setup were valid for 2010 or later and may represent the period of 2000–2005, we hypothesized that this finding could partially explain the poor model performance during validation.

We applied SWAT in the SRC to spatially quantify NO_3-N and TP emission from various pollution sources. The purpose of this spatial quantification was to identify CSAs in which the buffer zones that mitigate pollutant emissions to the surface water could be implemented. In Table 2, we specified the mean annual TN and TP loads that originated from sewage treatment plants, septic tank effluents and fish pond releases. In Figure 9 we presented the spatially variable NO_3-N and TP loads that predominantly originated from cultivated land and, to a smaller extent, from septic tank effluent. Integration of the sub-basin data from Figure 9 to calculate the total catchment load and subtracting the load assigned to septic tank effluent provided a rough estimate of the mean annual diffuse pollution load in the SRC, which reached 1,240,000 kg NO_3-N and 60,700 kg TP. Although these values also include emissions from urban (very small percentage) and forest (very low emission) runoff, this estimate confirms the initial hypothesis that diffuse agricultural pollution is largely the dominant source of pollution in the SRC. Although the SRC, or more widely, the Pilica catchment, have recently attracted the attention of a number of researchers studying pollution emissions and transport [32,61,62], this finding is new and has certain implications regarding water management. Particularly, regarding the fact that Poland has been sent to court by the EU Commission for failing to guarantee that they are addressing water pollution by nitrates effectively [63]. However, no Nitrate Vulnerable Zone (NVZ) has been included in the SRC under current legislation. However, Figure 9 indicates that some portions of the catchment could easily be designed as NVZs.

Strong evidence for the contributions of agriculture to pollutant emissions to streams also strengthens the basis for scenarios that assume the application of buffer zones in the identified CSAs. Previous modeling attempts of buffer zones in Poland using SWAT [51] have shown its limitations (*i.e.*, the fact that the VFS sub-model only accounts for the trapping effect in surface runoff (*cf.* Section 2.2.2). Consequently, the efficiency of applying buffer zones described by Piniewski *et al.* [51] when measured at the catchment outlet was negligible for NO_3-N and small for PO_4-P. In this study, we used buffer zone field monitoring data from the studied catchment to improve the mechanisms by which SWAT reduces pollutant losses. The modeled reduction rates were spatially variable, but higher than those in the study of Piniewski *et al.* [51]. In addition, the results showed that, the average contributions of the "shallow groundwater" mechanism to total reduction reach 28% for NO_3-N and 54% for TP. This demonstrates that the "surface water" trapping mechanism by VFS in SWAT is not sufficient (*i.e.*, it overlooks an important pathway by which both NO_3-N and TP compounds can reach the stream network). The efficiencies of buffer zones critically depend on the mechanisms by which N and P are transported from the land to the stream [64]. Although we have only empirically tested SWAT in the SRC, it is likely that this limitation would affect other areas, particularly areas of the vast Polish Plain, which are characterized by physiographic conditions that are similar to those of the SRC. However, the approach we used to consider the field measurements of the buffer zones in SWAT was fairly simplistic and based purely on parameter modification. In the future, larger field monitoring samples (in space and time) should allow for the development of a new SWAT sub-routine that would better reflect the pollutant pathways from field areas through buffer zones to streams under variable hydrological conditions.

To assess the effects of buffer zones on the nutrient loads that enter the Sulejów Reservoir, we summed the mean annual loads from all eight reaches with their outlets at the reservoir shoreline (*cf.* Figure 2). The results showed that applying buffer zones in the selected CSAs (occupying 20% of the arable land area and culminating in 12.4% of the land in the catchment) would contribute to the reduction of the pollutant loads entering the reservoir by 7% for NO_3-N and 16% for TP. This outcome is particularly important for TP, which is mainly responsible for reservoir eutrophication and for intensity of toxic algal blooms [34,65].

The estimated buffer zone efficiency can be considered as substantial. However, it is clear that other conservation practices are important for obtaining more pronounced reductions in pollutant runoff. Particularly, the activities should focus on reducing the inputs of nutrients to the landscape in the form of mineral and organic fertilizers by convincing the farmers to use fertilization plans more widely. Examples of other measures include extension of the closed period for spreading organic fertilizers, elimination of soil cultivation during the autumn, the cultivation of catch crops, and the construction of wetlands. Spatially-explicit indications of CSAs provide an opportunity for selecting effective measures. In the second step, their precise and cost-efficient application substantially increases the chance of improving the water quality in the catchment.

4. Conclusions

This study demonstrated that diffuse agricultural pollution is the main contributor of elevated NO_3-N and TP concentrations in the surface waters of the SRC relative to point source pollution from sewage treatment plants, septic tank effluents and fish pond releases. The application of a semi-distributed water quality model and performing a comprehensive spatial calibration and validation allowed us to spatially quantify the emission rates at the HRU and sub-basin level, which helped identify Critical Source Areas. These CSAs were selected to test the efficiencies of riparian buffer zones. The default SWAT sub-model designed for simulating the effects of buffer zones only accounts for nutrient trapping in surface runoff and overlooks an important sub-surface pathway in which nutrients can be trapped. The monitoring data from the SRC showed that the mean field-level reductions in the concentrations in the shallow groundwater near the buffer zone average 56% for NO_3-N and 76% for TP. These empirical reduction rates were used to enhance the capability of SWAT for representing the effects of the buffer zone. The scenario results showed that the efficiency of the buffer zones at the catchment level is lower than that at the field level but still significantly contributes to reductions in pollutant emission to the nearest streams and to reductions of the total pollutant load entering the Sulejów Reservoir (by 7% for NO_3-N and 16% for TP). Only using the default SWAT function of the simulating buffer zones would lead to an underestimation of buffer efficiency, particularly for phosphorus (54%). Thus, we argue that empirical data are important for improving existing models that monitoring more samples in the future should allow us to develop new SWAT routines for simulating the sub-surface trapping effects of the buffer zones.

The poor model performance of the nutrient load simulation during the validation period indicates that the nutrient load estimates from the SWAT model of the SRC are highly uncertain. However, it can be argued that simulated percent reductions in pollutant emission due to the application of buffer zones are more reliable, because of known model bias.

The implications from this study are valuable for water managers and other decision-makers. The use of water quality mathematical models to address contemporary water management problems is still limited in many countries, including Poland. Our study shows how the SWAT model is useful for the (1) quantification of point and diffuse pollution sources; (2) identification of high emission areas (CSAs) where measure implementation should be prioritized; and (3) quantification of the efficiency of conservation practices. All these three aspects are vital for the development of medium and long-term water quality improvement strategies by river basin managers. Further progress can be achieved by including the economic functions representing implementation costs of different conservation practices.

Acknowledgments: We thank two anonymous reviewers for providing constructive comments on the manuscript. This paper is an outcome of the EKOROB project: Ecotones for reducing diffuse pollution (LIFE08 ENV/PL/000519). This project was supported by the LIFE+ Environment Policy and Governance Programme, National Fund for Environmental Protection and Water Management, and funding dedicated for science in the period 2012–2014, and granted for implementation by the co-financed international project No. 2539/LIFE+2007-2013/2012/2 (www.en.ekorob.pl). We would like to acknowledge the Institute of Meteorology and Water Management—National Research Institute (IMGW-PIB) for providing hydrometeorological data. The first author benefited from the START 2014 stipend from the Foundation for Polish Science and from the Humboldt Research Fellowship for Postdoctoral Researchers from the Alexander von Humboldt Foundation in the Potsdam Institute for Climate Impact Research during the manuscript preparation phase.

Author Contributions: Mikołaj Piniewski, Ignacy Kardel, Katarzyna Izydorczyk and Wojciech Frątczak developed the methodological framework. Ignacy Kardel and Wojciech Frątczak processed field monitoring data. Mikołaj Piniewski, Paweł Marcinkowski, Ignacy Kardel and Marek Giełczewski developed the model setup. Mikołaj Piniewski and Paweł Marcinkowski performed model calibration. Mikołaj Piniewski, Paweł Marcinkowski and Ignacy Kardel run the model scenarios. Mikołaj Piniewski wrote the manuscript with inputs from Paweł Marcinkowski, Katarzyna Izydorczyk and Wojciech Frątczak, Marek Giełczewski and Paweł Marcinkowski created the artwork.

Conflicts of Interest: The authors declare no conflicts of interest.

References

1. European Commission. Directive 2000/60/EC of 23 October 2000 of the European Parliament and of the Council establishing a framework for community action in the field of water policy. *Off. J. Eur. Communities* **2000**, *L327*, 1–72.
2. European Commission. Council Directive 91/676/EEC of 12 December 1991 concerning the protection of waters against pollution caused by nitrates from agricultural sources. *Off. J. Eur. Communities* **1991**, *L375*, 1–8.
3. Zalewski, M.; Janauer, G.A.; Jolankai, G. Conceptual background. In *Ecohydrology: A New Paradigm for the Sustainable Use of Aquatic Resources*; Zalewski, M., Janauer, G.A., Jolankai, G., Eds.; International Hydrobiological Programme UNESCO: Paris, France, 1997.
4. Zalewski, M. Ecohydrology for implementation of the EU water framework directive. *Proc. Instit. Civil Eng. Water Manag.* **2011**, *164*, 375–385.
5. Zalewski, M. Ecohydrology: Process-oriented thinking towards sustainable river basins. *Ecohydrol. Hydrobiol.* **2013**, *13*, 97–103.
6. Zalewski, M. Ecohydrology and hydrologic engineering: Regulation of hydrology-biota interactions for sustainability. *J. Hydrol. Eng.* **2015**, *20*, doi: 10.1061/(ASCE)HE.1943-5584.0000999.
7. Arnold, J.G.; Srinivasan, R.; Muttiah, R.S.; Williams, J.R. Large-area hydrologic modeling and assessment: Part I. Model development. *J. Am. Water Resour. Assoc.* **1998**, *34*, 73–89.
8. Gassman, P.W.; Sadeghi, A.M.; Srinivasan, R. Applications of the SWAT Model Special Section: Overview and Insights. *J. Environ. Qual.* **2014**, *43*, 1–8.
9. Niraula, R.; Kalin, L.; Srivastava, P.; Anderson, J.A. Identifying critical source areas of nonpoint source pollution with SWAT and GWLF. *Ecol. Model.* **2013**, *268*, 123–133.
10. Panagopoulos, Y.; Makropoulos, C.; Baltas, E.; Mimikou, M. SWAT parameterization for the identification of critical diffuse pollution source areas under data limitations. *Ecol. Model.* **2011**, *222*, 3500–3512.
11. Wu, Y.; Chen, J. Investigating the effects of point source and nonpoint source pollution on the water quality of the East River (Dongjiang) in South China. *Ecol. Indic.* **2013**, *32*, 294–304.
12. Wu, Y.; Liu, S. Modeling of land use and reservoir effects on nonpoint source pollution in a highly agricultural basin. *J. Environ. Monit.* **2012**, *14*, 2350–2361.
13. Lowrance, R.; Todd, R.L.; Fail, J., Jr.; Hendrickson, O., Jr.; Leonard, R.; Asmussen, L. Riparian forests as nutrient filters in agricultural watersheds. *BioScience* **1984**, *34*, 374–377.
14. Naiman, R.J.; Decamps, H.; Fournier, F. The role of land/water ecotones in landscape management and restoration: A proposal for collaborative research. In *MAB Digest4*; Naiman, R.J., Decamps, H., Fournier, F., Eds.; UNESCO: Paris, France, 1989.
15. Mander, U.; Hayakawa, Y.; Kuusemets, V. Purification processes, ecological functions, planning and design of riparian buffer zones in agricultural watersheds. *Ecol. Eng.* **2005**, *24*, 421–432.

16. Dosskey, M.G.; Vidon, P.; Gurwick, N.P.; Allan, C.J.; Duval, T.P.; Lowrance, R. The role of riparian vegetation in protecting and improving chemical water quality in streams. *J. Am. Water Resour. Assoc.* **2010**, *46*, 261–277.

17. Dosskey, M.G. Toward quantifying water pollution abatement in response to installing buffer on crop land. *J. Environ. Manag.* **2001**, *28*, 577–598.

18. Vought, L.B.M.; Pinay, G.; Fuglsang, A.; Ruffinoni, C. Structure and function of buffer strips from a water quality perspective in agricultural landscapes. *Lands. Urban Plan.* **1995**, *31*, 323–331.

19. Dillaha, T.A.; Sherrard, J.H.; Lee, D.; Mostaghimi, S.; Shanholtz, V.O. Evolution of vegetative filters strips as a best managements practice for feed lots. *J. Water Pollut. Control Fed.* **1988**, *60*, 1231–1238.

20. Dunn, A.M.; Julien, G.; Ernst, W.R.; Cook, A.; Doe, K.G.; Jackman, P.M. Evaluation of buffer zone effectiveness in mitigating the risks associated with agricultural runoff in Prince Edward Island. *Sci. Total Environ.* **2011**, *409*, 868–882.

21. Syversen, N. Effect and design of buffer zones in the Nordic climate: The influence of width, amount of surface runoff, seasonal variation and vegetation type on retention efficiency for nutrient and particle runoff. *Ecol. Eng.* **2005**, *24*, 483–490.

22. Mayer, P.M.; Reynolds, S.K., Jr.; McCutchen, M.D.; Canfield, T.J. Meta-analysis of nitrogen removal in riparian buffers. *J. Environ. Qual.* **2007**, *36*, 1172–1180.

23. Carlyle, G.C.; Hill, A.R. Groundwater phosphate dynamics in a river riparian zone: Effects of hydrologic flowpaths, lithology and redox chemistry. *J. Hydrol.* **2001**, *247*, 151–168.

24. Hoffmann, C.C.; Berg, P.; Dahl, M.; Lasen, S.E.; Andersen, H.E.; Andersen, B. Groundwater flow and transport of nutrients through a riparian meadow -field data and modelling. *J. Hydrol.* **2006**, *331*, 315–335.

25. Weller, D.E.; Jordan, T.E.; Correll, D.L. Heuristic models for material discharge from landscapes with riparian buffers. *Ecol. Appl.* **1998**, *8*, 1156–1169.

26. Larose, M.; Heathman, G.C.; Norton, D.; Smith, D. Impacts of conservation buffers and grasslands on total phosphorus loads using hydrological modeling and remote sensing techniques. *Catena* **2011**, *86*, 121–129.

27. Parajuli, P.B.; Mankin, K.R.; Barnes, P.L. Applicability of targeting vegetative filter strips to abate fecal bacteria and sediment yield using SWAT. *Agric. Water Manag.* **2008**, *95*, 1189–1200.

28. Park, Y.S.; Park, J.H.; Jang, W.S.; Ryu, J.C.; Kang, H.; Choi, J.; Lim, K.J. Hydrologic Response Unit Routing in SWAT to Simulate Effects of Vegetated Filter Strip for South-Korean Conditions Based on VFSMOD. *Water* **2011**, *3*, 819–842.

29. Shan, N.; Ruan, X.H.; Pan, Z.R. Estimating the optimal width of buffer strip for nonpoint source pollution control in the Three Gorges Reservoir Area, China. *Ecol. Model.* **2014**, *276*, 51–63.

30. White, M.J.; Storm, D.E.; Busteed, P.R.; Stoodley, S.H.; Phillips, S.J. Evaluating nonpoint source critical source area contributions at the watershed scale. *J. Environ. Qual.* **2009**, *38*, 1654–1663.

31. Bloschl, G.   Scale and scaling in hydrology.   In *Wiener Mitteilungen, Wasser-Abwasser-Gewasser*; Technical University of Vienna: Vienna, Austria, 1996.

32. Wagner, I.; Izydorczyk, K.; Kiedrzyńska, E.; Mankiewicz-Boczek, J.; Jurczak, T.; Bednarek, A.; Wojtal-Frankiewicz, A.; Frankiewicz, P.; Ratajski, S.; Kaczkowski, Z.; *et al.* Ecohydrological system solutions to enhance ecosystem services: The Pilica River Demonstration Project. *Ecohydrol. Hydrobiol.* **2009**, *9*, 13–39.

33. Tarczyńska, M.; Romanowska-Duda, Z.; Jurczak, T.; Zalewski, M. Toxic cyanobacterial blooms in drinking water reservoir-causes, consequences and management strategy. *Water Sci. Technol.* **2001**, *1*, 237–246.

34. Izydorczyk, K.; Jurczak, T.; Wojtal-Frankiewicz, A.; Skowron, A.; Mankiewicz-Boczek, J.; Tarczyńska, M.   Influence of abiotic and biotic factors on microcystin content in Microcystis aeruginosa cells in a eutrophic temperate reservoir. *J. Plankton Res.* **2008**, *30*, 393–400.

35. Gągała, I.; Izydorczyk, K.; Jurczak, T.; Pawełczyk, J.; Dziadek, J.; Wojtal-Frankiewicz, A.; Jóźwik, A.; Jaskulska, A.; Mankiewicz-Boczek, J. Role of Environmental Factors and Toxic Genotypes in The Regulation of Microcystins-Producing Cyanobacterial Blooms. *Microb. Ecol.* **2014**, *67*, 465–479.

36. Williams, J.R. The Erosion-Productivity Impact Calculator (EPIC) Model: A Case History. *Philos. Trans. B* **1990**, *329*, 421–428.

37. Brown, L.C.; Barnwell, J.T.O.   *The Enhanced Stream Water Quality Models QUAL2E and QUAL2E-UNCAS: Documentation and User Manual*; EPA-600/3-87/007; U.S. Environmental Protetction Agency: Athens, GA, USA, 1987.

38. White, M.J.; Arnold, J.G. Development of a simplistic vegetative filter strip model for sediment and nutrient retention at the field scale. *Hydrol. Proc.* **2009**, *23*, 1602–1616.

39. Munoz-Carpena, R.; Parsons, J.E.; Gilliam, J.W. Modeling hydrology and sediment transport in vegetative filter strips. *J. Hydrol.* **1999**, *214*, 111–129.

40. Williams, J.R. Sediment-yield prediction with universal equation using runoff energy factor. In *Proceedings of the Sediment-Yield Workshop*; USDA Sedimentation Laboratory: Oxford, MS, USA, 1975.

41. Szcześniak, M.; Piniewski, M.   Improvement of hydrological simulations through applying daily precipitation interpolation schemes in meso-scale catchments. *Water* **2015**, *7*, 747–779.

42. Kowalczak, P.; Kundzewicz, Z.W. Water-related conflilcts in urban areas in Poland. *Hydrol. Sci. J. Spec. Issues* **2011**, *56*, 588–596.

43. Siegriest, R.L.; McCray, J.; Weintraub, L.; Chen, C.; Bagdol, J.; Lemonds, P.; van Cuyk, S.; Lowe, K.; Goldstein, R.; Rada, J. *Quantifying Site-Scale Processes and Watershed-Scale Cumulative Effects of Decentralized Wastewater Systems*; Colorado School of Mines: Golden, CO, USA, 2005.

44. Arnold, J.G.; Kiniry, J.R.; Srinivasan, R.; Williams, J.R.; Haney, E.B.; Neitsch, S.L. Soil and Water Assessment Tool Input Output Documentation, Version 2012; Texas Water Resources Institute TR-439; Texas Water Resources Institute: College Station, TX, USA, 2012.

45. Barszczewski, J.; Kaca, E. Water retention in ponds and the improvement of its quality during carp production. *J. Water Land Dev.* **2012**, *17*, 31–38.
46. Vsetickova, L.; Adamek, Z.; Rozkosny, M.; Sedlacek, P. Effects of semi-intensive carp pond farming on discharged water quality. *Acta Ichthyo. Piscat.* **2012**, *42*, 223–231.
47. Abbaspour, K. *SWAT-CUP2: SWAT Calibration and Uncertainty Programs—A User Manual*; Swiss Federal Institute of Aquatic Science and Technology (Eawag): Duebendorf, Switzerland, 2008; p. 95.
48. Abbaspour, K.C.; Johnson, C.A.; van Genuchten, M.T. Estimating uncertain flow and transport parameters using a sequential uncertainty fitting procedure. *Vadose Zone J.* **2004**, *3*, 1340–1352.
49. Moriasi, D.N.; Arnold, J.G.; van Liew, M.W.; Bingner, R.L.; Harmel, R.D.; Veith, T.L. Model evaluation guidelines for systematic quantification of accuracy in watershed simulations. *Trans. ASABE* **2007**, *50*, 885–900.
50. Marcinkowski, P.; Piniewski, M.; Kardel, I.; Giełczewski, M.; Okruszko, T. Modelling of discharge, nitrate and phosphate loads from the Reda catchment to the Puck Lagoon using SWAT. *Ann. Warsaw Univ. Life Sci. SGGW Land Reclam.* **2013**, *45*, 125–141.
51. Piniewski, M.; Kardel, I.; Giełczewski, M.; Marcinkowski, P.; Okruszko, T. Climate change and agricultural development: Adapting Polish agriculture to reduce future nutrient loads in a coastal watershed. *Ambio* **2014**, *43*, 644–660.
52. Santhi, C.; Kannan, N.; Arnold, J.G.; di Luzio, M. Spatial calibration and temporal validation of flow for regional scale hydrologic modelling. *J. Am. Water Resour. Assoc.* **2008**, *44*, 829–846.
53. Qi, C.; Grunwald, S. GIS-based hydrologic modelling in the Sandusky watershed using SWAT. *Trans. ASAE* **2005**, *48*, 169–180.
54. Piniewski, M.; Okruszko, T. Multi-site calibration and validation of the hydrological component of SWAT in a large lowland catchment. In *Modelling of Hydrological Processes in the Narew Catchment, Geoplanet: Earth and Planetary Sciences*; Świątek, D., Okruszko, T., Eds.; Springer: Berlin, Germany, 2011; pp. 15–41.
55. Van Liew, M.W.; Garbrecht, J. Hydrologic simulation of the Little Washita river experimental watershed using SWAT. *J. Am. Water Resour. Assoc.* **2003**, *39*, 413–426.
56. Vaze, J.; Post, D.A.; Chiew, F.H.S.; Perraud, J.M.; Viney, N.R.; Teng, J. Climate non-stationarity—Validity of calibrated rainfall-runoff models for use in climate change studies. *J. Hydrol.* **2010**, *394*, 447–457.
57. Zhu, Q.; Zhang, X.; Ma, C.; Gao, X.; Xu, Y. Investigating the uncertainty and transferability of parameters in SWAT model under climate change. *Hydrol. Sci. J.* **2015**, doi: 10.1080/02626667.2014.1000915.
58. Ekstrand, S.; Wallenberg, P.; Djodjic, F. Process based modelling of phosphorus losses from arable land. *Ambio* **2010**, *39*, 100–115.
59. Engel, B.; Storm, D.; White, M.; Arnold, J.G. A hydrologic/water quality model application protocol. *J. Am. Water Resour. Assoc.* **2007**, *43*, 1223–1236.

60. Kowalkowski, T.; Pastuszak, M.; Igras, J.; Buszewski, B. Differences in emission of nitrogen and phosphorus into the Vistula and Oder basins in 1995–2008—Natural and anthropogenic causes (MONERIS model). *J. Mar. Syst.* **2012**, *89*, 48–60.

61. Wagner, I.; Zalewski, M. Effect of hydrological patterns of tributaries on biotic processes in a lowland reservoir - consequences for restoration. *Ecol. Eng.* **2000**, *16*, 79–90.

62. Kiedrzyńska, E.; Kiedrzyński, M.; Urbaniak, M.; Magnuszewski, A.; Skłodowski, M.; Wyrwicka, A.; Zalewski, M. Point sources of nutrient pollution in the lowland river catchment in the context of the Baltic Sea eutrophication. *Ecol. Eng.* **2014**, *70*, 337–348.

63. Environment: Commission Takes Poland to Court over Nitrates and Water Pollution. Available online: http://europa.eu/rapid/press-release_IP-13-48_en.htm (accessed on 30 January 2015)

64. Heathwaite, A.L.; Johnes, P.J. Contribution of nitrogen species and phosphorus fractions to stream water quality in agricultural catchments. *Hydrol. Proc.* **1996**, *10*, 971–983.

65. Izydorczyk, K.; Skowron, A.; Wojtal, A.; Jurczak, T. The stream inlet to a shallow bay of a drinking water reservoir a "Hot-Spot" for *Microcystis* blooms initiation. *Int. Rev. Hydrobiol.* **2008**, *93*, 257–268.

Model Calibration Criteria for Estimating Ecological Flow Characteristics

Marc Vis, Rodney Knight, Sandra Pool, William Wolfe and Jan Seibert

Abstract: Quantification of streamflow characteristics in ungauged catchments remains a challenge. Hydrological modeling is often used to derive flow time series and to calculate streamflow characteristics for subsequent applications that may differ from those envisioned by the modelers. While the estimation of model parameters for ungauged catchments is a challenging research task in itself, it is important to evaluate whether simulated time series preserve critical aspects of the streamflow hydrograph. To address this question, seven calibration objective functions were evaluated for their ability to preserve ecologically relevant streamflow characteristics of the average annual hydrograph using a runoff model, HBV-light, at 27 catchments in the southeastern United States. Calibration trials were repeated 100 times to reduce parameter uncertainty effects on the results, and 12 ecological flow characteristics were computed for comparison. Our results showed that the most suitable calibration strategy varied according to streamflow characteristic. Combined objective functions generally gave the best results, though a clear underprediction bias was observed. The occurrence of low prediction errors for certain combinations of objective function and flow characteristic suggests that (1) incorporating multiple ecological flow characteristics into a single objective function would increase model accuracy, potentially benefitting decision-making processes; and (2) there may be a need to have different objective functions available to address specific applications of the predicted time series.

Reprinted from *Water*. Cite as: Vis, M.; Knight, R.; Pool, S.; Wolfe, W.; Seibert, J. Model Calibration Criteria for Estimating Ecological Flow Characteristics. *Water* **2015**, *7*, 2358–2381.

1. Introduction

The interactions between streamflow and aquatic ecosystems have occupied researchers across a range of disciplines for more than 50 years. Beginning with studies as early as Rantz [1] and continuing through Tennant [2] to the present day, numerous individual streamflow characteristics have been associated with various ecological responses [3]. More recently, studies have emphasized the importance of multiple streamflow characteristics operating simultaneously or interacting to influence ecological outcomes [4]. These streamflow characteristics are used to quantify relations between flow and ecological responses. At sites where streamflow records are available, the ecologically relevant streamflow characteristics (SFCs) can be derived directly from streamflow observations. However, many, probably most, sites of biological interest have few if any observed streamflow records.

Where streamflow records are unavailable, hydrological modeling is commonly used to derive flow time series, and these simulated time series are then used to derive streamflow characteristics. The basic assumption is that if a model is capable of reproducing observed streamflow with some accuracy, the simulated time series are also suitable to derive ecologically relevant flow characteristics. However, one has to note that flow simulations are never perfect and that they generally depend on the model and its parameterization. Therefore, the suitability of simulated flow series as a basis for the estimation of streamflow characteristics might vary considerably. Key issues that must be addressed include which aspects of the stream hydrograph (SFCs) should be estimated and which modeling approaches are best suited for estimating them.

At least two broad approaches to hydrologic modeling have been applied to ecological flow problems. Regional statistics have been used to predict ecologically relevant streamflow characteristics at ungauged sites to support the development of ecological response functions, with streamflow as the controlling variable [5–7]. Such statistical models depend on prior definition of the streamflow characteristics of interest and thus are of limited flexibility should other flow characteristics later emerge as important [8]. An alternative approach is the use of runoff models, which simulate an entire hydrograph for some period of interest from which any number of streamflow characteristics can subsequently be calculated [8]. Runoff models have been recommended by some authors as the tool of choice for ecological flow studies [4], while others have expressed reservations about their suitability for such applications [8,9].

There are two main criticisms related to using runoff models for application to ecological-flow studies. The first is the difficulty in transferring the calibrated model parameters from a gauged basin, where the model can be calibrated and verified, to an ungauged basin where model performance cannot be evaluated directly. This issue of predictions in ungauged catchments is an area of active research and can be addressed by different regionalization approaches [10]. However, even with perfectly estimated parameter values (*i.e.*, the estimated parameters for an ungauged catchment correspond to what had been achieved with local model calibration) a second issue remains. This is that the models are generally calibrated on some measure of overall model performance such as the model efficiency [8,9], while biological responses to streamflow are commonly associated with specific aspects of the hydrograph, such as the long-term mean or, often more important, high- or low-flow extremes [6,11–14]. This observation raises the question: Can alternative approaches to the design and calibration of runoff models improve their ability to estimate ecologically relevant flow characteristics with a level of accuracy and

257

precision needed to provide useful insights to the interaction between streamflow and ecosystems?

In this study, we used the HBV-light model [15–19] for runoff simulations. This model is an example of a multi-tank catchment model, with 10–15 parameters which are typically estimated by calibration. Several objective functions, each focusing on a different aspect of the hydrograph, were used to calibrate HBV-light. The aim of this study was to evaluate different objective functions for their ability to produce simulated time series that adequately preserve ecologically important flow characteristics.

2. Materials and Methods

2.1. Study Catchments

The 27 catchments used in this analysis represent parts of four Level 3 Ecoregions [20], listed east to west: Blue Ridge, Ridge and Valley, Central Appalachians, and Appalachian (Cumberland) Plateau (Figure 1). The catchments have average basin area of 829 square kilometers (km^2) (range 104–4799 km^2) and average elevation of 491 m above the North American Vertical Datum of 1988 (NAVD 88) (range 174–937 m) (Table 1). Hardwood forest and pasture are the dominant land cover in the study area. Soils are deep in the Blue Ridge ecoregion which leads to increased baseflow in comparison to the relatively thinner soils of the Appalachian Plateau and Ridge and Valley ecoregions [20] Generally, topographic slope and regolith thickness decreases from east to west, while karst development is most prominent in the Ridge and Valley [21]. Combined, these catchment characteristics produce noticeable and documented regional variations in hydrologic response and streamflow regimes [21–24].

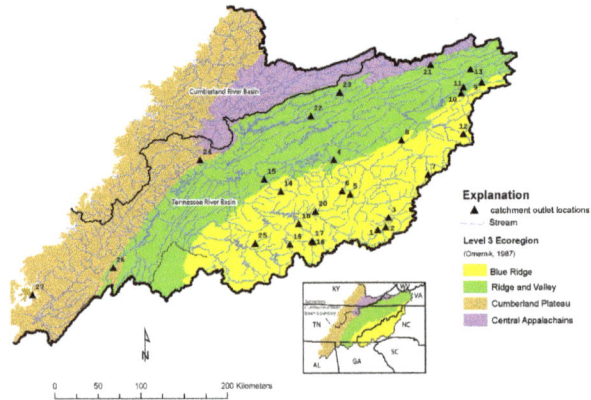

Figure 1. Catchment outlet locations for 27 basins modelled using 7 calibration schemes for HBV-light.

Table 1. U.S. Geological Survey (USGS) stream gaging sites used for model calibration and error evaluation. Latitude and longitude represent the basin outlet; ecoregion defined as the Level 3 ecoregion with the majority of the basin area; km², square kilometers; horizontal reference is North American Datum 1983; vertical reference is North American Vertical Datum 1988.

Map Number (Figure 1)	USGS Station Number	Latitude	Longitude	Average Elevation (m)	Primary Ecoregion (Omernik, 1987)	Basin Area (km²)
1	03441000	35.2731	−82.7058	645	Blue Ridge	104
2	03443000	35.2992	−82.6239	628	Blue Ridge	766
3	03446000	35.3981	−82.5950	637	Blue Ridge	173
4	03455000	35.9816	−83.1611	308	Blue Ridge	4799
5	03459500	35.6350	−82.9900	712	Blue Ridge	906
6	03460000	35.6675	−83.0736	749	Blue Ridge	127
7	03463300	35.8314	−82.1842	810	Blue Ridge	112
8	03465500	36.1765	−82.4574	463	Blue Ridge	2082
9	03471500	36.7604	−81.6312	642	Blue Ridge	198
10	03473000	36.6518	−81.8440	546	Blue Ridge	785
11	03475000	36.7132	−81.8187	555	Ridge and Valley	534
12	03479000	36.2392	−81.8222	795	Blue Ridge	236
13	03488000	36.8968	−81.7462	519	Ridge and Valley	578
14	03497300	35.6645	−83.7113	337	Blue Ridge	271
15	03498500	35.7856	−83.8846	259	Blue Ridge	697
16	03500000	35.1500	−83.3797	612	Blue Ridge	361
17	03500240	35.1589	−83.3942	615	Blue Ridge	146
18	03503000	35.3364	−83.5269	537	Blue Ridge	1130
19	03504000	35.1275	−83.6186	937	Blue Ridge	135
20	03512000	35.4614	−83.3536	562	Blue Ridge	476
21	03524000	36.9448	−82.1549	457	Ridge and Valley	1382
22	03528000	36.4251	−83.3982	323	Ridge and Valley	3816
23	03531500	36.6620	−83.0949	384	Central Appalachians	828
24	03540500	35.9831	−84.5580	232	Cumberland Plateau	1815
25	03550000	35.1389	−83.9806	474	Blue Ridge	268
26	03568933	34.8975	−85.4631	202	Ridge and Valley	379
27	03574500	34.6243	−86.3064	174	Cumberland Plateau	814

Temperature and precipitation in the study area vary with longitude and elevation. Average annual temperature in the area is 13.9 degrees Celsius (°C). The warmest months of the year are July and August, and the coldest are typically January and February [25]. The Blue Ridge averages about 1350 millimeters per year (mm/y) of precipitation annually, compared to 1450 mm/y in the Cumberland Plateau and Ridge and Valley [26]. Locally, precipitation in the Blue Ridge can exceed 2000 mm/y at the highest elevations. Less than 2 percent of the precipitation comes as snow (based on 1:10 ratio of rain to snow). The streamflow regime in the study area is characterized by peak runoff typically between December and April as the result of frozen or saturated soils and low evapotranspiration rates. Summer months typically have lower streamflows because of increased temperatures and evapotranspiration rates, though occasional convective or tropical storm systems may produce locally severe flooding. Lowest flows occur in the late-summer through the fall coinciding with continuing high temperatures and evapotranspiration rates combined with decreased precipitation (October is the driest month generally). Annual runoff for the study area varies from approximately 450 to more than 760 mm [27].

The Tennessee and Cumberland River basins (considered as one aquatic ecoregion by Abell *et al.* [28]) have the highest level of freshwater diversity in North America and possibly the most diversity for any temperate freshwater ecoregion in the world [29,30]. Included in this measure are 231 fish species (with 67 (29 percent) being endemic) along with a globally outstanding unionid mussel and crayfish fauna. Many of these species are restricted to the Tennessee and Cumberland River basins [28] (pp. 212–213). A wide range of human activities threaten these populations, including urbanization, mining, logging, agriculture, and other forms of land disturbance that alter hydrologic response [28]. In addition, the entire main channels of the Tennessee and Cumberland Rivers, together with many of their tributaries, have been impounded. Flow alteration as a result of these activities has degraded or destroyed stream habitat according to Abell *et al.* [28], with more than 57 fish species and 47 mussel species at risk in the Tennessee–Cumberland aquatic ecoregion [31] (cited in Abell *et al.* [28], p. 213).

2.2. HBV Model

The HBV model [15,16] is a simple multi-tank-type model for simulating runoff. Rainfall and air temperature data [32] as well as estimated potential evaporation data based on the American Society of Civil Engineers Penman–Monteith method [33–36] are inputs to the model, which consists of four commonly used routines: (1) snow; (2) soil moisture; (3) response; and (4) routing. These routines, or slight modifications, are commonly used in other similar models (for example PRMS; Leavesley, Lichty, Troutman, and Saindon, 1983). In the snow routine, snow accumulation and snow melt are calculated by a degree-day method [37]. The soil moisture routine represents soil–water storage, which is used in conjunction with temperature and precipitation to drive evaporation and groundwater recharge. Evaporation from the soil tank equals the potential evaporation if the relative soil moisture storage is above a certain fraction, while below that fraction a linear reduction is applied. The response routine consists of connected shallow and deep groundwater storage terms and simulates runoff by summing up three linear outflow equations representing peak, intermediate and base flow. The routing routine delivers simulated runoff to the catchment outlet based on a triangular weighting function in the routing routine.

Catchments can be separated into different elevation and vegetation zones as well as into subbasins in HBV. In this study, however, catchments were disaggregated using only different elevation zones to reduce problems of over-parameterization. Calculations were performed separately for each elevation zone according to catchment for the snow and soil-moisture routines. Groundwater storage was treated as a lumped representation for each catchment. The version of HBV used in this study, HBV-light [18], corresponds to a slightly modified version of HBV-6. HBV-light uses a warming-up period of normally one year to set state variable values according to the preceding meteorological conditions and parameter sets. A more detailed description of HBV-light can be found in [18].

2.3. Calibration

The HBV-light model was applied to the 27 catchments using a daily time step. Each catchment was separated into elevation zones of 200 m, which cover at least 5 percent of the area of their respective catchment. Elevation zones covering less than 5 percent of the catchment area were merged with neighboring elevation zones. Rainfall and temperature data were compiled for the different elevation zones with a lapse rate of 10 percent/100 m and 0.6 °C/100 m, respectively. The long-term monthly potential evaporation data were linearly interpolated to daily values and corrected by using the deviations of the temperature to its long-term mean.

For all catchments, the first three years of input data measurements were used for the "warming-up" of the model to estimate the initial state variables. The rest of the data were divided into two equal time periods (14 years) covering the hydrological years (1 October through 30 September) from 1983 to 1996 and from 1996 to 2009. Each time period served both as calibration and validation period; when using the first time period for calibration the second time period was used for validation, and vice versa. This approach to calibration, validation, and parameterization allows us to consider distributions of parameter values derived from multiple independent realizations of the model, providing a generally robust evaluation. To address parameter uncertainty and equifinality [38], each calibration was repeated 100 times (here called calibration trials), which because of the random elements of the Genetic Algorithm and Powell optimization (GAP, [39]) used for calibration, resulted in 100 different parameterizations. The feasible parameter value ranges were defined based on previous studies (Table 2) [40].

We considered seven different objective functions for calibration, which consisted of either single or combined statistical criteria evaluating the fit between observed and simulated values (Tables 3 and 4) to assess the influence of an objective function on the value of the simulated ecological indicators. The objective functions were chosen to represent different statistical aspects of streamflow. The combinations of criteria were defined to evaluate different aspects simultaneously; for example, combination 2 (C2) included Reff, MARE, Spearman, and Volume Error (see Table 3 for a description of the criteria). Reff and MARE are sensitive to peaks and low flows, respectively, and therefore help evaluate performance with respect to extreme discharge values. Volume Error expresses how well the model predicts overall runoff volume for the simulation period, whereas the Spearman rank coefficient reflects the model's success in replicating the overall timing and magnitude of discharge. Each objective function was used to calibrate the model for each time period, resulting in 14 simulated time series (seven objective functions for two different calibration periods) of streamflow for each catchment modeled.

Table 2. Parameter ranges used during the Genetic Algorithm and Powell optimization (GAP) calibrations within HBV-light. (°C, degrees Celsius; mm, millimeter; D, day).

Parameter	Explanation	Minimum	Maximum	Unit
Snow Routine				
TT	Threshold temperature	−2	2.5	°C
CFMAX	Degree-day factor	0.5	10	$mm \cdot °C^{-1} \cdot D^{-1}$
SFCF	Snowfall correction factor	0.5	1.2	-
CFR	Refreezing coefficient	0	0.1	-
CWH	Water holding capacity	0	0.2	-
Soil Routine				
FC	Maximum storage in soil box	100	550	mm
LP	Threshold for reduction of evaporation (relative storage in the soil box)	0.3	1	-
BETA	Shape coefficient	1	5	-
Response Routine				
PERC	Maximal flow from upper to lower box	0	4	$mm \cdot D^{-1}$
UZL	Maximal storage in the soil upper zone	0	70	mm
K0	Recession coefficient (upper box, upper outflow)	0.1	0.5	D^{-1}
K1	Recession coefficient (upper box, lower outflow)	0.01	0.2	D^{-1}
K2	Recession coefficient (lower box)	0.00005	0.1	D^{-1}
Routing Routine				
MAXBAS	Routing, length of weighting function	1	5	D

Table 3. Definitions criteria used in objective functions for the automatic calibration trials using the Genetic Algorithm and Powell optimization (GAP) algorithm.

Criterion	Description	Definition
Reff	Model efficiency	$1 - \dfrac{\Sigma (Q_{obs} - Q_{sim})^2}{\Sigma (Q_{obs} - \overline{Q}_{obs})^2}$
LogReff	Efficiency for log(Q)	$1 - \dfrac{\Sigma (\ln Q_{obs} - \ln Q_{sim})^2}{\Sigma (\ln Q_{obs} - \ln \overline{Q}_{obs})^2}$
Lindström	Lindström measure	$Reff - 0.1 \dfrac{\left| \Sigma (Q_{obs} - Q_{sim}) \right|}{\Sigma (Q_{obs})}$
MARE	Measure based on the Mean Absolute Relative Error [1]	$1 - \dfrac{1}{n} \Sigma \dfrac{\left| Q_{obs} - Q_{sim} \right|}{Q_{obs}}$
Spearman	Spearman rank correlation [2]	$\dfrac{\Sigma (R_{obs} - \overline{R}_{obs})(S_{sim} - \overline{S}_{sim})}{\sqrt{\Sigma (R_{obs} - \overline{R}_{obs})^2} \sqrt{\Sigma (S_{sim} - \overline{S}_{sim})^2}}$
VolumeError	Volume error	$1 - \dfrac{\left| \Sigma (Q_{obs} - Q_{sim}) \right|}{\Sigma (Q_{obs})}$

[1] Where n is the number of days; [2] Where R_{obs} and S_{sim} are the ranks of Q_{obs} and Q_{sim}, respectively.

Table 4. The three combination objective functions used during the Genetic Algorithm and Powell optimization (GAP) calibrations within HBV-light. The criteria were weighted equally in each case. See Table 3 for a more detailed specification of each of the criteria.

Combined Objective Function	Criteria
C1	Reff, LogReff, VolumeError
C2	Reff, MARE, Spearman, VolumeError
C3	Spearman, VolumeError

2.4. Evaluation

The choice of the SFCs is based on studies of Knight *et al.* [6], which identified 12 specific streamflow characteristics, from a larger suite identified in Knight *et al.* [41], as most appropriate indicators for fish species richness in the study area (Table 5). All SFCs were computed using the simulated runoff of each catchment that was calibrated with one of the seven objective functions and for the two different calibration and validation time periods. The value of each streamflow characteristic was determined for both time periods based on the measurement data. All indices were computed using the free EflowStats R-Package [42].

For each objective function, 100 calibration trials were accomplished per catchment for both periods (1983–1996 and 1996–2009), producing 100 independently optimized parameter sets per catchment per simulation period. For each objective function and streamflow characteristic, the sources of uncertainty in the results were analyzed. The spread reflects both differences in behavior among the 27 catchments and uncertainty among the parameter sets, but the relative importance of these two sources of variability is not uniform. The variability because of differences between catchments was analyzed by computing the medians of the streamflow characteristics over the 100 runs per catchment. To be able to compare the median values, normalization was carried out by dividing the median values by the corresponding observed flow characteristic value. For analyzing the spread resulting from parameter uncertainty, the ranges over 100 runs per catchment were divided by the range over the median values of the different catchments. The spread because of parameter uncertainty was compared to the variation between the different catchments.

To quantify the performance of objective functions in representing the different flow characteristics, Spearman rank correlation coefficients and Nash-Sutcliffe efficiencies (NSEs) were computed between the (median) simulated and observed flow characteristic values of the 27 different catchments. Where NSE of 1.0 corresponds to identical flow characteristic values between simulated and observed runoff time series for each catchment, a Spearman rank correlation coefficient of 1.0 only requires the order of observed and simulated flow characteristic values to be the same.

Table 5. Definition of streamflow characteristics used in this study (adapted and modified from Knight *et al.*, 2014 and Thomson and Archfield, 2014) (mm/day, millimeters per day; -, no units; %, percent).

Streamflow Characteristic	Abbreviation	Description	Units
Magnitude			
Mean annual runoff	MA41	Annual mean daily streamflow	mm/day
Maximum October runoff	MH10	Mean maximum October streamflow across the period of record	mm/day
Lowest 15% of daily runoff	Flowperc	85% exceedance of daily mean streamflow for the period of record	mm/day
Rate of streamflow recession	RA7	Median change in log of streamflow for days in which the change is negative across the period of record	mm/day
Ratio			
Average 30-day maximum runoff	DH13	Mean annual maximum of a 30-day moving average streamflow divided by the median for the entire record	–
Stability of runoff	TA1	Measure of the constancy of a flow regime by dividing daily flows into predetermined flow classes	–
Frequency			
Frequency of moderate floods	FH6	Average number of high-flow events per year that are equal to or greater than three times the median annual flow for the period of record	number/year
Frequency of moderate floods	FH7	Average number of high-flow events per year that are equal to or greater than three times the median annual flow for the period of record	number/year
Variability			
Variability of March runoff	MA26	Standard deviation for March streamflow divided by the mean streamflow for March	–
Variability in high-flow pulse duration	DH16	100 times the standard deviation for the yearly average high-flow pulse durations (daily flow greater than the 75th percentile) divided by the mean of the yearly average high pulse durations	%
Variability of low-flow pulse count	FL2	100 times the standard deviation for the average number of yearly low-flow pulses (daily flow less than the 25th percentile) divided by the mean low-flow pulse counts	%
Date			
Timing of annual minimum runoff	TL1	Julian date of annual minimum flow occurrence	Julian day

3. Results

The model efficiencies that could be achieved for the different catchments varied from 0.64 to 0.91 (calibration) and 0.61 to 0.90 (validation), indicating reasonably good runoff simulation with the calibrated HBV-light model. As an example of the performance of the simulations with regard to the streamflow characteristics, the results for two indices (DH16 (variability in high-flow pulse duration) and MA41 (mean annual runoff)) for one catchment (03455000) are shown in Figure 2. Each plot contains 28 boxplots (one for each combination of an objective function, time period and calibration or validation). Each of the boxplots is based on 100 streamflow characteristic values obtained by using the 100 different parameter sets per catchment for the simulations. In both cases, there were clear deviations of the flow characteristics computed from the simulated time series compared to the observed runoff series as indicated by the red lines (red line represents observed SFC value). The streamflow characteristic DH16 was largely underestimated, especially for period 1 (1983–1996) (Figure 2a). The spread among the 100 different simulations was considerably larger for period 2 (1996–2009) than for period 1. For SFCs such as MA41 (Figure 2b), the performance differences in predicting the streamflow characteristic were prominent between the four combinations of calibration and validation periods.

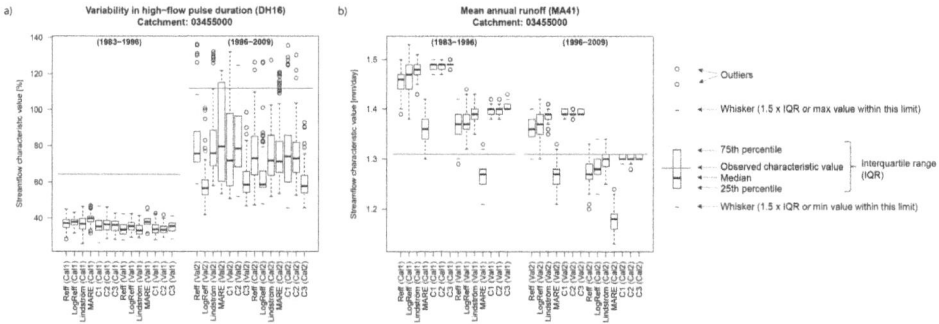

Figure 2. Boxplots for catchment 4 (03455000) and (**a**) streamflow characteristic DH16 (Variability in high-flow pulse duration); (**b**) streamflow characteristic MA41 (Mean annual runoff). Cal1 and Cal2 are calibration of period 1, respectively period 2, whereas Val1 and Val2 are validation of period 1, respectively period 2.

The agreement between observed and simulated flow characteristics varied considerably among the different catchments (Figure 3). Each plot contains 28 boxplots (one for each combination of an objective function, time period and calibration or validation). Each boxplot is based on 27 values (one value per catchment), which were normalized by dividing the median streamflow characteristic value based on simulated runoff by the corresponding streamflow characteristic value computed based on the observed runoff time series. The spread between the different catchments is much smaller for the streamflow characteristic MA41 (mean annual runoff) than for the other flow characteristics. Except for the criteria LogReff and MARE, MA41 was reproduced well for both calibration periods, whereas values were slightly underestimated when being validated on period 1 and slightly overestimated when validated on period 2. Both MA41 (mean annual runoff) and MH10 (maximum October runoff) were reproduced less well for parameter sets derived by calibration based on the criteria LogReff and MARE, both of which are more sensitive to low flow conditions than the other criteria.

The distribution of the 27 relative ranges (per catchment—Dividing the range over the 100 runs per catchment by the range over the 27 median catchment values) is a measure for the consistency over the different catchments (Figure 4). While for some cases there was a low variation (indicated by narrow distributions of relative range), for many cases a considerable variation was observed. For calibrations based on the Nash-Sutcliffe efficiency, for instance, the median relative range varied from around 0.1 for MA41 (mean annual runoff) to above 1 for FL2 (variability of low-flow pulse count).

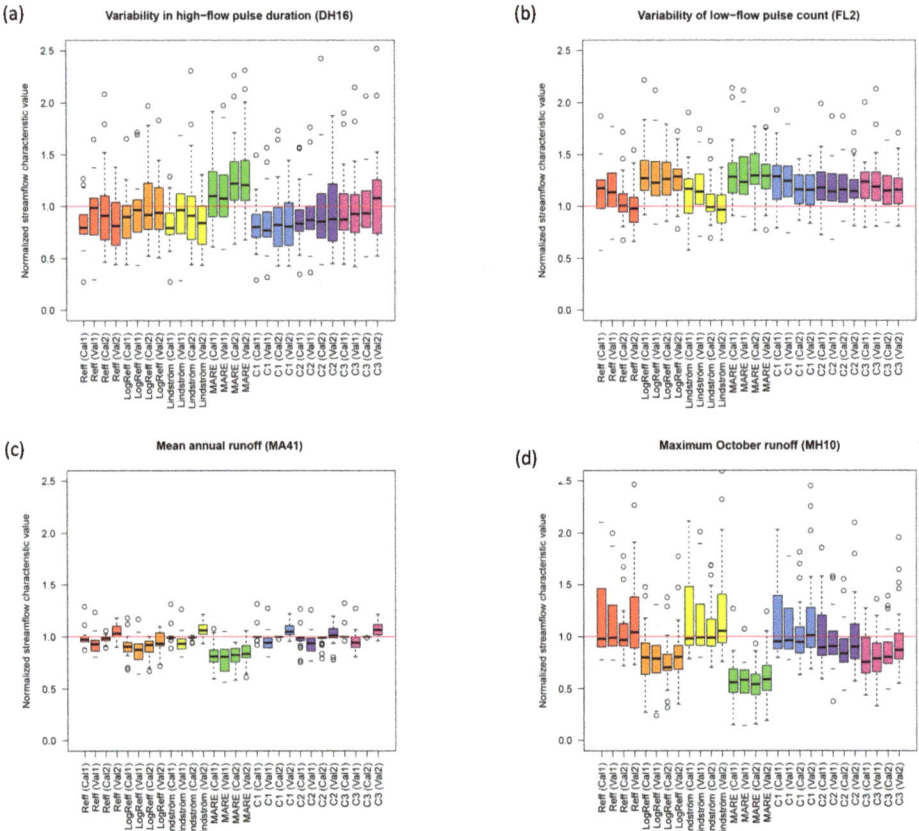

Figure 3. *Cont.*

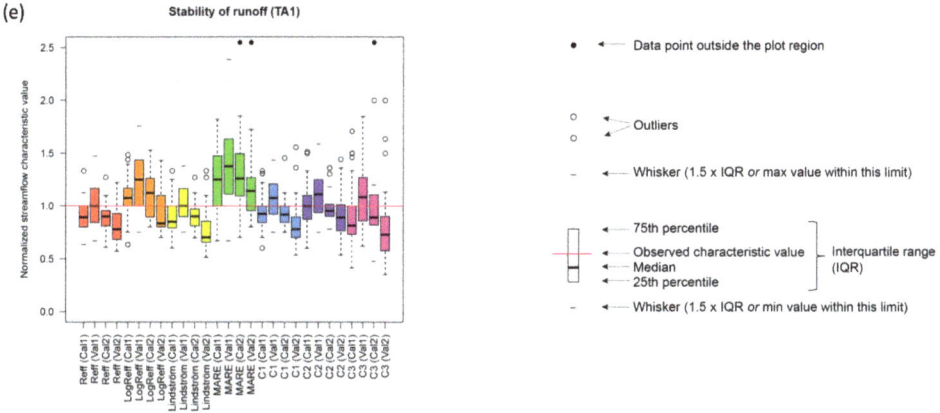

Figure 3. Normalized median flow characteristic values for five different flow characteristics: (**a**) DH16 (Variability in high-flow pulse duration); (**b**) FL2 (Variability of low-flow pulse count); (**c**) MA41 (Mean annual runoff); (**d**) MH10 (Maximum October runoff) and (**e**) TA1 (Stability of runoff). Each color corresponds to an objective function. Per objective function, the four boxplots represent (from left to right) calibration period 1 (Cal1), validation period 1 (Val1), calibration period 2 (Cal2) and validation period 2 (Val2). Each boxplot is based on 27 normalized median flow characteristic values, one value for each of the 27 catchments. Medians were computed over 100 runs per catchment. Normalization was carried out by dividing the median values by the corresponding observed flow characteristic value.

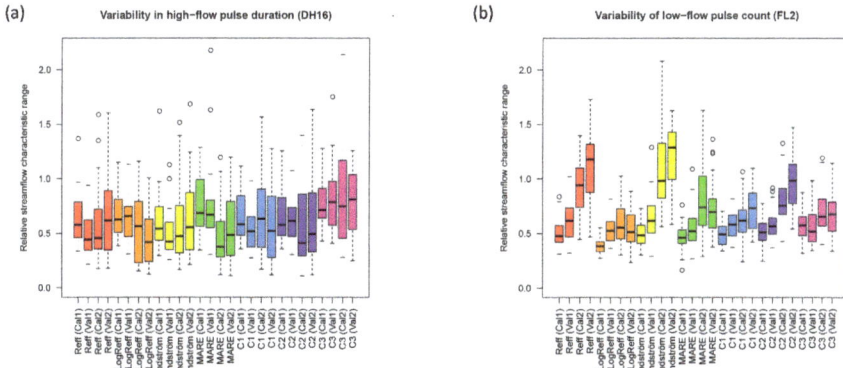

Figure 4. *Cont.*

267

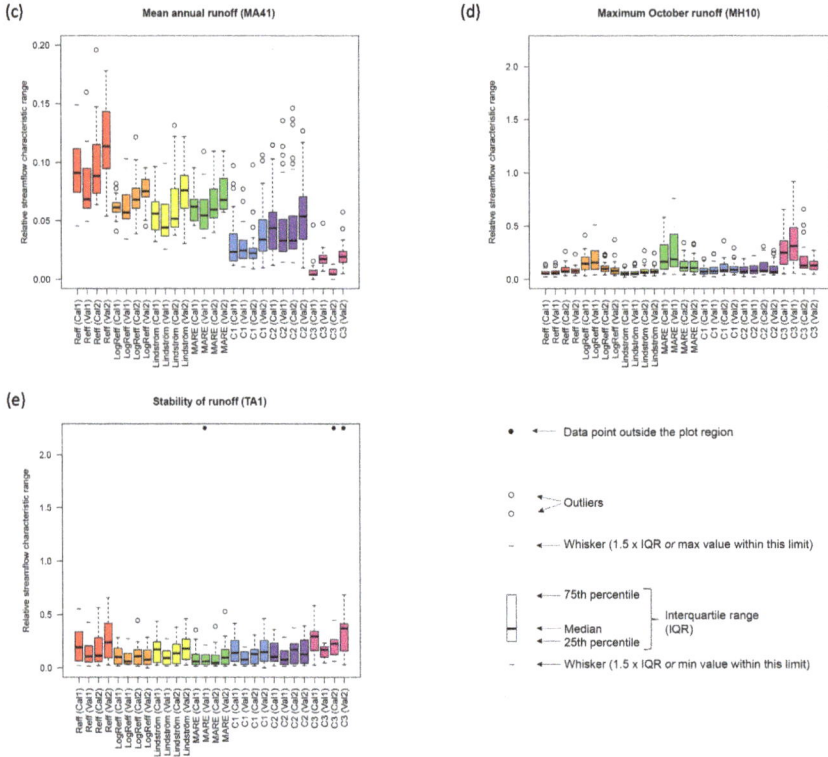

Figure 4. Relative ranges as a measure for parameter uncertainty for streamflow characteristics (**a**) DH16 (Variability in high-flow pulse duration); (**b**) FL2 (Variability of low-flow pulse count); (**c**) MA41 (Mean annual runoff); (**d**) MH10 (Maximum October runoff) and (**e**) TA1 (Stability of runoff). Each color corresponds to an objective function. Per objective function, the four boxplots represent (from left to right) calibration period 1 (Cal1), validation period 1 (Val1), calibration period 2 (Cal2) and validation period 2 (Val2). Each boxplot is based on 27 values, one value for each of the 27 catchments. Relative ranges were computed by dividing the range over the 100 runs per catchment by the range over the 27 median catchment values. Note that the Mean annual runoff (MA41) has been plotted on a different scale.

Agreement among the different streamflow characteristics and the different objective functions varied considerably (Figure 5). Comparison of streamflow characteristics based on observed runoff series against the medians of those obtained from simulated time series allows evaluating the agreement in relation to the variation between catchments. These scatter plots show that the agreement varied considerably among both the different streamflow characteristics and the different objective functions. While only plots with flow characteristics calculated for the first calibration period are shown, results were similar for the other calibration and validation periods. The performance for all streamflow characteristics and all combinations of calibration/validation periods were evaluated using the Spearman rank correlation coefficients (Table 6), which evaluates how well the relative ranking of the indices between the catchments is captured, and the model efficiencies (Table 7), which evaluate how well the exact values were predicted. Typically, the values were similar for periods 1 and 2, when the parameterizations obtained by calibration for the respective period were used, resulting in a median difference of 0.015 for the Spearman Rank correlation and 0.0855 for NSE. In general, results are expected to be poorer for the validation period in comparison to the calibration period; however, for the respective validation periods the values were only slightly lower (median difference of −0.0215 (Spearman) and −0.029 (NSE)). This indicates that results were similar for the two periods and were similar when looking at the validation periods. The average median percent error for estimated streamflow characteristics was almost always less than zero, indicating that the objective functions used for model calibration typically underestimated each of the 12 streamflow characteristics being evaluated (Table 8).

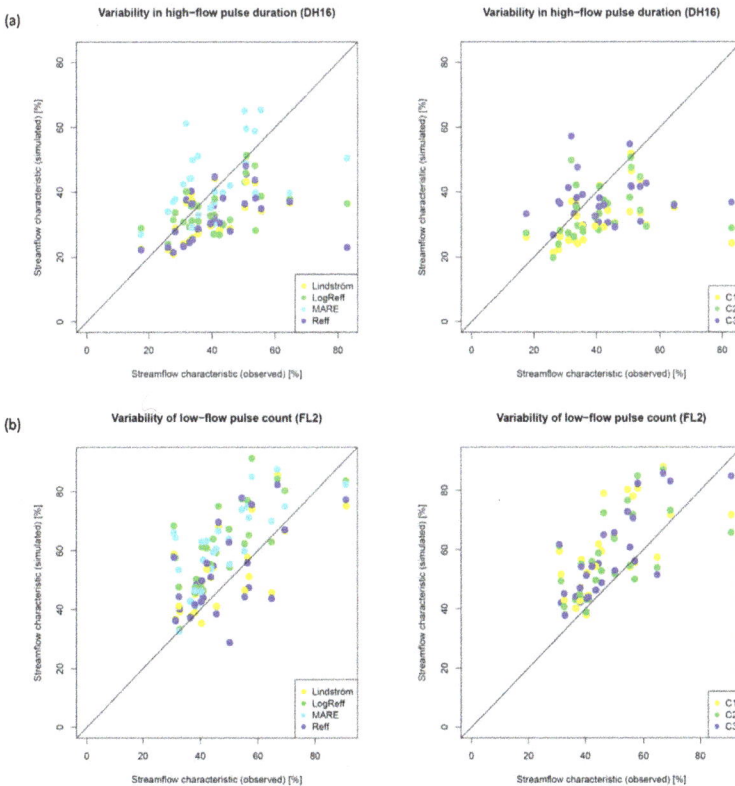

Figure 5. *Cont.*

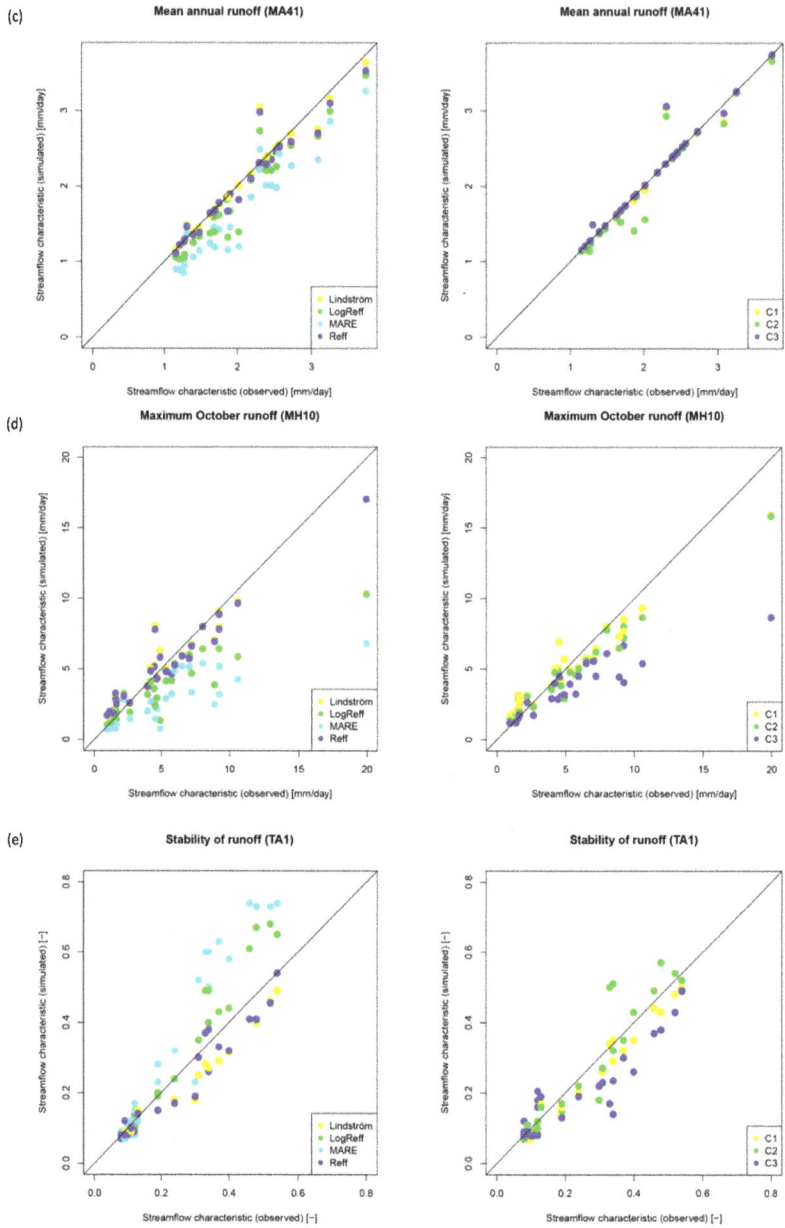

Figure 5. Scatterplots for the streamflow characteristics (**a**) DH16 (Variability in high-flow pulse duration); (**b**) FL2 (Variability of low-flow pulse count); (**c**) MA41 (Mean annual runoff); (**d**) MH10 (Maximum October runoff) and (**e**) TA1 (Stability of runoff) for calibration period 1. The points represent the median value of all 100 calibration trials in each catchment based on single criteria objective functions (**left column**) and multi-criteria objective functions (**right column**).

Table 6. Spearman rank correlation coefficients between objective functions (horizontal) and streamflow characteristics (vertical) based on observed respective simulated streamflow (for each group of four values: upper − left = calibration period 1 (Cal1), upper − right = validation period 2 (Val2), lower − left = validation period 1 (Val1), lower − right = calibration period 2 (Cal2)). Colors are ranging from white (for a Spearman rank correlation of 0) to dark green (for a Spearman rank correlation of 1).

	Reff		LogReff		Lindström		MARE		C1		C2		C3	
MA41	0.973	0.978	0.930	0.927	0.980	0.983	0.919	0.918	0.980	0.981	0.947	0.928	0.981	0.986
	0.957	0.991	0.929	0.947	0.961	0.998	0.926	0.950	0.961	1.000	0.952	0.979	0.962	1.000
MH10	0.930	0.831	0.874	0.853	0.916	0.837	0.834	0.829	0.941	0.837	0.958	0.874	0.918	0.898
	0.960	0.940	0.862	0.868	0.958	0.934	0.822	0.829	0.957	0.918	0.942	0.903	0.885	0.933
Flowperc	0.796	0.978	0.810	0.986	0.790	0.961	0.814	0.979	0.808	0.980	0.810	0.983	0.685	0.867
	0.778	0.985	0.808	0.996	0.781	0.980	0.804	0.996	0.803	0.995	0.806	0.996	0.683	0.897
RA7	0.736	0.724	0.877	0.885	0.726	0.735	0.888	0.896	0.870	0.873	0.851	0.892	0.696	0.797
	0.756	0.836	0.930	0.930	0.719	0.775	0.848	0.902	0.878	0.919	0.880	0.917	0.744	0.789
DH13	0.977	0.938	0.974	0.948	0.971	0.908	0.960	0.960	0.981	0.945	0.976	0.945	0.926	0.691
	0.955	0.866	0.976	0.937	0.955	0.877	0.964	0.957	0.971	0.910	0.978	0.885	0.871	0.573
TA1	0.972	0.929	0.968	0.943	0.977	0.906	0.947	0.974	0.968	0.884	0.960	0.899	0.875	0.766
	0.936	0.956	0.933	0.966	0.952	0.942	0.884	0.936	0.958	0.948	0.942	0.964	0.904	0.924
FH6	0.943	0.851	0.916	0.906	0.935	0.875	0.728	0.863	0.953	0.916	0.900	0.921	0.569	0.663
	0.926	0.888	0.853	0.931	0.931	0.898	0.634	0.855	0.942	0.930	0.901	0.919	0.498	0.613
FH7	0.948	0.933	0.881	0.889	0.949	0.935	0.810	0.887	0.967	0.945	0.965	0.952	0.688	0.563
	0.927	0.951	0.842	0.889	0.941	0.960	0.763	0.805	0.945	0.967	0.944	0.967	0.480	0.520
MA26	0.849	0.917	0.789	0.906	0.855	0.920	0.704	0.858	0.894	0.923	0.903	0.915	0.631	0.856
	0.752	0.932	0.699	0.894	0.782	0.935	0.672	0.829	0.821	0.933	0.831	0.928	0.381	0.769
DH16	0.534	0.645	0.443	0.662	0.503	0.673	0.402	0.471	0.510	0.745	0.525	0.683	0.145	0.482
	0.429	0.549	0.421	0.654	0.410	0.514	0.346	0.645	0.526	0.659	0.511	0.650	0.094	0.518
FL2	0.521	0.443	0.740	0.628	0.609	0.449	0.734	0.703	0.709	0.602	0.684	0.668	0.755	0.594
	0.548	0.617	0.659	0.604	0.579	0.659	0.641	0.626	0.672	0.711	0.620	0.695	0.616	0.628
TL1	0.477	0.394	0.643	0.520	0.471	0.347	0.612	0.753	0.603	0.330	0.531	0.428	0.574	0.418
	0.407	0.112	0.646	0.546	0.418	0.065	0.623	0.777	0.497	0.362	0.531	0.201	0.600	0.280

4. Discussion

In the absence of observed data, environmental flow studies necessarily rely on some form of streamflow estimation to model the response of aquatic ecology to alteration of the streamflow regime. Knight *et al.* [23] and Murphy *et al.* [8] raised the question of validity and began evaluation of model accuracies for predicting known ecologically-relevant streamflow characteristics. Murphy *et al.* [8] and Shrestha *et al.* [9] highlight that typical calibration approaches, often focused on daily, monthly, or annual mean values, are inadequate when predicting more subtle aspects of the flow regime. An increasing body of work is making use of statistical modeling approaches to address hydrologic and hydro-ecological questions [5,7,43–45]. However, as already stated by Murphy *et al.* [8] and Shrestha *et al.* [9], runoff models have advantages as well as limitations, particularly in regard to developing streamflow time series reflecting land cover, human population, or climatic projections. As such, runoff models should be closely evaluated to better understand if the calibration approaches and predictive accuracies yield results amenable to their end use.

Table 7. Nash-Sutcliffe efficiencies between objective functions (horizontal) and streamflow characteristics (vertical) based on observed respective simulated streamflow (for each group of four values: upper − left = calibration period 1 (Cal1), upper − right = validation period 2 (Val2), lower − left = validation period 1 (Val1), lower − right = calibration period 2 (Cal2)). Colors are ranging from white (for Nash-Sutcliffe efficiencies of 0 or lower to dark green (for a Nash-Sutcliffe efficiency of 1).

		Reff		LogReff		Lindström		MARE		C1		C2		C3	
MA41	upper	0.917	0.936	0.840	0.881	0.936	0.933	0.584	0.626	0.946	0.939	0.922	0.927	0.949	0.930
	lower	0.858	0.967	0.746	0.835	0.900	0.993	0.490	0.554	0.914	0.999	0.875	0.965	0.916	1.000
MH10	upper	0.848	0.820	−0.627	0.570	0.841	0.796	−3.942	−1.220	0.820	0.871	0.796	0.879	−1.630	0.663
	lower	0.859	0.934	−0.931	0.332	0.874	0.926	−5.692	−2.258	0.848	0.926	0.756	0.850	−1.367	0.667
Flowperc	upper	0.416	0.749	0.611	0.837	0.356	0.660	0.647	0.960	0.463	0.680	0.614	0.804	0.170	0.477
	lower	0.484	0.868	0.569	0.967	0.491	0.820	0.465	0.966	0.538	0.848	0.591	0.939	0.373	0.669
RA7	upper	0.209	0.281	0.071	0.193	0.279	0.370	−0.420	−0.284	−0.043	−0.063	−0.229	−0.197	−9.226	−7.224
	lower	−0.628	−0.230	0.369	0.385	−0.608	−0.277	0.156	0.186	0.276	0.252	0.190	0.231	−5.173	−4.088
DH13	upper	0.372	−0.164	0.884	0.472	−0.601	−1.895	0.910	0.858	0.797	0.522	0.770	0.874	−7.603	−20.044
	lower	0.638	0.427	0.919	0.748	0.437	−0.030	0.814	0.914	0.902	0.813	0.672	0.817	−4.235	−14.891
TA1	upper	0.898	0.432	0.856	0.882	0.829	0.108	0.672	0.803	0.918	0.477	0.886	0.749	0.502	−1.020
	lower	0.863	0.912	0.718	0.845	0.892	0.926	0.548	0.685	0.881	0.974	0.839	0.953	0.806	0.705
FH6	upper	0.709	0.628	−1.354	−0.967	0.660	0.559	−7.331	−4.461	0.513	0.502	0.210	0.282	−3.781	−5.629
	lower	0.714	0.622	−0.788	−0.465	0.717	0.612	−4.768	−3.426	0.736	0.680	0.533	0.522	−2.536	−4.020
FH7	upper	0.746	0.756	−0.440	−1.246	0.585	0.600	−0.752	−1.837	0.769	0.725	0.842	0.820	−13.413	−22.837
	lower	0.813	0.826	0.290	−0.242	0.801	0.820	−0.260	−0.612	0.912	0.930	0.932	0.954	−9.425	−11.728
MA26	upper	0.618	0.849	0.080	0.033	0.582	0.832	−0.418	−1.114	0.789	0.882	0.848	0.872	−4.116	−4.256
	lower	0.331	0.862	0.184	0.320	0.324	0.886	0.178	−0.513	0.500	0.894	0.564	0.878	−1.898	−2.343
DH16	upper	−3.044	−0.329	−3.375	0.050	−3.323	−0.307	−0.463	−0.371	−3.727	−0.006	−2.768	0.192	−3.474	−0.562
	lower	−0.937	−0.182	−2.056	0.186	−1.012	−0.234	−1.025	0.006	−1.535	−0.092	−1.562	0.119	−2.785	−0.309
FL2	upper	0.118	−1.176	−0.469	−1.557	0.201	−0.931	−0.556	−1.448	−0.266	−0.827	−0.167	−1.773	0.139	−0.948
	lower	−0.040	−1.198	−0.530	−1.841	0.056	−1.123	−0.759	−1.703	−0.203	−0.409	−0.132	−1.246	−0.104	−1.018
TL1	upper	−0.376	−4.676	−0.211	−3.016	−0.310	−5.502	−0.361	−2.672	−0.017	−4.483	−0.196	−4.053	−0.023	−2.708
	lower	−0.505	−4.322	−0.250	−3.892	−0.518	−4.338	−0.557	−2.218	−0.400	−4.503	−0.489	−5.932	0.021	−3.529

Table 8. Median percent error for streamflow characteristics by model objective function for calibration period 1 (Cal1).

Objective Function	MA41	MH10	RA7	TA1	DH13	FH7	FH6	FL2	MA26	DH16	TL1	E85	Average Median Error (Percent)
Lindström	−0.6	−1.8	−25.0	−15.2	−18.1	−23.0	−12.0	16.8	9.1	−20.8	3.7	19.1	−5.6
LogReff	−9.5	−20.0	−50.0	7.7	−9.5	−37.5	−27.0	26.9	−7.3	−10.0	4.8	15.2	−9.7
MARE	−18.9	−44.0	−57.1	25.0	−7.4	−44.4	−41.4	28.2	−19.6	9.9	5.5	−7.3	−14.3
Reff	−2.5	−2.1	−18.2	−10.8	−14.7	−20.0	−12.0	17.5	9.8	−20.2	4.2	9.8	−4.9
C1	0.0	−4.8	−50.0	−7.7	−13.1	−19.0	−14.1	28.6	4.9	−19.7	3.4	29.9	−5.1
C2	−0.8	−10.6	−42.9	0.0	−7.5	−14.0	−18.2	17.7	2.2	−16.4	4.0	13.2	−6.1
C3	0.0	−24.5	−44.4	−18.9	−18.9	−69.3	−37.6	23.6	−28.1	−12.5	3.4	24.1	−16.9
Average Median Percent Error	−4.6	−15.4	−41.1	−2.8	−12.7	−32.5	−23.2	22.8	−4.2	−12.8	4.1	14.9	−

While the HBV-light model was used in this study, there is little reason to assume that results would be discernibly different if another calibrated runoff model were used. Partly this reflects the fact that most mechanistic runoff models are fundamentally similar in concept and application, using more or less the same or similar routines. Fundamentally, if calibration is used, the simulated series are fitted to the observed series according to some objective function, and regardless of the specific model being used, this fit does not ensure agreement in all possible aspects of the hydrograph shape.

The accuracy of prediction and appropriateness of calibration is important in the context of environmental flow application as error of predicting flow-regime components will be translated and probably amplified as error in estimating ecological response. A given approach to model calibration will lead to accurate prediction of the runoff with regard to the used objective function measure, however accurate prediction of other aspects may be lacking. For example, Knight *et al.* [41] (Figure 2) published linear functions representing the 80th quantile upper-bound relationship of specialized insectivore scores to three streamflow characteristics (TA1, FH6, and RA7; see Table 5 for definitions). Following Murphy *et al.* [8], we use these relations to evaluate the accuracy of streamflow characteristic predictions as well as predicted ecological response based on the seven calibration approaches discussed herein for a single model (catchment 03488000). Using the equations from Knight *et al.* [41] and simulated streamflow presented in this paper, values of insectivore scores varied from 0.49 to 0.87 for RA7, 0.53 to 0.8 for TA1, and 0.58 to 0.84 for FH6 (Table 9; Figure 6). While median percent difference error for estimated specialized insectivore score for RA7 was a modest 8.2 percent under the estimate using observed data, individual departures from the observed values ranged from −19.7 to 42.6 percent for RA7, −13.1 to 31.1 percent for TA1, and −10.8 to 29.2 percent for FH6. Model results in this example are similar to those for a regional regression model reported by Murphy *et al.* [8] (9 percent difference for streamflow characteristic and 16 percent over estimation for insectivore score using HBV-light. Results presented here are considerably different than those for a rainfall-runoff model example from Murphy *et al.* [8], showing 90 percent overestimated for the same ecological score.

The objective functions used for model calibration resulted overall in an underprediction of the 12 streamflow characteristics being evaluated (Table 8). The general underprediction of the flow characteristics is a result similar to that seen in Murphy *et al.* [8] where a TOPMODEL application calibrated on mean annual flow was evaluated in the context of predicting the same streamflow characteristics. The median errors presented here are within plus-or-minus 30 percent of observed values, proposed by Kennard *et al.* [46] as an acceptable band of uncertainty, for 8 to 12 streamflow characteristics (out of 12) depending on the objective function (Figure 7, Table 8). This is in stark contrast to the rainfall runoff model evaluated in Murphy *et al.* [8]) where 13 of 19 streamflow characteristics were outside this band. While similar patterns are seen in overall model results, the calibration approaches evaluated in this paper appear to have provided more accurate estimates across the flow regime as defined by these characteristics. These results can be attributed both to the use of 100 parameter sets, which resulted in more robust flow characteristic estimations, and the use of different objective functions. Parameter uncertainty was substantial for many streamflow characteristics depending on which objective function was used. Despite this, high model efficiencies could still be achieved in many cases when using the median of 100 calibration trials as a more robust prediction for streamflow characteristics.

Table 9. Comparison of selected streamflow characteristics based on simulated and observed streamflow time series for a single model location (site 13 (03488000)) and calibration period 1 (Cal1). (TA1, RA7, and FH6, defined in Table 5; values in parentheses represent the specialized insectivore score using the associated streamflow characteristic value based on linear equations presented in Knight *et al.* [41], Figure 2; hydro, percent error for streamflow characteristic derived from simulated and observed streamflow time series; eco, percent error for specialized insectivore score based on streamflow characteristic derived from simulated and observed streamflow time series).

Objective Function (see Table 3 for Definitions)	RA7		Percent Error	TA1		Percent Error	FH6		Percent Error
	Simulated	Observed	Hydro/Eco	Simulated	Observed	Hydro/Eco	Simulated	Observed	Hydro/Eco
Lindström	0.14 (0.49)		27.3/−19.7	0.4 (0.55)		−16.7/−9.8	13 (0.59)		13.4/−9.2
LogReff	0.1 (0.66)		−9.1/8.2	0.67 (0.75)		39.6/23	10.08 (0.7)		−12/7.7
MARE	0.06 (0.83)		−45.5/36.1	0.73 (0.8)		52.1/31.1	6.62 (0.84)		−42.2/29.2
Reff	0.125 (0.55)	0.11 (0.61)	13.6/−9.8	0.41 (0.56)	0.48 (0.61)	−14.6/−8.2	13.38 (0.58)	11.46 (0.65)	16.8/−10.8
C1	0.12 (0.57)		9.1/−6.6	0.43 (0.57)		−10.4/−6.6	12.92 (0.59)		12.7/−9.2
C2	0.09 (0.7)		−18.2/14.8	0.57 (0.68)		18.8/11.5	12.38 (0.62)		8/−4.6
C3	0.05 (0.87)		−54.5/42.6	0.38 (0.53)		−20.8/−13.1	6.54 (0.84)		−42.9/29.2

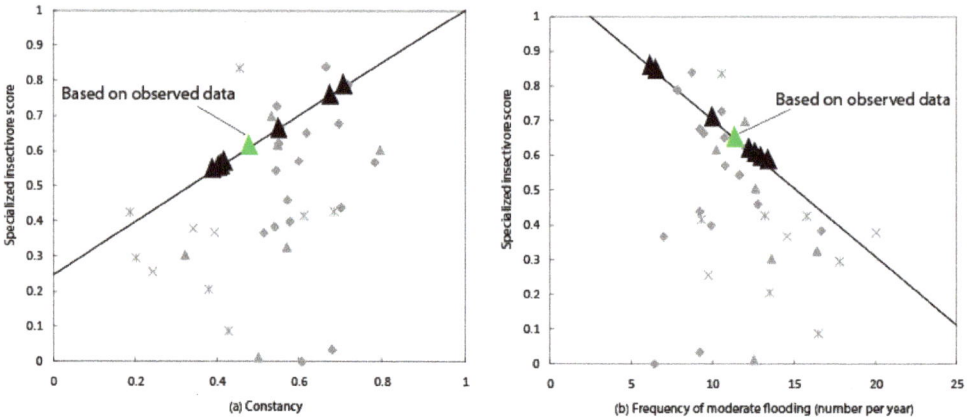

Figure 6. *Cont.*

274

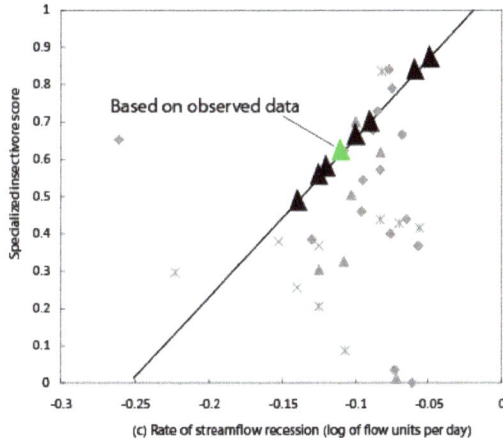

(c) Rate of streamflow recession (log of flow units per day)

Figure 6. Example of an ecological flow application by comparison of estimated values for three streamflow characteristics for site 13 (03488000) (Table 1, Figure 1) and calibration period 1 (Cal1). (**a**) Constancy; (**b**) Frequency of moderate flooding (number per year) and (**c**) Rate of streamflow recession (log of flow units per day). Black triangles represent model estimated values based on the seven objective functions. Green triangle represents streamflow characteristics based on observed data. Values for RA7 (Rate of streamflow recession) were multiplied by negative 1 to convert values to those in the original analysis. Thin black lines represent 80th percentile quantile regression lines based on the 33 data point (grayed) in the background used by Knight *et al.* [41]. (Figure modified from Knight *et al.* [41]).

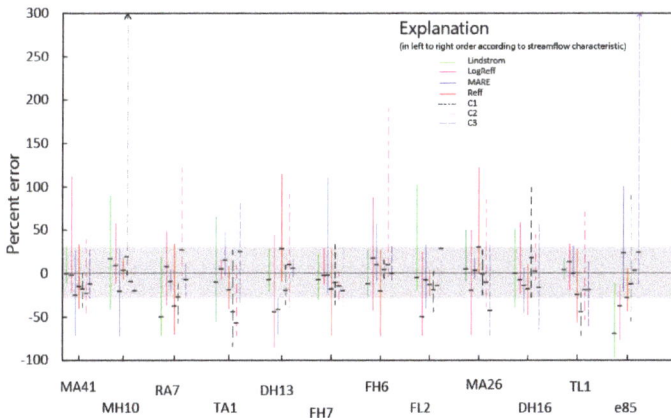

Figure 7. Minimum, maximum, and median percent errors according to objective function and streamflow characteristic for calibration period 1 (Cal1). Each vertical bar is based on the median error for the 27 catchments. The gray band in the center of the figure represents ±30 percent difference [46] Vertical bars with arrows indicate the maximum percent error exceeded the axis scale.

275

While the low average median percentage error would indicate a good performance with regard to the estimated flow characteristics, the scatter plots and computed Nash-Sutcliffe efficiencies and Spearman rank correlations reveal a slightly different picture. Spearman rank correlations were rather high for many of the objective functions and streamflow characteristics. For many of those objective function and flow characteristic combinations, however, Nash-Sutcliffe efficiencies were much lower. This shows that, although a clear bias might be observed in the predicted streamflow characteristic values, the order between the catchments was preserved quite well. In practice it might be more important to determine how well the flow characteristics are reproduced relative to the variation among catchments in the region than to determine the relative error value. When evaluating the scatter plots (Figure 5), low values of the Nash-Sutcliffe efficiencies indicated that the represented variability was relatively low, and the low Spearman rank correlations indicated that some flow characteristics that were not similar on a ranking scale were estimated correctly for the different catchments.

Considering individual streamflow characteristics, a pattern in predictive accuracy is evident. Most notably, streamflow characteristics that reflect average conditions (MA41, MA26, TA1, and TL1) were predicted quite well, with average median percent errors ranging from 2.8 to 4.6 percent absolute (Table 8). However, for some of these characteristics, especially TL1, the relative variation of the simulated values among the catchments were rather poor (Tables 6 and 7). Aspects of the hydrograph representative of high-flow conditions (MH10, FH7, FH6, DH13, DH16, and RA7) were underpredicted consistently (between 12.7 and 41.1 percent), with individual model calibrations underpredicting values up to 70 percent under observed. Low-flow characteristics were overpredicted (FL2 and E85) by 22.8 and 14.9 percent respectively. This appears to indicate that the model, regardless of calibration, may be retaining water during high-flow periods and allowing it to release during low-flow periods. The considerable underprediction of RA7 (rate of streamflow recession) indicates that higher flow events receded at a slower rate, which is suggestive of water stored in groundwater, and subsequently abundant groundwater discharge. The underprediction of RA7 and overprediction of low-flow characteristics are complementary.

MA41 (mean annual runoff) was predicted extremely well, particularly when using those calibrations where the objective function included the volume error as criterion, which is expected as this criterion is equivalent to the mean annual runoff. Predictions of MA41 also performed quite well when calibrated using the Nash-Sutcliffe efficiency. This performance might be attributed to the sensitivity of the Nash-Sutcliffe efficiency for high flows, which could reduce the error in the estimation of mean annual runoff. As noted by Murphy et al. [8], inclusion of ecological flow characteristics as criteria in calibrations may yield better simulations.

5. Conclusions

The accuracy of simulated runoff resulting from seven objective functions was evaluated in this paper by comparing streamflow characteristics based on observed and predicted streamflow time series. While the ultimate goal is to produce the most accurate simulated streamflow time series at ungauged catchments based on the transfer of calibrated parameter sets from gauged to ungauged catchments, the comparison in this study addresses an important part of the total uncertainty, namely the uncertainty related to the prediction accuracy specific streamflow characteristics that were not part of the calibration routine. The primary conclusion is that good model performance in terms of objective functions, such as the frequently used Nash-Sutcliffe model efficiency, does not ensure that all flow characteristics computed from these simulations will correspond to those derived from observed runoff. This is an important consideration that is often overlooked by users of model output who use simulated time series for various analyses, supporting resource allocation decisions, or establishing flow policy. While expecting simulated runoff series to agree with the observed in all possible aspects is unreasonable, this analysis serves as a further reminder of the substantial errors possible, using ecological flow characteristics as the example.

Two novel approaches were used in this study. First, we evaluated the effectiveness of seven objective functions for simulating streamflow time series and subsequent streamflow characteristic calculations. This allowed for critical examination of the importance of the objective function choice, as results differed substantially among objective functions. Results indicate there was no single best calibration strategy, but not surprisingly, different strategies provided better predictions for different streamflow characteristics. However, there was some indication that the combined objective functions, which evaluate the runoff simulations in different aspects, might be generally more suitable across a range of flow characteristics. Second, parameter uncertainty was explicitly considered by using the combination of 100 different equally possible parameter sets for each calibration trial instead of the typical single optimal calibrated parameter set. Our results confirmed the value of this approach by showing that different parameter sets can be similar with respect to the objective function used (similarity between the Nash-Sutcliffe for example) but differ greatly with respect to other characteristics. We demonstrated that using only one parameter set could result in substantial uncertainties, which can be reduced by using the values based on several parameter sets as more robust estimation.

More research is needed to determine which objective functions are most useful to ensure acceptable simulations of ecological flow characteristics, or other regime-defining characteristics. One suitable approach beyond the objective functions used in this paper might be to include streamflow characteristics of particular interest as objective functions in the calibration. This corresponds to the suggestion to include various hydrological signatures as diagnostic tools [47]. The fact that simulation-based flow characteristics varied largely depending upon which objective functions were used indicates that there is a considerable potential to improve model calibrations by considering specific flow characteristics when evaluating model performance during calibration. While it can be expected that performances improve when a certain streamflow characteristic is explicitly included in the objective function, it is less clear which criteria should be included to ensure acceptable simulations for calculation of streamflow characteristics in general. Further research is therefore motivated to explore which criteria to include in the objective function to obtain streamflow simulations that preserve as many streamflow characteristics as possible.

Acknowledgments: This paper is a product of discussions and activities that took place at the U.S. Geological Survey John Wesley Powell Center for Analysis and Synthesis as part of the workgroup focusing on Water Availability for Ungauged Rivers (https://powellcenter.usgs.gov/). Funding for this research was provided by the Tennessee Wildlife Resources Agency, the National Park Service, the U.S. Geological Survey Cooperative Water Program, and the University of Zurich. The use of trade, firm, or product names is for descriptive purposes only and does not imply endorsement by the U.S. Government.

Author Contributions: Rodney Knight and Jan Seibert conceived the initial ideas for this study; Marc Vis performed the simulations; Jan Seibert, Marc Vis, Sandra Pool and Rodney Knight analyzed the results; all authors contributed to writing of the paper.

Conflicts of Interest: The authors declare no conflict of interest.

References

1. Rantz, S.E. *Stream Hydrology Related to the Optimum Discharge for King Salmon Spawning in the Northern California Coast Ranges*; U.S. Geological Survey Water-Supply Paper 1779-AA: Washington, DC, USA, 1964; p. 15.
2. Tennant, D.L. Instream Flow Regimens for Fish, Wildlife, Recreation and Related Environmental Resources. *Fisheries* **1976**, *1*, 6–10.
3. Olden, J.D.; Poff, N.L. Redundancy and the choice of hydrologic indices for characterizing streamflow regimes. *River Res. Appl.* **2003**, *19*, 101–121.
4. Poff, N.L.; Richter, B.D.; Arthington, A.H.; Bunn, S.E.; Naiman, R.J.; Kendy, E.; Acreman, M.; Apse, C.; Bledsoe, B.P.; Freeman, M.C.; *et al.* The ecological limits of hydrologic alteration (ELOHA): A new framework for developing regional environmental flow standards. *Freshw. Biol.* **2010**, *55*, 147–170.
5. Carlisle, D.M.; Wolock, D.M.; Meador, M.R. Alteration of streamflow magnitudes and potential ecological consequences: A multiregional assessment. *Front. Ecol. Environ.* **2011**, *9*, 264–270.
6. Knight, R.R.; Murphy, J.C.; Wolfe, W.J.; Saylor, C.F.; Wales, A.K. Ecological limit functions relating fish community response to hydrologic departures of the ecological flow regime in the Tennessee River basin, United States. *Ecohydrology* **2014**, *7*, 1262–1280.

7. Sanborn, S.C.; Bledsoe, B.P. Predicting streamflow regime metrics for ungauged streamsin Colorado, Washington, and Oregon. *J. Hydrol.* **2006**, *325*, 241–261.

8. Murphy, J.C.; Knight, R.R.; Wolfe, W.J.; Gain, W.S. Predicting Ecological Flow Regime at Ungaged Sites: A Comparison of Methods. *River Res. Appl.* **2013**, *29*, 660–669.

9. Shrestha, R.R.; Peters, D.L.; Schnorbus, M.A. Evaluating the ability of a hydrologic model to replicate hydro-ecologically relevant indicators. *Hydrol. Process.* **2014**, *28*, 4294–4310.

10. Hrachowitz, M.; Savenije, H.H.G.; Blöschl, G.; McDonnell, J.J.; Sivapalan, M.; Pomeroy, J.W.; Arheimer, B.; Blume, T.; Clark, M.P.; Ehret, U.; *et al.* A decade of Predictions in Ungauged Basins (PUB)—A review. *Hydrol. Sci. J.* **2013**, *58*, 1198–1255.

11. Clausen, B.; Biggs, B. Relationships between benthic biota and hydrological indices in New Zealand streams. *Freshw. Biol.* **1997**, *38*, 327–342.

12. Clausen, B.; Biggs, B.J. Flow variables for ecological studies in temperate streams: Groupings based on covariance. *J. Hydrol.* **2000**, *237*, 184–197.

13. Poff, N.L.; Ward, J.V. Implications of Streamflow Variability and Predictability for Lotic Community Structure: A Regional Analysis of Streamflow Patterns. *Can. J. Fish. Aquat. Sci.* **1989**, *46*, 1805–1818.

14. Puckridge, J.T.; Walker, K.F.; Costelloe, J.F. Hydrological persistence and the ecology of dryland rivers. *Regul. Rivers Res. Manag.* **2000**, *16*, 385–402.

15. Bergström, S. *Development and Application of a Conceptual Runoff Model for Scandinavian Catchments*; SMHI: Norrköping, Sweden, 1976; No. RHO 7; p. 134.

16. Bergström, S. *The HBV Model: Its Structure and Applications*; SMHI Hydrology: Norrköping, Sweden, 1992; p. 35.

17. Lindström, G.; Johansson, B.; Persson, M.; Gardelin, M.; Bergström, S. Development and test of the distributed HBV-96 hydrological model. *J. Hydrol.* **1997**, *201*, 272–288.

18. Seibert, J.; Vis, M.J.P. Teaching hydrological modeling with a user-friendly catchment-runoff-model software package. *Hydrol. Earth Syst. Sci.* **2012**, *16*, 3315–3325.

19. Singh, V.P. *Computer Models of Watershed Hydrology*; Water Resources Publications: Highlands Ranch, CO, USA, 1995.

20. Omernik, J.M. Ecoregions of the Conterminous United States. *Ann. Assoc. Am. Geogr.* **1987**, *77*, 118–125.

21. Wolfe, W.; Haugh, C.; Webbers, A.; Diehl, T. Preliminary Conceptual Models of the Occurrence, Fate, and Transport of Chlorinated Solvents in Karst Regions of Tennessee. Department of Interior, US Geological Survey, Water Resources Investigations Report 97-4097. Available online: http://pubs.usgs.gov/wri/wri974097/new4097.pdf (accessed on 3 April 2015).

22. Hoos, A.B. Recharge Rates and Aquifer Hydraulic Characteristics for Selected Drainage Basins in Middle and East Tennessee. Department of the Interior, US Geological Survey, Water Resources Investigations Report 90-4015, 34. Available online: http://pubs.water. usgs.gov/wri904015/ (accessed on 24 June 2010).

23. Knight, R.R.; Gain, W.S.; Wolfe, W.J. Modelling ecological flow regime: An example from the Tennessee and Cumberland River basins. *Ecohydrology* **2012**, *5*, 613–627.

24. Law, G.S.; Tasker, G.D.; Ladd, D.E. Streamflow-Characteristic Estimation Methods for Unregulated Streams of Tennessee. *US Geological Survey, Scientific Investigations Report 2009–5159, 212. p, 1 Plate*; Available online: http://pubs.usgs.gov/sir/2009/5159/ (accessed on 16 June 2010).

25. U.S. Department of Commerce Climatography of the United States No. 85 Divisional Normals and Standard Deviations of Temperature, Precipitation, and Heating and Cooling Degree Days 1971–2000 (And Previous Normals Periods) Section 1: Temperature. United States Department of Commerce: Washington, DC, USA, 1971.

26. U.S. Department of Commerce Climatography of the United States No. 85 Divisional Normals and Standard Deviations of Temperature, Precipitation, and Heating and Cooling Degree Days 1971–2000 (And Previous Normals Periods) Section 2: Precipitation. United States Department of Commerce: Washington, DC, USA, 1971.

27. Moody, D.W.; Chase, E.B.; Aronson, D.A. *National Water Summary 1985—Hydrologic Events and Surface-Water Resources*; United States Geological Survey Water-Supply Paper 2300: Chapter on Tennessee Surface-Water Resources; United States Geological Survey: Reston, VA, USA, 1986; pp. 425–429.

28. Abell, R.A.; Olson, D.M.; Dinerstein, E.; Hurley, P.T.; Diggs, J.T.; Eichbaum, W.; Walters, S.; Wettengel, W.; Allnutt, T.; Loucks, C.J.; *et al. Freshwater Ecoregions of North America: A Conservation Assessment*; Island Press: Washington, DC, USA, 2000.

29. Olson, D.M.; Dinerstein, E. The Global 200: A Representation Approach to Conserving the Earth's Most Biologically Valuable Ecoregions. *Conserv. Biol.* **1998**, *12*, 502–515.

30. Etnier, D.A.; Starnes, W.C. *The fishes of Tennessee*; The University of Tennessee Press: Knoxville, TN, USA, 1993.

31. Master, L.L.; Flack, S.R.; Stein, B.A.; Conservancy, N. *Rivers of Life: Critical Watersheds for Protecting Freshwater Biodiversity*; Nature Conservancy: Arlington, VA, USA, 1998.

32. Thornton, P.E.; Thornton, M.M.; Mayer, B.W.; Wilhelmi, N.; Wei, Y.; Devarakonda, R.; Cook, R.B. *Daymet: Daily Surface Weather Data on a 1-km Grid for North America, Version 2*; Oak Ridge National Laboratory: Oak Ridge, TN, USA, 2014.

33. Monteith, J.L. Evaporation and Environment. In *The State and Movement of Water in Living Organism 19th Symposia of the Society Experimental Biology*; University Press: Cambridge, UK, 1965; pp. 205–234.

34. Walter, I.; Allen, R.; Elliott, R.; Itenfisu, D.; Brown, P.; Jensen, M.; Mecham, B.; Howell, T.; Snyder, R.; Eching, S.; *et al.* The ASCE Standardized Reference Evapotranspiration Equation. PREPARED BY Task Committee on Standardization of Reference Evapotranspiration of the Environmental and Water Resources Institute. Available online: http://kimberly.uidaho.edu/water/asceewri/ascestzdetmain2005.pdf (accessed on 3 April 2015).

35. Rotstayn, L.D.; Roderick, M.L.; Farquhar, G.D. A simple pan-evaporation model for analysis of climate simulations: Evaluation over Australia. *Geophys. Res. Lett.* **2006**, *33*, L7715.

36. Hobbins, M.; Wood, A.; Streubel, D.; Werner, K. What Drives the Variability of Evaporative Demand across the Conterminous United States? *J. Hydrometeorol.* **2012**, *13*, 1195–1214.

37. Rango, A.; Martinec, J. Revisiting the degree-day method for snowmelt computations. *J. Am. Water Resour. Assoc.* **1995**, *31*, 657–669.

38. Beven, K.; Freer, J. Equifinality, data assimilation, and uncertainty estimation in mechanistic modelling of complex environmental systems using the GLUE methodology. *J. Hydrol.* **2001**, *249*, 11–29.

39. Seibert, J. Multi-Criteria calibration of a conceptual runoff model using a genetic algorithm. *Hydrol. Earth Syst. Sci.* **2000**, *4*, 215–224.

40. Seibert, J. Regionalisation of parameters for a conceptual rainfall-runoff model. *Agric. For. Meteorol.* **1999**, *98–99*, 279–293.

41. Knight, R.R.; Brian Gregory, M.; Wales, A.K. Relating streamflow characteristics to specialized insectivores in the Tennessee River Valley: A regional approach. *Ecohydrology* **2008**, *1*, 394–407.

42. Thompson, J.; Archfield, S. *The EflowStats R package Introduction to EflowStats*; United States Geological Survey: Reston, VA, USA, 2014.

43. Castellarin, A.; Camorani, G.; Brath, A. Predicting annual and long-term flow-duration curves in ungauged basins. *Adv. Water Resour.* **2007**, *30*, 937–953.

44. McManamay, R.A. Quantifying and generalizing hydrologic responses to dam regulation using a statistical modeling approach. *J. Hydrol.* **2014**, *519*, 1278–1296.

45. Zhu, Y.; Day, R.L. Regression modeling of streamflow, baseflow, and runoff using geographic information systems. *J. Environ. Manag.* **2009**, *90*, 946–953.

46. Kennard, M.J.; Mackay, S.J.; Pusey, B.J.; Olden, J.D.; Marsh, N. Quantifying uncertainty in estimation of hydrologic metrics for ecohydrological studies. *River Res. Appl.* **2010**, *26*, 137–156.

47. Gupta, H.V.; Wagener, T.; Liu, Y. Reconciling theory with observations: Elements of a diagnostic approach to model evaluation. *Hydrol. Process.* **2008**, *22*, 3802–3813.

Modeling of Soil Water and Salt Dynamics and Its Effects on Root Water Uptake in Heihe Arid Wetland, Gansu, China

Huijie Li, Jun Yi, Jianguo Zhang, Ying Zhao, Bingcheng Si, Robert Lee Hill, Lele Cui and Xiaoyu Liu

Abstract: In the Heihe River basin, China, increased salinity and water shortages present serious threats to the sustainability of arid wetlands. It is critical to understand the interactions between soil water and salts (from saline shallow groundwater and the river) and their effects on plant growth under the influence of shallow groundwater and irrigation. In this study, the Hydrus-1D model was used in an arid wetland of the Middle Heihe River to investigate the effects of the dynamics of soil water, soil salinization, and depth to water table (DWT) as well as groundwater salinity on *Chinese tamarisk* root water uptake. The modeled soil water and electrical conductivity of soil solution (EC_{sw}) are in good agreement with the observations, as indicated by *RMSE* values (0.031 and 0.046 $cm^3 \cdot cm^{-3}$ for soil water content, 0.037 and 0.035 $dS \cdot m^{-1}$ for EC_{sw}, during the model calibration and validation periods, respectively). The calibrated model was used in scenario analyses considering different DWTs, salinity levels and the introduction of preseason irrigation. The results showed that (I) *Chinese tamarisk* root distribution was greatly affected by soil water and salt distribution in the soil profile, with about 73.8% of the roots being distributed in the 20–60 cm layer; (II) root water uptake accounted for 91.0% of the potential maximal value when water stress was considered, and for 41.6% when both water and salt stress were considered; (III) root water uptake was very sensitive to fluctuations of the water table, and was greatly reduced when the DWT was either dropped or raised 60% of the 2012 reference depth; (IV) arid wetland vegetation exhibited a high level of groundwater dependence even though shallow groundwater resulted in increased soil salinization and (V) preseason irrigation could effectively increase root water uptake by leaching salts from the root zone. We concluded that a suitable water table and groundwater salinity coupled with proper irrigation are key factors to sustainable development of arid wetlands.

Reprinted from *Water*. Cite as: Li, H.; Yi, J.; Zhang, J.; Zhao, Y.; Si, B.; Hill, R.L.; Cui, L.; Liu, X. Modeling of Soil Water and Salt Dynamics and Its Effects on Root Water Uptake in Heihe Arid Wetland, Gansu, China. *Water* **2015**, *7*, 2382–2401.

1. Introduction

In arid and semi-arid wetlands, salinity and water scarcity are two serious and chronic environmental problems threatening the ecosystem [1]. In the Heihe River basin, China, wetlands are now often experiencing extended periods of high soil salinization levels and associated water availability problems due to the impacts of high evaporative conditions, poor surface drainage, human population pressures, and the associated changes in land use [2,3]. Soil salinity limits plant growth [4] and negatively influences soil quality [5,6], resulting in the change of community structure, density, and growth status. Wetland areas decreased by 42% from 1975 to 2010 within the middle Heihe River basin [7]. Reed marsh areas decreased from 597.8 ha in the late 1990s to the current 492 ha, and that the reed plant height and reed stem were reduced by 25% and 4%, respectively [8]. Owing to the changes in water availability, land desiccation, and salinization, the vegetation has shifted from hydrophytes towards halophytes, psammophytes, xerophytes and super-xerophytes [9]. Meanwhile, soil salinization has caused clogging of soil pores and channels for water flow that has resulted in a considerable reduction in soil permeability, soil porosity, and soil hydraulic conductivity [10]. Consequently, before developing reliable countermeasures, it is important to evaluate the interactions between soil water and salts and their impacts on plant water use, based on factors such as the groundwater quality, the water table, and the plant tolerance to salinity.

In an arid climate where rainfall is very limited, shallow groundwater plays a key role in ecosystem functions. Therefore, it is particularly important to understand the effects of the depth to the water table (DWT) and groundwater quality on root zone water contents, salinity, and plant water use. Jolly *et al.* [1] concluded that in arid/semiarid environments, where the surface water regime was vulnerable to rainfall variability, the persistence of wetlands can be dependent, either completely or partially, on the contributions from groundwater. Ayars *et al.* [11] reported that the potential for meeting crop water needs from shallow groundwater ranged up to 50% of the total irrigation. Crops like alfalfa and forage grasses exhibit more continuous water uptake patterns for their long growing seasons and robust root systems [12]. However, most of these previous studies have focused on farmland with few of them giving consideration to wetlands, especially in arid environments [1].

The interactions between the soil water, salt, shallow groundwater, and root water uptake are complex and influenced by numerous factors. Evaluating the interactions of these factors through field research is difficult and time-consuming. In addition, salt variation is often small and is not easily detectable in a single growth season. Simulation models that integrate the soil water movement, solute transport, and plant water uptake provide information that otherwise cannot be obtained from field experiment [13]. Since the 1970s, many numerical solutions were developed to describe water and solute transport [14–16]. Most of these models are based on the solutions to the Richards equation for water flow and the convection-dispersion equation (CDE) for solute transport [17]. However, accurate predictions of these models rely on precise measurement of hydraulic characteristics [18]. For some parameters (e.g., α, n and l in the van Genuchten-Mualem function [19,20]), however, it is difficult to measure hydraulic characteristics at the plot scale. Numerous studies have indicated that the laboratory measured hydraulic parameters and/or parameters derived from pedotransfer functions, in combination with inverse optimization algorithms are an effective approach for improving the description of unsaturated hydraulic functions [4,17,21].

In this study, a widely used model Hydrus-1D [22], which simulates one-dimensional transport of water, heat, and multiple solutes in variably saturated media, was adopted to simulate the soil water flow, solute transport and root water uptake in an arid wetland with shallow saline groundwater. Our objectives were: (I) to test the feasibility of the Hydrus-1D model approach in simulating water flow and solute transport with observed data; (II) to characterize the interactions between soil water, salt, and groundwater and their effects on *Chinese tamarisk* root water uptake and root distribution, and (III) to conduct scenario analyses for the soil water, salt dynamics and root water uptake under different conditions. In addition, the long term salinity trends under different DWTs and groundwater salinities (EC_{gw}) were investigated.

2. Materials and Methods

2.1. Study Area

The study area is located at Pingchuan town, in the Middle Heihe River basin, Gansu, China (39°20′ N, 100°06′ E). The landform is representative of a riparian wetland covered with *Chinese tamarisk*, which is the dominant plant community in this area and in the study area served as a shelter forest. Soil electrical conductivities (soil/water ratio of 1/5, $EC_{1:5}$) ranged between 1.51 and 26.72 dS· m^{-1}. The DWT ranged from 0.47 m in the rainy fall to 1.44 m in the dry winter. The salinity of the shallow groundwater varied between 2.0 and 4.0 dS· m^{-1}. The climate is a continental arid temperate zone with annual average precipitation of 116.7 mm from 1965 to 2010 with about 60% of the precipitation received during July–September. The annual average potential evaporation is 2366 mm· year^{-1}. The annual average temperature is 7.6 °C, and the lowest and highest temperatures are about -27 °C in January and 39.1 °C in July, respectively [23]. Soil was formed from alluvial deposits with a silty loam texture (Table 1).

Table 1. Soil physical properties and calibrated parameters at the study area.

Soil Layer (cm)	0–15	15–25	25–55	55–65	65–100	100–150
Texture	Silt Loam	Silt Loam	Silt Loam	Silt Loam	Silt Loam	Coarse Sand
Clay (%)	11	10	11	13	16	-
Silt (%)	54	65	66	64	56	-
Sand (%)	35	25	23	23	28	100
Bulk density(g·cm^{-3})	1.2	1.35	1.42	1.44	1.44	1.42
θ_r (cm^3·cm^{-3})	0.093	0.08	0.075	0.071	0.079	0.051
θ_s(cm^3·cm^{-3})	0.543	0.493	0.502	0.462	0.534	0.376
α (cm^{-1})	0.019	0.028	0.032	0.033	0.035	0.034
n	2.06	1.70	1.50	1.36	1.71	4.43
l	0.5	0.5	0.5	0.5	0.5	0.5
K_s (cm·d^{-1})	2.5	10.6	24.7	6.7	34.0	1428.0
λ (cm)	16.8	19.5	48.2	29.1	10.8	124.0

Note: The particle size limits were 0.05 to 2 mm for sand, 0.05–0.002 mm for silt and <0.002 mm for clay. θ_r, residual water content; θ_s, saturated water content; α, reciprocal value of air entry pressure; n, the smoothness of pore size distribution; l, pore connectivity parameter; K_s, saturated hydraulic conductivity; and λ, dispersivity.

2.2. Measurements

Field data were collected during two growing seasons of *Chinese tamarisk* (2 May to 27 October 2012 and 2013). Soil water content was measured using a Neutron Moisture Meter (NMM, L520-D, Nanjing, China) every 5 days with a depth interval of 10 cm down to 100 cm. Soil salinity based on the soil diluted extract method (soil/water ratio of 1/5, $EC_{1:5}$) was measured every 15 days with a measurement interval of 10 cm down to 90 cm. A shallow monitoring well was installed in the vicinity of the neutron probe to measure DWT every 5 days. The groundwater electrical conductivity (EC_{gw}) was measured every 15 days. At the beginning of the experiment, undisturbed soil samples (diameter, 5 cm, height 5 cm) from five representative layers were collected for the laboratory measurement of soil bulk density (BD), texture, saturated hydraulic conductivity (K_s) and water content (θ_s). The BD was calculated from the volume-mass relationship for each core sample. Soil texture was determined using the pipette sampling method [24]. K_s values of the undisturbed soil cores were determined using a falling head method [25]. The soil cores were first saturated from the bottom and then submerged in water for 24 h. After weighing, the saturated soil samples were dried at 105 °C to constant mass, and their mass-based saturated soil water content was determined. θ_s values were determined by multiplying saturated mass-based soil water content with BD. In addition, root distribution was measured using a root auger (Eijkelkamp, The Netherlands), and soil cores were sampled in 10 cm depth increments. Root biomass was obtained by washing away soil particles, oven-dried and weighed (Figure 1).

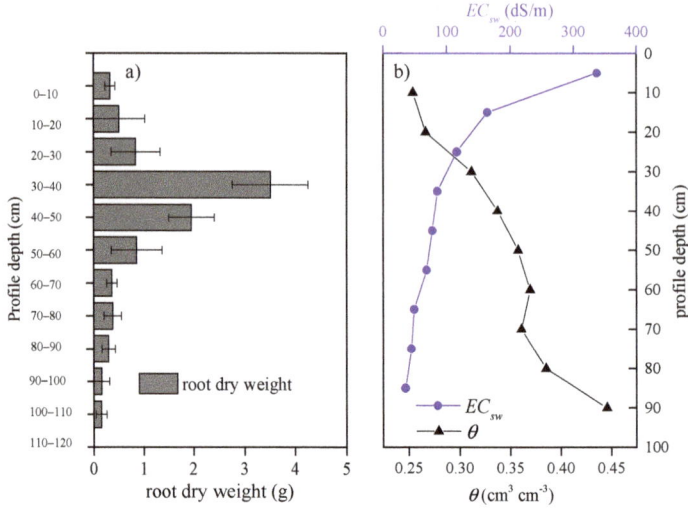

Figure 1. Graphical representation of (**a**) the root distribution of *Chinese tamarisk* measured at the start of the 2012 growing season and (**b**) the average value of soil water (θ) and EC_{sw} during the 2012 growing season.

2.3. Model Simulation

2.3.1. Soil Water Flow and Solute Transport

Simulations of soil water flow and solute transport were performed with Hydrus-1D [22]. This program numerically solves the Richards equation for water flow and uses advection-dispersion equations (CDE) for heat and solute transport in variably saturated porous media. Variably-saturated water flow is described using the Richards' equation:

$$\frac{\partial \theta}{\partial t} = \frac{\partial}{\partial z}\left[K(h)\left(\frac{\partial h}{\partial z}+1\right)\right] - S(z,t) \tag{1}$$

where θ is soil water content ($L^3 \cdot L^{-3}$), t is time (T), z is the vertical space coordinate (L), K is the hydraulic conductivity ($L \cdot T^{-1}$), h is the pressure head (L), S is the sink term accounting for root water uptake ($L^3 \cdot L^{-3} \cdot T^{-1}$). The unsaturated soil hydraulic properties were described using the van Genuchten-Mualem functional relationships [19,20]:

$$\theta(h) = \begin{cases} \theta_r + \dfrac{\theta_s - \theta_r}{\left(1+|\alpha h|^n\right)^m} & h < 0 \\ \theta_s & h \geqslant 0 \end{cases} \tag{2}$$

$$K(h) = K_s S_e^l\left[1-\left(1-S_e^{1/m}\right)^m\right]^2 \tag{3}$$

286

$$S_e = \frac{\theta - \theta_r}{\theta_s - \theta_r} \tag{4}$$

where θ_r and θ_s are the residual and saturated water contents ($L^3 \cdot L^{-3}$), respectively. K_s ($L \cdot T^{-1}$) is the saturated hydraulic conductivity, α, (L^{-1}) and n represent the empirical shape parameters, $m = 1 - 1/n$; l is the pore connectivity parameter, which is taken as 0.5 [20]. S_e is the effective saturation.

2.3.2. Solute Transport

The partial differential equations governing equilibrium one-dimensional solute transport under transient flow in variably-saturated medium is defined in Hydrus-1D as:

$$\frac{\partial \theta C}{\partial t} = \frac{\partial}{\partial z}\left(\theta D \frac{\partial C}{\partial z}\right) - \frac{\partial v \theta C}{\partial Z} \tag{5}$$

where C is the solute concentration of the liquid phase ($M \cdot L^{-3}$). D is the dispersion coefficient ($L^2 \cdot T^{-1}$), and v is the average pore water velocity ($L \cdot T^{-1}$). The dispersion coefficient is defined as (ignoring molecular diffusion):

$$D = \lambda v \tag{6}$$

where λ is dispersivity (L). The dispersivity is viewed as a material constant independent of the flow rate. Since v is obtained from the numerical solution of the water flow model (the water flux q divided by θ), dispersivity is the only solute transport parameter needed for solving the CDE equation.

2.3.3. Root Water Uptake

The potential transpiration rate, T_p ($L \cdot T^{-1}$), is spread in the root zone according to the normalized root density distribution function, β (z, t) (L^{-1}). The actual root water uptake, S, is obtained from the potential root water uptake (i.e., potential transpiration) S_p, through multiplication with a stress response function $\alpha(h, h_\varphi, z, t)$ accounting for water and osmotic stresses [26,27] as follows:

$$S(h, h_\varphi, z, t) = \alpha(h, h_\varphi, z, t)\, S_p(z, t) = \alpha(h, h_\varphi, z, t)\, \beta(z, t)\, T_p(t) \tag{7}$$

where stress response function $\alpha(h, h_\varphi, z, t)$ is a dimensionless function of the soil water (h) and osmotic (h_φ) pressure heads ($0 \leqslant \alpha \leqslant 1$). $S_p(z, t)$ and $S(h, h_\varphi, z, t)$ are the potential and actual volumes of water removed from a unit volume of soil per unit of time ($L^3 \cdot L^{-3} \cdot T^{-1}$), respectively. The actual transpiration rate, T_a ($L \cdot T^{-1}$), is then obtained by integrating Equation (7) over the root domain L_R:

$$T_a = \int_{L_R} S\left(h, h_\varphi, z, t\right) dz = T_p \int_{L_R} \alpha\left(h, h_\varphi, z, t\right) \beta\left(z, t\right) dz \tag{8}$$

We further assumed that the effects of water and salinity were multiplicative [28]: $\alpha(h, h_\varphi) = \alpha(h)\,\alpha(h_\varphi)$, so that different stress response functions could be used. Root water uptake due to water stress was described using the model introduced by Feddes *et al.* [26]:

$$\alpha_h = \begin{cases} 0, & h > h_1,\ h \leqslant h_4 \\ \frac{h-h_1}{h_2-h_1}, & h_2 < h \leqslant h_4 \\ 1, & h_3 < h \leqslant h_2 \\ \frac{h-h_4}{h_3-h_4}, & h_4 < h \leqslant h_3 \end{cases} \tag{9}$$

where root water uptake is assumed to be zero close to saturation (*i.e.*, wetter than the "anaerobiosis point", h_1). For $h < h_4$ (the wilting point pressure head), root water uptake is also completely stressed. Root water uptake is considered to be at the potential rate when pressure heads range from h_2 to h_3, and when pressure head values are between h_1 and h_2 (or h_3 and h_4), root water uptake increases (or decreases) linearly with h.

Hydrus-1D assumes h_3 is a function of T_p and allows users to specify two different T_p (T_{p1} and T_{p2}) and h_3 (h_{3-1} and h_{3-2}), respectively. The calculation equations are:

$$h_3 = \begin{cases} h_{3-1} + \frac{(h_{3-2}-h_{3-1})}{(T_{p1}-T_{p2})(T_{p1}-T_p)} & T_{p2} < T_p < T_{p1} \\ h_{3-2} & T_p < T_{p2} \\ h_{3-1} & T_p > T_{p1} \end{cases} \tag{10}$$

Root water uptake due to osmotic stress was described with an S-shaped function developed by van Genuchten, 1987 [28]:

$$\alpha_\varphi = \frac{1}{1 + \left(\frac{h_\varphi}{h_{\varphi 50}}\right)^p} \tag{11}$$

where p represents experimental constants. The exponent p was found to be approximately 3 when only salinity stress data was applied [28]. The parameter $h_{\varphi 50}$ represents the pressure head at which the water extraction rate is reduced by 50% during conditions of negligible water stress.

288

In Hydrus-1D, EC is expressed as electrical conductivity of soil solution (EC_{sw}). Alternatively, we measured $EC_{1:5}$ and converted it into the saturated paste extracts (EC_e) using the following relationship where $EC_e = (2.46 + 3.03\,\theta_{sp}^{-1})\,EC_{1:5}$, where θ_{sp} is the water content of the saturated paste (θ_{sp} Kg·Kg^{-1}) [29]. Then EC_e values are converted into EC_{sw} by the following equation [30]:

$$EC_{sw} = EC_e \frac{BD \cdot SP}{100 \cdot \theta} = EC_e \frac{\theta_s}{\theta} \tag{12}$$

where SP is the saturation percentage (the water content of the saturated soil-paste, expressed on a dry-weight basis), BD is the bulk density (g·cm^{-3}) and θ_s is the saturated soil water content (L^3·L^{-3}).

Furthermore, Hydrus-1D uses the following relationship to convert EC_{sw} to osmotic pressure head (cm):

$$h_\varphi = -3.8106 EC_{sw} + 0.5072 \tag{13}$$

The equation is very similar to the relationship reported by the US Salinity Laboratory Staff (1954) [31] for estimating the osmotic pressure of soil solutions from EC measurements ($h_\varphi = -3.7188\ EC_{sw}$).

2.4. Initial and Boundary Condition

Initial conditions were set in the model with measured soil water contents and electrical conductivities on 2 May 2012. At the soil surface, an atmospheric boundary condition was specified using the daily data of precipitation and reference crop evapotranspiration (ET_0) obtained from the Linze Station (2 km away from the site). Daily values of the ET_p were calculated using the reference evapotranspiration (ET_0) via Penman-Monteith method [32] multiplying by a crop coefficient of the investigated *Chinese tamarisk*. The crop coefficient was estimated from fraction of ground cover and plant height [33], that is, crop coefficient for the middle season is 1.05, increased from 0.55 to 1.05 linearly in the first 10 days of plant development and decreased from 1.05 to 0.55 linearly in the last seven days of plant defoliation. Then, ET_p was divided into potential evaporation (E_p) and potential transpiration (T_p) according to Beer's Law:

$$E_p\,(t) = ET_p\,(t) \cdot exp^{-k \cdot LAI(t)} \tag{14}$$

$$T_p\,(t) = ET_p\,(t) - E_p\,(t) \tag{15}$$

where k is an extinction coefficient set to be 0.463 and LAI is the leaf area index. The model was used to directly calculate actual E and T given the soil moisture conditions and the root water uptake functions. LAI values were measured during different stages of the growing season using a LI-COR area meter (Model LI-3100C, LI-COR Environmental and Biotechnology Research Systems, Lincoln, Nebraska), and were linearly interpolated between the measurement dates. At the bottom, variable pressure head and concentration boundary conditions were specified for water flow and solute transport using the measured water table depths and groundwater EC, respectively. For solute transport, we assumed that the rain water was free of solutes and implemented a no flux boundary condition at the soil surface. The boundary conditions used in the calibration and validation processes are shown in Figure 2.

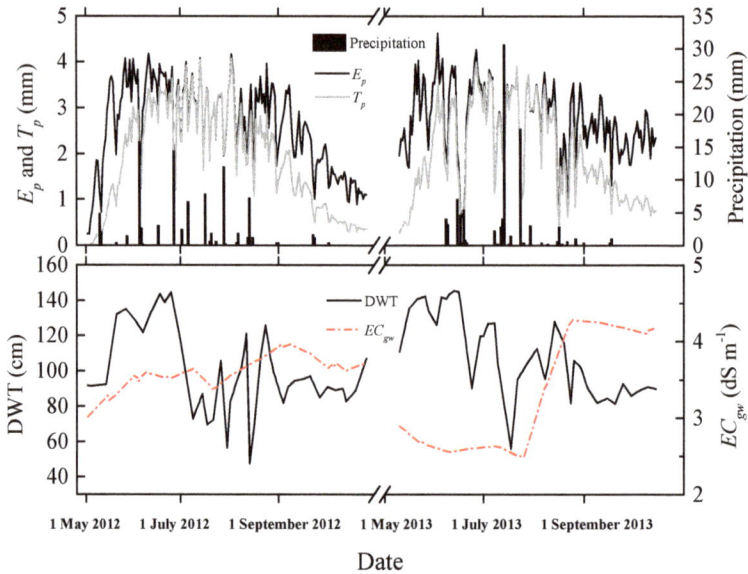

Figure 2. Dynamics of potential evapotranspiration, precipitation, depth to water table (DWT) and groundwater electric conductivity (EC_{gw}).

2.5. Model Calibration and Validation

In this study, the Hydrus-1D model was calibrated using site-specific boundary conditions and measured water contents, θ, and electrical conductivities, EC_{sw} values during 2012. Saturated water content (θ_s) and hydraulic conductivity (K_s) were determined from the soil cores taken as stated above. The other van Genuchten-Mualem parameters were estimated via Rosetta pedotransfer functions [34] using the particle size distribution and bulk density dataset. For initial values of solute transport parameters in the root zone (0–100 cm), the dispersivity (λ) was set to an average value (8.9) based on 67 soils with silt loam textures according to Vanderborght and Vereecken [35]. Based on aquifer materials, thickness, and

290

hydraulic conductivity, the dispersivity of the aquifer was fixed as 124 cm according to Gelhar *et al.* [36].

The parameters of the Feddes model were synthesized based on Moayyad [37] with Grinevskii [38]: $h_1 = -0.1$ cm, $h_2 = -2$ cm, $h_{3-1} = -80$ cm, $h_{3-2} = -250$ cm, $h_4 = -15,000$ cm. Without consideration of the water stress (*i.e.*, reset the Feddes model parameters to make water stress vanish), parameters of the S-shaped function developed by van Genuchten were fitted as: $h_{\varphi 50} = 326.4$ cm, $p = 3$. Root distribution was specified according to measured root dry weight distribution along the soil profile (Figure 1).

In Hydrus, inverse parameter estimation employed a relatively simple, gradient-based, local optimization approach based on the Marquardt-Levenberg method [39]. In this case, inverse solutions were used to optimize soil hydraulic and solute transport parameters simultaneously using the observed data, initial conditions, initial estimates, and boundary conditions. That is, α, n and λ in the five upper soil layers were fitted first since Hydrus could optimize 15 parameters at a time. θ_r was the last parameter estimated. Then, the model was validated with the observed data of the 2013 growing season without changing the calibrated parameters.

The agreement between the predicted and observed data was evaluated by root mean square error (*RMSE*) and coefficient of determination (R^2):

$$RMSE = \sqrt{\frac{\sum_{i=1}^{N} (O_i - P_i)^2}{N - 1}} \tag{16}$$

$$R^2 = 1 - \frac{\sum_{i=1}^{N} (P_i - O_i)^2}{\sum_{i-1}^{N} (O_i - \overline{O})^2} \tag{17}$$

where O_i and P_i are the *i*th values of observed and predicted values, respectively, and $\overline{O}$ is the average of observed values. N is the number of observations.

2.6. Simulation Scenarios

In order to understand the impacts of groundwater change on plant water use, we simulated root water uptake under different DWTs and EC_{gw} conditions. Taking the data of 2012 as the reference, eight DWT (*i.e.*, DWT would either raise or drop 15%, 30%, 45% and 60% based on the 2012 reference depth, respectively) and eight EC_{gw} (*i.e.*, EC_{gw} would either increase or decrease 15%, 30%, 45% and 60% based on the 2012 reference value, respectively) were assumed in this process (Table 2).

In addition, to evaluate long-term salinity trends, a long-term time series analysis was conducted considering the fluctuations of DWT and EC_{gw} in relation to 2012 base values. Firstly, using the LARS-WG weather generator [40] and historical meteorological data from 1954 to 2012, we generated 30 years of weather data which had the same statistical characteristics as the historical data. Then, soil salinization risks were assessed using the generated long-term time series data for three water levels (DWT is 60, 100 (average water level of 2012) and 140 cm) and three groundwater electrical conductivities (EC_{gw} is 1.75, 3.75 (average EC_{gw} of 2012) and 5.75 dS·m^{-1}). The initial value of EC_{sw} was taken as 3.75 dS·m^{-1} for different water tables and 107.25 dS·m^{-1} (average root zone EC_{sw} of 2012) for different EC_{gw}, respectively.

Salt normally accumulated before the plant germination in our studied area (*i.e.*, the dry and cold winter time). In order to elucidate the impacts of artificial watering on soil water, salt dynamics, and to determine how to alleviate salt stress on the arid wetlands, we evaluated the influence of a preseason irrigation event applied at the initial stages of plant growth starting on 2 May 2012. To account for irrigation amounts ranging from 1 to 80 cm, eighty Hydrus simulation scenarios were run. All simulation scenarios are listed in Table 2.

Table 2. Simulation scenarios performed in this study.

Main Scenarios	Scenarios in Detail
Root water uptake predictions	Depth to water table raise 15% of the 2012 reference depth, DWT+15%
	Depth to water table raise 30% of the 2012 reference depth, DWT+30%
	Depth to water table raise 45% of the 2012 reference depth, DWT+45%
	Depth to water table raise 60% of the 2012 reference depth, DWT+60%
	Depth to water table drop 15% of the 2012 reference depth, DWT-15%
	Depth to water table drop 30% of the 2012 reference depth, DWT-30%
	Depth to water table drop 45% of the 2012 reference depth, DWT-45%
	Depth to water table drop 60% of the 2012 reference depth, DWT-60%
	Groundwater electrical conductivity increase 15% of the 2012 reference value, EC_{gw} + 15%
	Groundwater electrical conductivity increase 30% of the 2012 reference value, EC_{gw} + 30%
	Groundwater electrical conductivity increase 45% of the 2012 reference value, EC_{gw} + 45%
	Groundwater electrical conductivity increase 60% of the 2012 reference value, EC_{gw} + 60%
	Groundwater electrical conductivity decrease 15% of the 2012 reference value, EC_{gw} − 15%
	Groundwater electrical conductivity decrease 30% of the 2012 reference value, EC_{gw} − 30%
	Groundwater electrical conductivity decrease 45% of the 2012 reference value, EC_{gw} − 45%
	Groundwater electrical conductivity decrease 60% of the 2012 reference value, EC_{gw} − 60%
Long term (30 years) salinity trends	Depth to water table is 60 cm, DWT = 60 cm
	Depth to water table is 100 cm, DWT = 100 cm
	Depth to water table is 140 cm, DWT = 140 cm
	Groundwater electrical conductivity is 1.75 dS/m, EC_{gw} = 1.75 dS/m
	Groundwater electrical conductivity is 3.75 dS/m, EC_{gw} = 3.75 dS/m
	Groundwater electrical conductivity is 5.75 dS/m, EC_{gw} = 5.75 dS/m
Preseason irrigation strategy	A human irrigation (irrigation amount range from 1 to 80 cm) applied at the initial stages of plant growth (2 May 2012).

3. Results and Discussion

3.1. Model Calibration and Validation

There was good agreement between observed and simulated soil water contents and salt contents as indicated by the smaller $RMSE$ and higher R^2 values. Calibration periods resulted in $RMSE$ = 0.031 cm$^3 \cdot$ cm^{-3} and R^2 = 0.88 for soil water and $RMSE$ = 0.037 dS$\cdot$ m^{-1} and R^2 = 0.92 for EC_{sw}. Validation period resulted in $RMSE$ = 0.046 cm$^3 \cdot$ cm^{-3} and R^2 = 0.82 for soil water and $RMSE$ = 0.035 dS$\cdot$ m^{-1} and R^2 = 0.95 for EC_{sw}. These results demonstrated that despite the considerable demands on input data, Hydrus-1D was an effective tool for evaluating water and solute transport [41–43] and would be acceptable in performing scenario simulations. The calibrated parameters are shown in Table 1. There was generally very good agreement between the simulated and measured soil water contents, though there are some discrepancies. It is not possible to specifically identify the causes of the discrepancies, but they might be partially attributable to preferential flow caused by macropores and cracks [44,45], spatial heterogeneity and observation errors [46], and the locally occurring chemical processes, such as adsorption-desorption, and proportional root uptake [22], and precipitation/dissolution reactions in soils [47].

3.2. Soil Water and Salt Dynamics and Their Effects on Root Water Uptake of Tamarisk

The roots of *Chinese tamarisk* are primarily distributed in the 20–60 cm soil layer, accounting for 73.76% of total dry weight (Figure 1a) with the maximum values (35.55%) at the 30–40 cm soil depth. This distribution may be partially attributed to a large salt accumulation near the soil surface that is unfavorable to root growth (Figure 1b). The accumulation of salts has primarily been caused by high atmospheric demands that caused water movement towards the soil surface from the shallow saline groundwater. Since the groundwater has high salinity levels, salts are also transported with the water and accumulated in the root zone. In addition, scarce rainfall and poor surface drainage have also been shown to contribute to this process in arid regions [48]. Our results are consistent with the reports of Li *et al.* [49], who found that root growth of *tamarisk* seemed to be repressed when the salinity ($EC_{1:5}$) was greater than 6 dS$\cdot$ m^{-1}. Similarly, although the deeper soil layer contains little salt, shallow water table results in relatively high water contents and small values of aeration porosity. These conditions may limit root growth, respiration, and water uptake. Therefore, the optimum depth observed for plant growth was between 20 to 60 cm, because of the salt stress near the soil surface, and the saturation and anaerobic conditions below the 60 cm soil depth (Figure 3). Therefore, the long-term effects of water and salt stress caused *Chinese tamarisk* to develop its root system in the most suitable strata.

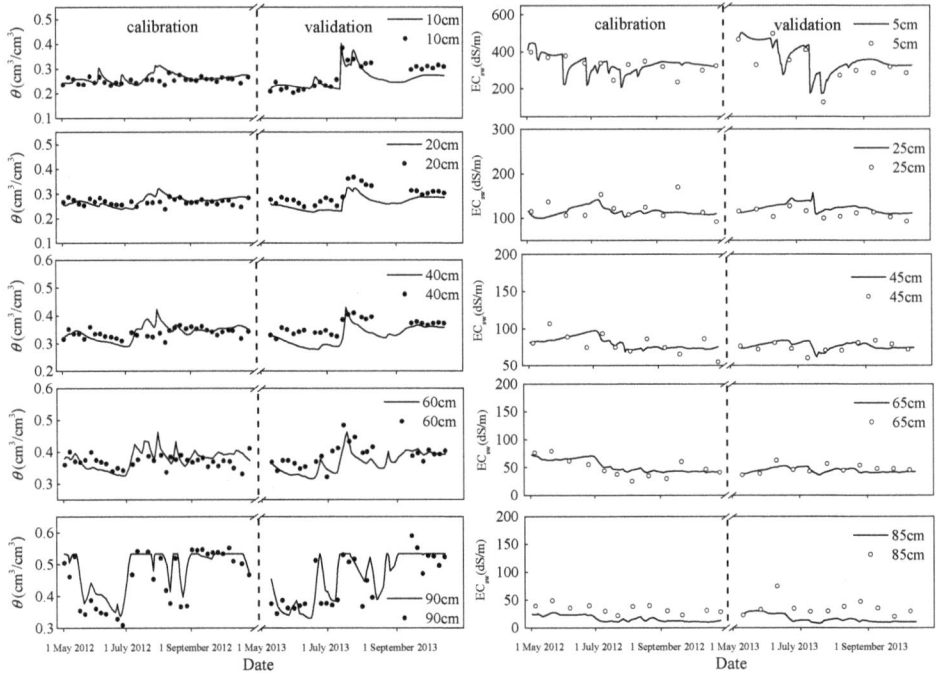

Figure 3. Measured and simulated soil water and EC_{sw} in both calibration and validation period during the growing season of *Chinese tamarisk*.

Dynamics of the profiled water contents were primarily attributable to the natural precipitation, evapotranspiration, and location of the water table. In general, soil water contents increased from the surface layer to the bottom layer because of a shallow water table. The soil moisture at the 20–60 cm strata remained relatively constant during the growing season and served as a soil moisture buffer layer (Figures 1 and 3). Because of the shallow groundwater tables, the soil moisture fluctuated dramatically for the 80–100 cm layer and ranged from 0.33 to 0.53 cm^3·cm^{-3} (Figure 3). Similarly, due to the relatively limited precipitation and large quantity of evaporation during 2012, the EC fluctuated intensively with EC_{sw} values ranging from 207.1 to 448.4 dS·m^{-1} in the surface layer. Meanwhile, EC decreased from the surface layer to the bottom layer because of the intensive evaporation, poor surface drainage, and negligible precipitation (Figures 1 and 3).

Because of the sparse vegetative cover that was effected by water and salinity stresses, cumulative evaporation reached 149 mm during the growing season. Accordingly, the migration of salt with intensive evapotranspiration was thought to be the main cause for soil salinization in this area. Infiltration was only 91 mm during the 2012 growing season that was less than both evaporation and transpiration (Figure 4). Furthermore, rainfall infiltration could dissolve large quantities of soluble salts from the upper layer. Though precipitation in this region is unable to provide sufficient water for plant growth, cumulative upward soil water flux attributable to groundwater charge reached 216 mm during the growing season of 2012. Further, compared with the infiltration water from the upper boundary, the recharged water from the groundwater has a low salt concentration and can be easily utilized by plant roots. Therefore, the groundwater plays a critical role in the maintenance of *Chinese tamarisk* growth and water supplements. These observations are in agreement with Morris and Collopy [50], who reported that more than half the tree water uptake (*Eucalyptus Camaldulensis* and *Casuarina cunninghamiana*) was drawn from the groundwater. Satchithanantham *et al.* [51] found that during the dry mid-season, when the *ET* was at its peak, the groundwater supplied up to 92% of the water for consumptive use by potatoes. Ayars *et al.* [11] observed that almost 100% of the consumptive use by alfalfa was supplied by contributions from the shallow groundwater.

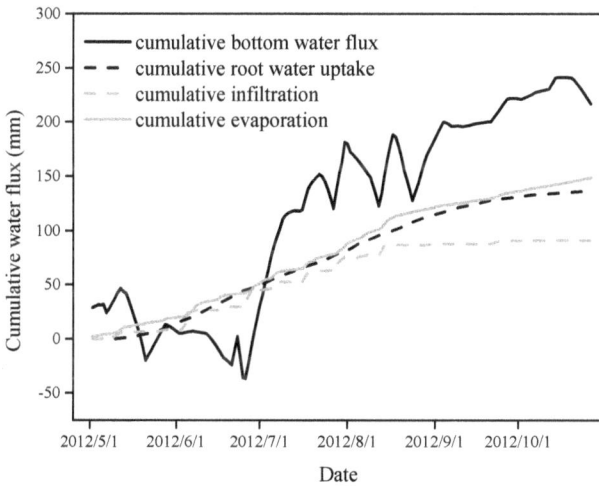

Figure 4. Cumulative water flux during the growing season of 2012.

Root water uptake reached 91.0% of its potential maximal value when water stress was considered and only 41.6% of that amount when both water and salt stress conditions were taken into consideration (Figure 5). This phenomenon has been attributed to the vast quantities of soluble salts that results in decreased solute potentials and increased ion toxicity [52]. These types of observations have resulted in assessments of salt stress being the dominating factor affecting root water uptake in arid riparian wetlands [1]. Therefore, there is a pressing need to develop appropriate management measures to reduce the impacts of water and salt stresses on *Chinese tamarisk*.

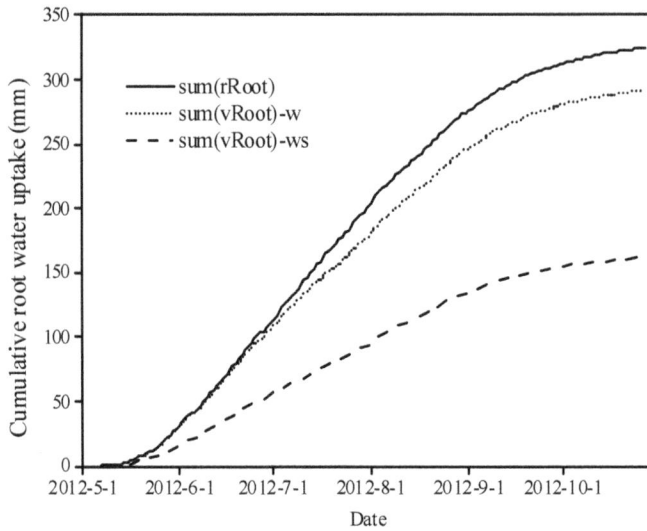

Figure 5. Cumulative root water uptake under different stress conditions. Note: sum(rRoot), potential cumulative root water uptake; sum(vRoot)-w, cumulative root water uptake under water stress only; sum(vRoot)-ws, cumulative root water uptake under coupled water and salt stress.

3.3. Scenario Simulations

3.3.1. Root Water Uptake Predictions

As indicated by Figure 6, cumulative root water uptake is more sensitive to fluctuations of water table than EC_{gw}. Root water uptake reached the maximum values of 136.6 mm when DWT was at the 2012 reference depth (CK), then decreased gradually as the water table rose. Cumulative root water uptake only reached 72.5 mm when DWT dropped 60% (DWT − 60%). This reduction in root water uptake was mainly because excessive shallow water table caused water stress in the root zone. Similarly, the cumulative root water uptake declined dramatically when the DWT increased from the 2012 reference depth to DWT + 60%. The cumulative root water uptake only reached 118.6 mm when DWT increased 60%. These simulation results suggest that either too shallow or too deep a water table will have dramatic impacts on the root water uptake. Contrary to the effects of groundwater table, root water uptake exhibited almost no change within the assumed range of EC_{gw} values. This relative lack of response may be partially attributed to a high degree of salt tolerance for this plant. For example, in many riparian systems of the southwestern United States, increased salinity caused by changes in water flow, have favored salt-tolerant *tamarisk* and greatly reduced the recruitment and growth of native salt-sensitive riparian species [53,54]. In addition, salt accumulation in the root zone is a slow process and the change of EC_{gw} may not result in obvious increases of root zone salinity within a single year which will be discussed in the following section (Section 3.3.2).

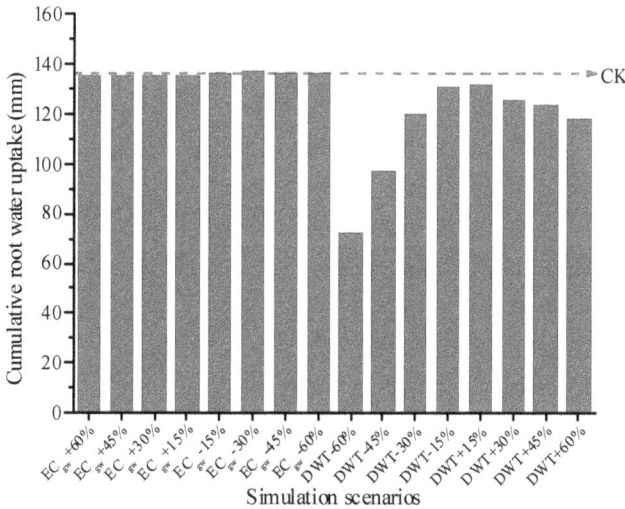

Figure 6. Cumulative root water uptake of *Chinese tamarisk* under different simulation conditions. CK is cumulative root water uptake of 2012 growing season.

3.3.2. The Long-Term Salinization Trends

To assess long-term salinity trends, Hydrus-1D was combined with a stochastic weather generator LARS-WG to evaluate the long term changes of soil salinity under different water tables and EC_{gw}. Figure 7a illustrates that soil salinization increased year by year with a saline shallow ground water (EC_{gw} = 3.75 dS·m^{-1}). In general, root zone EC_{sw} increased with the upward movement of water table whereas the amplitude of the EC_{sw} decreased with the elevation of the water table. Average root zone EC_{sw} after 30 years were 62.30, 57.34 and 47.15 dS·m^{-1} when DWT was at a depth of 60, 80 and 100 cm, respectively. These results indicated that a shallow water table contributed to increased soil salinization although the same conditions promoted root water uptake. Ibrakhimov *et al.* [55] found that elevated groundwater levels resulted in increased soil salinization by the annual addition of 3.5–14 t/haof salts depending on groundwater salinity. Xie *et al.* [4] reported that there is a contradiction between available water, salt stress, and reed water uptake with variations in DWT. Similarly, root zone EC_{sw} increased with the increased EC_{gw}, values of 142.08, 177.53, and 210.55 dS·m^{-1} when accompanying EC_{gw} values were 1.75, 3.75, and 5.75 ds·m^{-1} after 30 years, respectively (Figure 7b). The results indicated that soil salinization conditions will deteriorate continuously without human intervention and highlighted the importance of preventing human induced EC_{gw} increases that would occur from subsurface irrigation drainage from farmland.

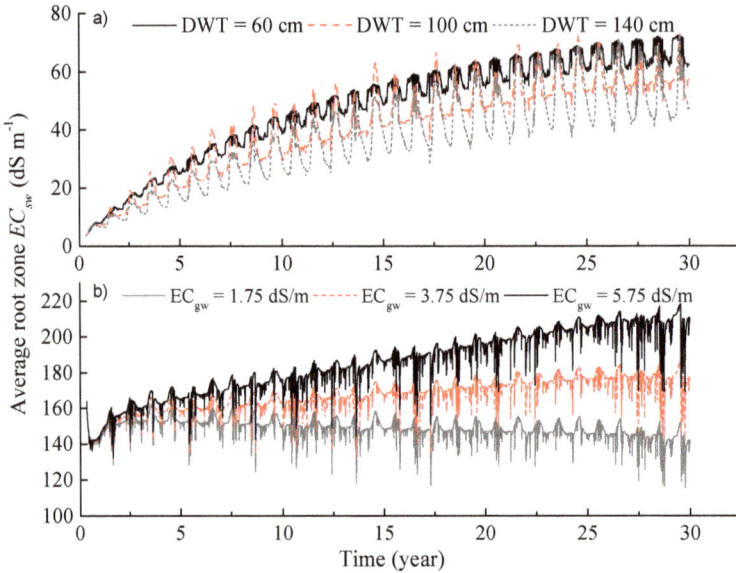

Figure 7. Temporal changes in soil salinity in averaged root zone as affected by (**a**) water table and (**b**) groundwater electrical conductivity.

3.3.3. Preseason Irrigation Strategy

Preseason irrigation increased root water uptake significantly. Compared with no irrigation, root water uptake increased 4.5%, 40.2%, 79.3%, 100.6% and 115.4% when the irrigation amounts were 10, 20, 30, 40 and 50 cm, respectively (Figure 8). These results indicated that root water uptake generally increased with increased irrigation quantities when irrigation amounts ranged from 10 to 50 cm. However, root water uptake increased only 2 mm (from 294 to 296 mm) when the irrigation quantities increased from 50 to 80 cm, and demonstrated that irrigation quantities less than 50 cm were sufficient to promote *tamarisk* root water uptake (Figure 9b). Note that cumulative root water uptake displayed a decreased trend when the irrigation quantities were less than 6 cm and indicated that although a small quantity of irrigation increased the root zone water content to some extent, it was not sufficient for the salt to be effectively leached out of the root zone. Furthermore, the irrigated water cannot be easily used by the plant roots because of the water contained a large amount of soluble salts that were dissolved from the upper soil layer. This observation implied that the precipitation in the region is not beneficial for plants and even threatens plant growth, since single precipitation events are normally less than 6 cm. The increased root water uptake resulting from fresh water irrigation was mainly because sufficient quantities of water from the upper boundary layers can leach salts below the root zone and effectively alleviate salt stress (Figure 9b). Because these salts can re-accumulate as a result of evaporation (Figure 9a), repeated irrigation is needed. The average root zone EC_{sw} dropped rapidly below 52.5, 27.7, and 13.7 dS·m^{-1} in 15 days when the irrigation quantities were 20, 40, and 60 cm, respectively. Therefore, irrigation before the growing season is essential, but too small or too large a quantity of irrigation is not advisable. In our case, 30–40 cm of preseason irrigation was reasonable.

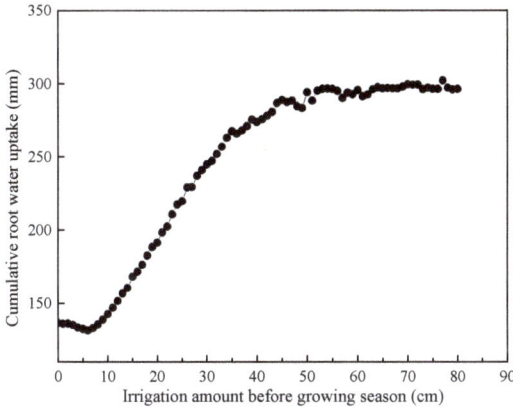

Figure 8. Cumulative root water uptake of *Chinese tamarisk* with different irrigation amounts.

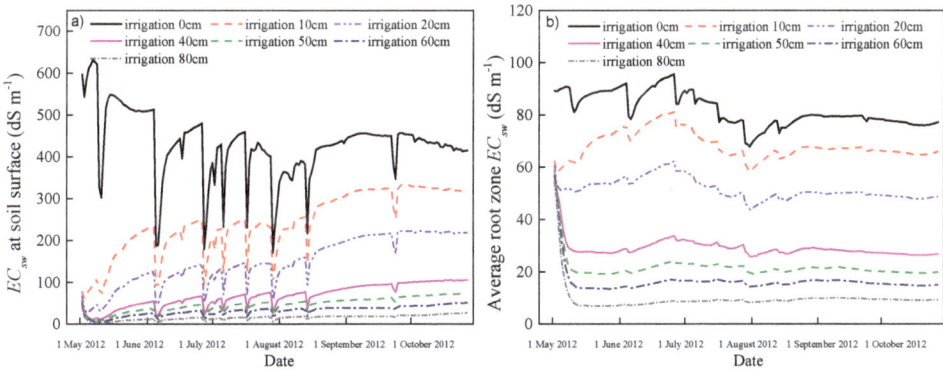

Figure 9. Temporal changes of soil salinity at the soil surface (**a**) and the averaged root zone electrical conductivity as affected by irrigation amount (**b**).

The advantages of artificial irrigation to maintain plant growth has also been addressed by other researchers. Holland *et al.* [56] observed that a two-fold to five-fold increase in plant water potential and a three-fold to six-fold increase in *Eucalyptus camaldulensis* water consumption in three to four months after watering in the riparian region. Xie *et al.* [4] reported that irrigation clearly increased reed water use, especially when irrigation quantities were higher than 3 cm·d^{-1}. Askri *et al.* [57] demonstrated that in order to increase date palm water use, high irrigation frequencies and shallow groundwater are needed to maintain high water content and low salinity in the root zone. Therefore, we would suggest it would be beneficial to have artificial irrigation to maintain the sustainable development of the arid riparian wetlands.

4. Conclusions

In this study, soil water and salt dynamics and their effects on *Tamarisk* root water uptake were characterized by coupling measured data with simulation scenarios in the Heihe riparian wetland, China. The Hydrus-1D model simulations of soil water and salt dynamics matched the observed data fairly well during both the calibration and validation periods as indicated by smaller *RMSE* and higher R^2 values, which demonstrated the feasibility of using the model under different simulation scenarios.

Chinese tamarisk extends its root system into the most suitable strata with 73.6% of the total root system distributed in the 20–60 cm soil layer because of the long-term effects of water and salt stress. Groundwater is the main water source for *Chinese tamarisk* in the study area. Cumulative root water uptake only accounted for 41.6% of the potential value under the joint influences of water and salt stress. This result indicated the necessity of human interventions to alleviate water and salt stress. Furthermore, root water uptake was most sensitive to the fluctuations of water table levels. Too deep or too shallow a groundwater table was found to severely repress root water uptake. Shallow groundwater was found to result in increased soil salinization, especially when the groundwater contains a large amount of salts (high EC_{sw}). Preseason irrigation has the potential to leach salt out of the root zone and maintain the EC_{sw} at a reduced level during the growing season which would result in increased water uptake. Cumulative root water uptake increased when irrigation quantities were initially increased. Further increases in irrigation quantities diminished the increased rate of root water uptake.

Irrigation before the growing season is necessary, but the irrigation quantities should be taken into consideration. This study provided insights into soil water and salt redistribution and their effects on plant water use, and should help in the establishment of improved management practices for arid riparian wetlands.

Acknowledgments: This study was financially supported by National Natural Science Foundation of China (91025018, 41371233, 41371234, 41201279), Northwest Agriculture and Forestry University fund (Z111021308), and the "111" Project (B12007). We thank our colleagues for the joined field work and data analysis. Special thanks go to the staff of the Linze CERN station.

Author Contributions: All authors contributed to the design and development of this manuscript. Huijie Li and Jun Yi, carried out the data analysis and prepared the first draft of the manuscript; Ying Zhao is the graduate advisor of Huijie Li and contributed many ideas to the study; Jianguo Zhang, Bingcheng Si, Robert Hill, Lele Cui and Xiaoyu Liu provided some important advices on the concept of methodology and writing of the manuscript.

Conflicts of Interest: The authors declare no conflict of interest.

References

1. Jolly, I.D.; McEwan, K.L.; Holland, K.L. A review of groundwater-surface water interactions in arid/semi-arid wetlands and the consequences of salinity for wetland ecology. *Ecohydrology* **2008**, *1*, 43–58.
2. Niu, Y.; Liu, X.D.; Zhang, H.B.; Meng, H.J. The ecological function evaluation of wetland in upper and middle reaches of heihe basin. *Wetl. Sci.* **2007**, *3*, 215–220. (In Chinese)
3. Lei, K.; Zhang, M.X. The wetland resources in china and the conservation advices. *Wetl. Sci.* **2005**, *3*, 81–86. (In Chinese)
4. Xie, T.; Liu, X.; Sun, T. The effects of groundwater table and flood irrigation strategies on soil water and salt dynamics and reed water use in the yellow river delta, china. *Ecol. Model.* **2011**, *222*, 241–252.
5. Ahmad, M.U.D.; Bastiaanssen, W.G.M.; Feddes, R.A. Sustainable use of groundwater for irrigation: A numerical analysis of the subsoil water fluxes. *Irrig. Drain.* **2002**, *51*, 227–241.
6. Gowing, J.W.; Rose, D.A.; Ghamarnia, H. The effect of salinity on water productivity of wheat under deficit irrigation above shallow groundwater. *Agric. Water Manag.* **2009**, *96*, 517–524.
7. Jiang, P.H.; Zhao, R.F.; Zhao, H.L.; Lu, L.P.; Xie, Z.L. Relationships of wetland landscape fragmentation with climate change in middle reaches of heihe river, China. *J. Appl. Ecol.* **2013**, *24*, 1661–1668. (In Chinese)
8. Zhang, H.B.; Meng, H.J.; Liu, X.D.; Zhao, W.J.; Wang, X.P. Vegetation characteristics and ecological restoration technology of typical degradation wetlands in the middle of heihe river basin, zhangye city of gansu province. *Wetl. Sci.* **2012**, *10*, 194–199. (In Chinese)
9. Feng, Q.; Liu, W.; Su, Y.H.; Zhang, Y.W.; Si, J.H. Distribution and evolution of water chemistry in heihe river basin. *Environ. Geol.* **2004**, *45*, 947–956.
10. Tedeschi, A.; Dell'Aquila, R. Effects of irrigation with saline waters, at different concentrations, on soil physical and chemical characteristics. *Agric. Water Manag.* **2005**, *77*, 308–322.
11. Ayars, J.E.; Shouse, P.; Lesch, S.M. *In situ* use of groundwater by alfalfa. *Agric. Water Manag.* **2009**, *96*, 1579–1586.
12. Ayars, J.E.; Christen, E.W.; Soppe, R.W.; Meyer, W.S. The resource potential of *in-situ* shallow ground water use in irrigated agriculture: A review. *Irrig. Sci.* **2005**, *24*, 147–160.

13. Singh, R.; Singh, J. Irrigation planning in cotton through simulation modeling. *Irrig. Sci.* **1996**, *17*, 31–36.

14. Feddes, R.A.; Kabat, P.; van Bakel, P.; Bronswijk, J.J.B.; Halbertsma, J. Modelling soil water dynamics in the unsaturated zone—State of the art. *J. Hydrol.* **1988**, *100*, 69–111.

15. de Jong, R.; Bootsma, A. Review of recent developments in soil water simulation models. *Can. J. Soil Sci.* **1996**, *76*, 263–273.

16. Saito, H.; Šimůnek, J.; Mohanty, B.P. Numerical analysis of coupled water, vapor, and heat transport in the vadose zone. *Vadose Zone J.* **2006**, *5*, 784–800.

17. Jacques, D.; Šimůnek, J.; Timmerman, A.; Feyen, J. Calibration of richards' and convection-dispersion equations to field-scale water flow and solute transport under rainfall conditions. *J. Hydrol.* **2002**, *259*, 15–31.

18. Wang, T.; Franz, T.E.; Zlotnik, V.A. Controls of soil hydraulic characteristics on modeling groundwater recharge under different climatic conditions. *J. Hydrol.* **2015**, *521*, 470–481.

19. Van Genuchten, M.T. A closed-form equation for predicting the hydraulic conductivity of unsaturated soils. *Soil Sci. Soc. Am. J.* **1980**, *44*, 892–898.

20. Mualem, Y. A new model for predicting the hydraulic conductivity of unsaturated porous media. *Water Resour. Res.* **1976**, *12*, 513–522.

21. Šimůnek, J.; van Genuchten, M.T.; Šejna, M. Hydrus: Model use, calibration, and validation. *Tansac Asabe* **2012**, *55*, 1261–1274.

22. Šimůnek, J.; van Genuchten, M.T.; Sejna, M. The hydrus-1d software package for simulating the one-dimensional movement of water, heat, and multiple solutes in variably-saturated media. *Univ. Calif. -Riverside Res. Rep.* **2005**, *3*, 1–240.

23. Zhao, W.; Liu, B.; Zhang, Z. Water requirements of maize in the middle heihe river basin, china. *Agric. Water Manag.* **2010**, *97*, 215–223.

24. Gee, G.W.; Or, D. 2.4 particle-size analysis. *Methods Soil Anal. Part* **2002**, *4*, 255–293.

25. Klute, A.; Dirksen, C. Hydraulic conductivity and diffusivity: Laboratory methods. In *Methods of Soil Analysis: Part 1—Physical and Mineralogical Methods*; American Society of Agronomy: Madison, WI, USA, 1986; pp. 687–734.

26. Feddes, R.A.; Kowalik, P.J.; Zaradny, H.X. *Simulation of Field Water Use and Crop Yield*; Centre for Agricultural Publishing and Documentation: Wageningen, The Netherlands, 1978.

27. Šimůnek, J.; Hopmans, J.W. Modeling compensated root water and nutrient uptake. *Ecol. Model.* **2009**, *220*, 505–521.

28. Van Genuchten, M.T. *A Numerical Model for Water and Solute Movement in and below the Root Zone*; United States Department of Agriculture Agricultural Research Service U.S. Salinity Laboratory: Riverside, CA, USA, 1987.

29. Slavich, P.G.; Petterson, G.H. Estimating the electrical conductivity of saturated paste extracts from 1: 5 soil, water suspensions and texture. *Soil Res.* **1993**, *31*, 73–81.

30. Corwin, D.L.; Lesch, S.M. Application of soil electrical conductivity to precision agriculture. *Agron. J.* **2003**, *95*, 455–471.

31. Richards, L.A. U.S. salinity laboratory staff. In *USDA Handbook No. 60. Diagnosis and Improvement of Saline and Alkali Soils*; USDA: Washington, DC, USA, 1954; p. 13.

32. Allen, R.G.; Pereira, L.S.; Raes, D.; Smith, M. *Crop Evapotranspiration-Guidelines for Computing Crop Water Requirements-FAO Irrigation and Drainage Paper 56*; FAO: Rome, Italy, 1998; Volume 300, p. 6541.

33. Allen, R.G.; Pereira, L.S. Estimating crop coefficients from fraction of ground cover and height. *Irrig. Sci.* **2009**, *28*, 17–34.

34. Schaap, M.G.; Leij, F.J.; van Genuchten, M.T. Rosetta: A computer program for estimating soil hydraulic parameters with hierarchical pedotransfer functions. *J. Hydrol.* **2001**, *251*, 163–176.

35. Vanderborght, J.; Vereecken, H. Review of dispersivities for transport modeling in soils. *Vadose Zone J.* **2007**, *6*, 29–52.

36. Gelhar, L.W.; Welty, C.; Rehfeldt, R.K. A critical review of data on field-scale dispersion in aquifers. *Water Resour. Res.* **1992**, *28*, 1955–1974.

37. Moayyad, B. *Importance of Groundwater Depth, Soil Texture and Rooting Depth on Arid Riparian Evapotranspiration*; New Mexico Institute of Mining and Technology: Socorro, NM, USA, 2001.

38. Grinevskii, S.O. Modeling root water uptake when calculating unsaturated flow in the vadose zone and groundwater recharge. *Mosc. Univ. Geol. Bull.* **2011**, *66*, 189–201.

39. Marquardt, D.W. An algorithm for least-squares estimation of nonlinear parameters. *J. Soc. Ind. Appl. Math.* **1963**, *11*, 431–441.

40. Semenov, M.A.; Barrow, E.M.; Lars-Wg, A. *A Stochastic Weather Generator for Use in Climate Impact Studies*; User Manual: Hertfordshire, UK, 2002.

41. Kandelous, M.M.; Kamai, T.; Vrugt, J.A.; Šimůnek, J.; Hanson, B.; Hopmans, J.W. Evaluation of subsurface drip irrigation design and management parameters for alfalfa. *Agric. Water Manag.* **2012**, *109*, 81–93.

42. Ramos, T.B.; Šimůnek, J.; Gonçalves, M.C.; Martins, J.C.; Prazeres, A.; Pereira, L.S. Two-dimensional modeling of water and nitrogen fate from sweet sorghum irrigated with fresh and blended saline waters. *Agric. Water Manag.* **2012**, *111*, 87–104.

43. Siyal, A.A.; Bristow, K.L.; Šimůnek, J. Minimizing nitrogen leaching from furrow irrigation through novel fertilizer placement and soil surface management strategies. *Agric. Water Manag.* **2012**, *115*, 242–251.

44. Garg, K.K.; Das, B.S.; Safeeq, M.; Bhadoria, P.B.S. Measurement and modeling of soil water regime in a lowland paddy field showing preferential transport. *Agric. Water Manag.* **2009**, *96*, 1705–1714.

45. Patil, M.D.; Das, B.S.; Bhadoria, P.B.S. A simple bund plugging technique for improving water productivity in wetland rice. *Soil Tillage Res.* **2011**, *112*, 66–75.

46. Vazifedoust, M.; van Dam, J.C.; Feddes, R.A.; Feizi, M. Increasing water productivity of irrigated crops under limited water supply at field scale. *Agric. Water Manag.* **2008**, *95*, 89–102.

47. Robarge, W.P. *Precipitation/Dissolution Reactions in Soils*; CRC Press: Boca Raton, FL, USA, 1999; p. 2.

48. Herczeg, A.L.; Dogramaci, S.S.; Leaney, F.W.J. Origin of dissolved salts in a large, semi-arid groundwater system: Murray basin, australia. *Mar. Freshw. Res.* **2001**, *52*, 41–52.

49. Li, C.; Lei, J.; Zhao, Y.; Xu, X.; Li, S. Effect of saline water irrigation on soil development and plant growth in the taklimakan desert highway shelterbelt. *Soil Tillage Res.* **2015**, *146*, 99–107.

50. Morris, J.D.; Collopy, J.J. Water use and salt accumulation by *Eucalyptus camaldulensis* and *Casuarina cunninghamiana* on a site with shallow saline groundwater. *Agric. Water Manag.* **1999**, *39*, 205–227.

51. Satchithanantham, S.; Krahn, V.; Sri Ranjan, R.; Sager, S. Shallow groundwater uptake and irrigation water redistribution within the potato root zone. *Agric. Water Manag.* **2014**, *132*, 101–110.

52. Bastias, E.; Alcaraz-Lopez, C.; Bonilla, I.; Martinez-Ballesta, M.C.; Bolanos, L.; Carvajal, M. Interactions between salinity and boron toxicity in tomato plants involve apoplastic calcium. *J. Plant Physiol.* **2010**, *167*, 54–60.

53. Brotherson, J.D.; Field, D. Tamarix: Impacts of a successful weed. *Ra ngelands* **1987**, *9*, 110–112.

54. Sala, A.; Smith, S.D.; Devitt, D.A. Water use by tamarix ramosissima and associated phreatophytes in a mojave desert floodplain. *Ecol. Appl.* **1996**, 888–898.

55. Ibrakhimov, M.; Khamzina, A.; Forkutsa, I.; Paluasheva, G.; Lamers, J.P.A.; Tischbein, B.; Vlek, P.L.G.; Martius, C. Groundwater table and salinity: Spatial and temporal distribution and influence on soil salinization in khorezm region (uzbekistan, aral sea basin). *Irrig. Drain. Syst.* **2007**, *21*, 219–236.

56. Holland, K.L.; Charles, A.H.; Jolly, I.D.; Overton, I.C.; Gehrig, S.; Simmons, C.T. Effectiveness of artificial watering of a semi-arid saline wetland for managing riparian vegetation health. *Hydrol. Process.* **2009**, *23*, 3474–3484.

57. Askri, B.; Ahmed, A.T.; Abichou, T.; Bouhlila, R. Effects of shallow water table, salinity and frequency of irrigation water on the date palm water use. *J. Hydrol.* **2014**, *513*, 81–90.

Development of a Prototype Web-Based Decision Support System for Watershed Management

Dejian Zhang, Xingwei Chen and Huaxia Yao

Abstract: Using distributed hydrological models to evaluate the effectiveness of reducing non-point source pollution by applying best management practices (BMPs) is an important support to decision making for watershed management. However, complex interfaces and time-consuming simulations of the models have largely hindered the applications of these models. We designed and developed a prototype web-based decision support system for watershed management (DSS-WMRJ), which is user friendly and supports quasi-real-time decision making. DSS-WMRJ is based on integrating an open-source Web-based Geographical Information Systems (Web GIS) tool (Geoserver), a modeling component (SWAT, Soil and Water Assessment Tool), a cloud computing platform (Hadoop) and other open source components and libraries. In addition, a private cloud is used in an innovative manner to parallelize model simulations, which are time consuming and computationally costly. Then, the prototype DSS-WMRJ was tested with a case study. Successful implementation and testing of the prototype DSS-WMRJ lay a good foundation to develop DSS-WMRJ into a fully-fledged tool for watershed management. DSS-WMRJ can be easily customized for use in other watersheds and is valuable for constructing other environmental decision support systems, because of its performance, flexibility, scalability and economy.

Reprinted from *Water*. Cite as: Zhang, D.; Chen, X.; Yao, H. Development of a Prototype Web-Based Decision Support System for Watershed Management. *Water* **2015**, *7*, 780–793.

1. Introduction

Climate change, population growth and unreasonable exploitation of water resources have caused environmental deterioration, the unavailability of freshwater and an imbalance between supply and demand to a global extent, thus seriously affecting the sustainable development and utilization of water resources. At present, more than 1.2 billion people and 60% of global basins lie at the edge of water resource shortage [1]. How to relieve or eliminate the deterioration of the water environment and realize the sustainable utilization of water resource have become common concerns of and challenges for humankind. Scientific and effective tools are urgently needed to fulfill the purpose of the sustainable utilization of water

305

resource. The decision support system for watershed management (DSS-WM) is one of the representative management tools and plays an important role in watershed management.

Driven by the latest advancements of information and communication technologies, hydrologic sciences and other disciplines, there is booming research on watershed management using hydrological models. For example, distributed hydrological and hydrodynamic models, such as SWAT (Soil and Water Assessment Tool), HSPF (Hydrological Simulation Program Fortran), AGNPS (agricultural non-point source pollution model) and WASP (Water Quality Analysis Simulation Program), are all geared with management modules for the simulation and evaluation of management effects on flow, sediment and nutrients [2–6]. Although great achievements have been gained by these models, the complex model structures and interfaces have impeded their applications by inexperienced users. Besides, the time-consuming and computationally costly procedures of model simulations have further hindered the application of these models, especially under circumstances where real-time or quasi-real-time support for decision-making is required.

To overcome the aforementioned shortcomings, hydrologists and environmental scientists have designed and invented many dedicated DSS-WMs to assist with watershed management. Similar to other environmental DSS, these DSS-WMs usually consist of a decision-making information database and user interfaces and models [7–9]. According to the operational environments, DSS-WMs can be divided into desktop-based and web-based. Desktop-based DSS-WMs usually provide intuitive wizard style interfaces, which eliminate the complexity of the models. For example, under the impetus of the MULINO (Multi-sectoral, Integrated and Operational DSS) project, Mysiak et al. [10] developed mDSS (a decision support system for water resource management that has been developed under the European research project, MULINO) for optimizing the management of water resources by integrating hydrological models with multiple-criteria evaluation procedures. Cau and Paniconi [11] linked SWAT and mDSS to assess four alternatives, including intensive agriculture and dairy farming and treated wastewater for irrigation. Hipel et al. [12] designed and developed the GMCR II (graph model for conflict resolution) for conflict resolution over multiple stakeholders in controlling water pollutions.

It is a general trend to turn to the Internet as a platform for software solutions, and so is DSS-MWs. Rao et al. [13] developed a prototype web-based DSS based on a commercial Web GIS (Web-based Geographical Information Systems) tool, ArcIMS (Arc Internet Map Server), and a hydrological model, SWAT. Additionally, the prototype was then applied to a small watershed, Panhandle in Oklahoma, targeted at aiding a better management plan. Model parallel simulation and a cloud computing platform were not attempted in their work. Zeng et al. [14] constructed a web-based decision-making system by integrating the ArcGIS Engine, the distributed hydrological model, Hydrologic Engineering Center's Hydrologic Modeling System (HEC-HMS), genetic algorithm (GA) and artificial neural network (ANN). HEC-HMS was applied to the prediction of runoff; ANN was used to predict the city water resources demand; GA was used to achieve the goal of distributing water resources among the regions of the city. Sun [15] migrated a web-based DSS to a public cloud, which is an extension of web-based solutions.

Throughout the development of DSSs for watershed management, great achievements have been made by integrating models and other technologies for better watershed management. However, there are still some inadequacies, such as: (1) Most integrated models are conceptual or empirical models, and distributed hydrological ones are few; (2) More DSSs for watershed management are desktop-based, while the web-based ones are still rare; and (3) The performance of the systems are not well explained or evaluated, which is a key factor to achieve the goal of real-time decision support.

Our objectives in this study are to design and develop a web-based decision support system for watershed management (DSS-WMRJ), which is user friendly and supports quasi-real-time decision making. We build the DSS-WMRJ by integrating an open source Web GIS tool (Geoserver), a modeling component (SWAT, Soil and Water Assessment Tool), a cloud computing platform (Hadoop) and other open source components and libraries. In addition, a private cloud is used in an innovative manner to parallelize the model simulations, which are time consuming and computationally costly. The successful implementation and testing of the prototype DSS-WMRJ shows that it is able to fulfill the goal of quasi-real-time decision support and provide intuitive interfaces.

2. System Design

2.1. Architecture of Decision Support System for Watershed Management (DSS-WMRJ)

To meet the requirements of availability, stability, interoperability and portability, a systematic architecture of four tiers, including the presentation, proxy, application and database and model, is considered (Figure 1).

The presentation tier provides a graphic user interface, which is accessible via the browsers of many devices, for users to perform system management, map operations, spatial and attribute information retrieval, watershed management, and so on. The map viewer is achieved by the Openlayers component, which communicates with map services to retrieve the grid or vector map through Asynchronous JavaScript and XML (AJAX) and to render the map in the browser. Thus, it provides operation experience approximate to a desktop GIS tool.

FusionCharts is the only commercial software used for presenting the watershed management results, due to its dynamic and excellent chart functionalities.

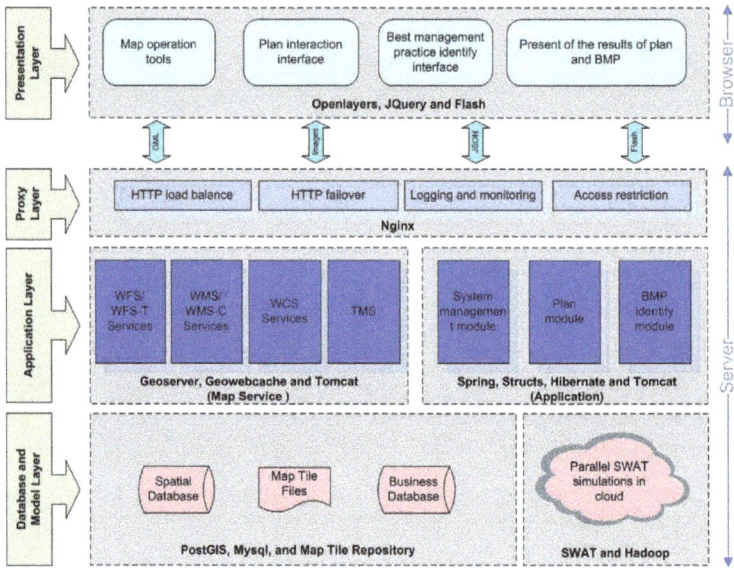

Figure 1. The system architecture of decision support system for watershed management (DSS-WMRJ). WFS/WFS-T, Web Feature Service/Web Feature Service-Transaction; WMS, Web Map Service; WCS, Web Coverage Service.

The open-source component Nginx is used as the proxy tier, which lies between the presentation and application tiers and acts as a communication agent for these two tiers. The deployment and configuration of Nginx is easy, while it provides useful functionalities, such as load balancing, failover, access control, logging, monitoring, *etc.* When the system exceeds the workload limit of the system, the system administrator can add more background services and a simple configuration of Nginx to scale up the system. Therefore, the proxy tier is very important for enhancing the performance and improving the stability of the system.

The application tier consists of two components: map service and watershed management services. Both components adopt the Service-Oriented Architecture (SOA). The map services component uses an open source Web GIS tool (Geoserver), which is in compliance with the Open Geospatial Consortium (OGC) standards, such as Web Map Service (WMS), Web Feature Service/Web Feature Service-Transaction (WFS/WFS-T) and Web Coverage Service (WCS). The watershed management services provide functionalities, such as planning and BMP identification. These components are standard compliant and service oriented, making them scalable and interoperable.

308

The database and model tier is located on the bottom of the architecture. This tier consists of databases and model simulation services. The databases store and manage attribute data, spatial data and map tiles via a spatial database, an object-relational database and a file system. The spatial data are stored in the PostGreSQL database with the use of the PostGIS library, which adds support for the use and management of geographic objects. Spatial and other regular indices are created for every map layer stored in the spatial database to increase the speed of retrieval. Map tiles are pre-generated and stored in the map tile repository. This will accelerate the mapping processes, as WMS can directly deliver the caching map tiles to the client when a map request is sent to it. In addition, the model simulation service is a key component of the DSS-WMRJ, which guarantees quasi-real-time decision making by parallelizing model simulations on a private cloud. A detailed description of the model simulation service is given in the next section.

2.2. Model Simulation Service

The decision making procedures usually require a great many model simulations, for example, when an uncertainty analysis is required in the decision making as the model input, the structure and parameters contain various degree of uncertainty or when evaluating the environmental effect of combinations of different management measures, which may themselves involve different configurations (making the decision making procedures very time consuming and computationally costly). Therefore, a fast model simulation is the key factor of DSS-WMRJ to achieve real-time or quasi-real-time decision making.

An open-source cloud computing platform (Hadoop) is used to parallelize model simulations in order to accelerate the simulation procedures. Hadoop is an implementation of the Google MapReduce algorithm [16,17]. It consists of two components: the Hadoop Distributed File System (HDFS) and the distributed computation framework (MapReduce). HDFS is a robust distributed file system, which is able to read and write data in parallel over a large number of machines and achieves much higher throughput than traditional technologies. This feature is very useful to process mass model simulation results. MapReduce is a distributed computing framework that consists of the JobTracker and the TaskTrackers. It provides two important application programming interfaces (APIs): Mapper and Reducer. With these interfaces, developers can quickly write efficient parallel codes. Hadoop parallelizes tasks as follows: (1) Clients submit a job to the JobTracker, which is the master of the MapReduce framework; (2) The JobTracker then divides the submitted job into task sets and distributes these task sets to TaskTrackers; and (3) the tasks in the assigned set are further distributed to Mapper or Reducer, which then executes the task.

To achieve SWAT parallel simulations on a Hadoop cluster, developers must implement the aforementioned APIs of MapReduce, and co-operation among the presentation tier, application service and model simulation service is needed. Figure 2 shows the procedures of paralleling SWAT simulations. These procedures are summarized as follows: (1) A user is prompted for certain specific inputs that pertain to management practices and submits these inputs to the application service; (2) The application service translates the inputs into parameter sets of the SWAT model and distributes these parameter sets to the model simulation service; (3) The model simulation service parallelizes the model simulations, which involves operations, such as model input file editing, model executing, simulation result extracting and saving results to the HDFS; (4) When the submitted job is finished, the application service gathers all simulation results in the HDFS and generates a statistic report, which is XML-based, and delivers it to the presentation tier; and (5) Finally, the presentation tier renders the report through its chart component.

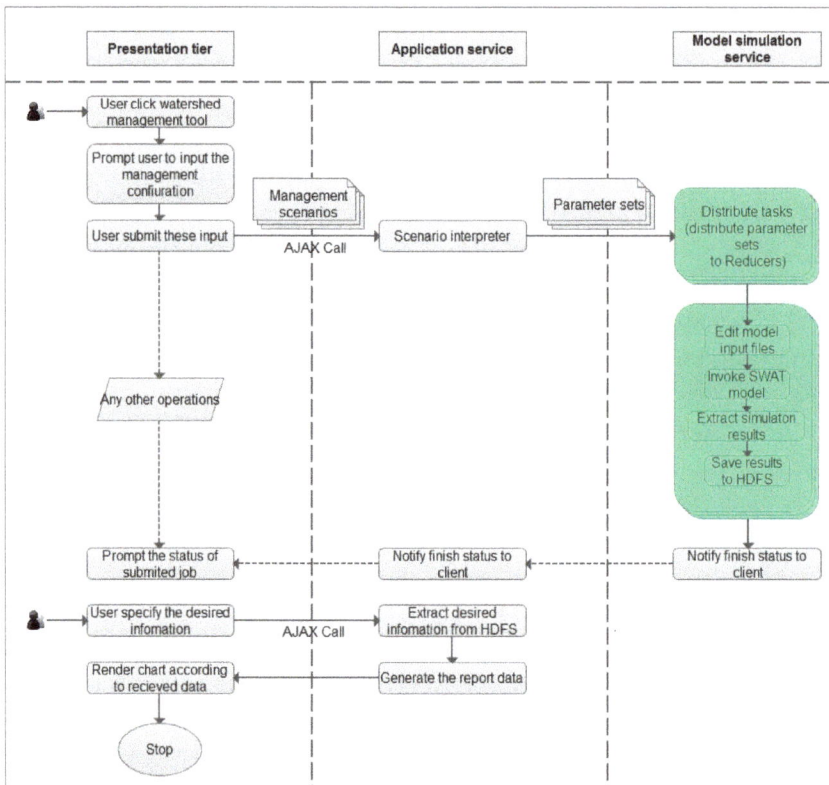

Figure 2. Parallelizing the simulation of management scenarios (the colored parts are the execution in parallelizing). AJAX, Asynchronous JavaScript and XML; HDFS, Hadoop Distributed File System; SWAT, Soil and Water Assessment Tool.

3. Case Study and DSS-WMRJ Test

3.1. Model Setup

The SWAT model [18,19] is a semi-distributed, continuous, watershed-scale hydrological model that was developed by the Agricultural Research Service of the United States Department of Agriculture (USDA-ARS) to simulate the quantity and quality of surface water and groundwater. It not only deeply depicts the physical hydrological cycle, but also considers the impact of human activities, such as land use change, water conservancy facilities, agriculture management practices and other environment protecting facilities (e.g., vegetation filter strips and grassed waterways) on the hydrological processes.

Jinjiang basin with an area of 5629 km^2 is selected as the test watershed for which to implement and evaluate the DSS-WMRJ. ArcSWAT, one of the graphical user interface procedures for SWAT, is used to delineate Jinjiang basin [20]. The basin is divided into 99 subbasins based on the DEM data and with a threshold area of 3000 ha. The subbasins are subdivided into HRUs, which represent homogeneous soil and land use according to the soil type, land use and topographic slope, with threshold values of 5%, 20% and 20%, respectively, resulting in 886 HRUs. Additionally, the watershed model is set to run in daily mode. The SWAT has been calibrated based on water discharge data, but not calibrated for sediment and nutrients, because of the insufficient monitoring data.

3.2. System Implementation

According to the design scheme of DSS-WMRJ, a prototype of DSS-WMRJ was established by incorporating the hydrological model of an experimental watershed. Figure 3 is the GUI of the prototype of DSS-WMRJ. The left column provides the functionalities of the system and layer management. The system functionalities control the privileges of users, and the layer management controls the switching on or off of layers. The right column provides watershed management functionalities and some general map-related functionalities, such as map roaming, zoom in/out, overview map, and so on.

For a prototype of DSS-WMRJ, we only developed a tool to evaluate the soil and water conservation effect of a vegetation filter strip (VFS-Tool), which is a widely-used conservation practice to remove agricultural and urban pollutants before they reach nearby water bodies by establishing a strip of dense vegetative filter around the upslope pollutant sources. The interfaces of the VFS-Tool (Figure 4) are intuitive and easy to use. Users just need to click the tool icon in the toolbar, enter or select certain specific inputs that pertain to VFS and submit these inputs to the server.

311

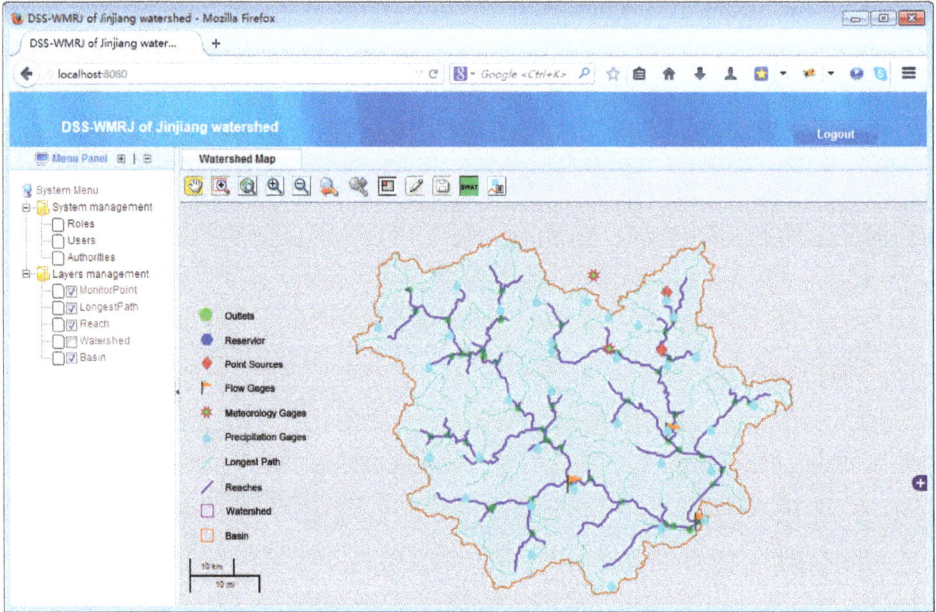

Figure 3. Main interface of DSS-WMRJ.

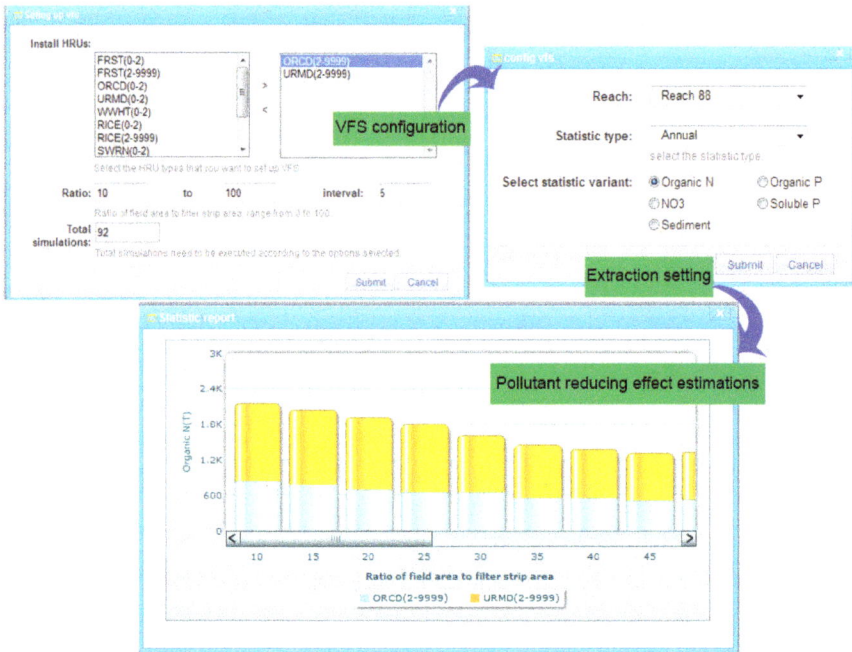

Figure 4. Interfaces of the vegetation filter strip (VFS)-Tool.

3.3. Performance Tests

The model simulation service is a key component of DSS-WMRJ that directly determines the achievement of real-time or quasi real-time decision making. The performance of this component is tested and evaluated. To perform the tests, one management scenario was used, which established VFSs around two kinds of HRUs (whose land use type is orchard or urban with a slope greater than two degrees) with varying ratios of field area to filter strip area (from 10 to 100 at an interval of five; Figure 4). The management scenario generated a total of 92 model simulations, which needed 110.4 min to finish if running the model in series, as each simulation took about 1.2 min. We will not go into the details about the pollution-reduction effect of VFSs, as our main objective here was to demonstrate the performance of the model simulation service. To evaluate the scalability of the model simulation service, the management scenario was performed in a Hadoop cluster (private cloud) with different number "of TaskTrackers (from one to eight), and each TaskTracker was allowed to perform four tasks simultaneously. These TaskTrackers are virtual machines on two physical servers. The configurations of the virtual machines are identical, and so are the physical ones, with the configuration details being listed in Table 1.

Table 1. The configurations of virtual and physical machines.

Virtual Machines	Physical Machines
4 logical processors Model name: Intel(R) Xeon(R) CPU E5520 @ 2.27 GHz CPU MHz: 2,261.060 RAM: 4 GB OS: 64-bit Red Hat Enterprise Linux 5.4	2 physical processors 4 cores for each physical processor 16 logical processors Model name: Intel(R) Xeon(R) CPU E5520 @ 2.27 GHz CPU MHz: 2,261.060 RAM: 16 GB VM management software: VMware ESXi 4.1

Figure 5 shows the results of the tests. The simulation time decreased as the number of TaskTrackers increased from one to eight. When eight TaskTrackers were used, the simulation time reached the lowest point (about 4.4 min). The number of TaskTrackers and the number of tasks in each TaskTracker are the major factors that affect the performance of the model simulation service (other factors are ignorable). As the task number in each TaskTracker was set to a constant value of four, the number of TaskTrackers becomes the only determinant factor that is negatively and nonlinearly proportional to the simulation time. Thus, we chose the inverse first order equation to generate a fit curve of the simulation time *vs.* the number of TaskTrackers (Figure 5). According to the trend of the fit curve, we believe that the lowest simulation time that could be achieved is about 2.14 min by using 23 TaskTrackers, as the simulation job cannot further parallelize beyond this number.

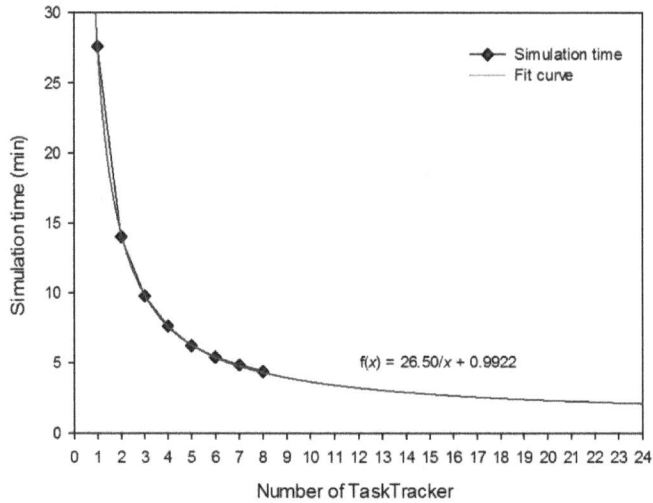

Figure 5. The performance of the model simulation service.

We also evaluated our model simulation service of DSS-WMRJ by comparing it with a widely-used SWAT auto-calibration tool (SWAT-CUP), which operates on a PC (Table 1). SWAT-CUP took 97.9 min to finish 92 simulations (not including the post-processing to gather information in order to generate a management report), while our service only took 4.4 min when running on eight TaskTrackers (each allowed four tasks running simultaneously). Although it is not a stringent comparison, as these two tools run on different environments, it still provided some convincing results that our model simulation service substantially reduced the execution time by parallelizing the model simulations on Hadoop clusters and, therefore, is able to support decision making with a reasonable amount of simulation time.

4. Discussion

The most outstanding features of DSS-WMRJ are its quasi-real-time decision making support, intuitive and wizard-style interfaces and excellent scalability. The implementation and test results showed that DSS-WMRJ can meet the goals of achieving intuitive and concise interfaces and supporting real-time or quasi real-time decision making. Besides, it is scalable, as the users just need to add more computing machines to the Hadoop cluster to scale up the system and achieve the goal of reducing the model simulation time. Other components of DSS-WMRJ, such as the map service and watershed management components, can also be scaled up by deploying machines and a simple configuration of Nginx.

Building on open source software and libraries is another valuable feature of DSS-WMRJ worthy of note (except the commercial chart component FusionCharts, due to its dynamic and excellent chart functionalities, but this component can be replaced by an open-source one). This feature makes it economic, as software license costs and other costs are not a factor, making it applicable to other watersheds. However, open-source software has some disadvantages, such as a lack of rapid building-up of tools and technology support, which is usually available for commercial software. Under joint efforts of the open-source community, the gap between open source and commercial software is increasingly narrowing.

Our DSS-WMRJ also has some other advantages. For example, it is accessible at any time and from anywhere by using a browser via the Internet or Intranet; and it is beneficial for information sharing and cooperation between individuals or institutions; thus, these will prompt users to participate in the decision making processes. It is easy to maintain and upgrade, as the system is deployed on the server. Besides, the web-based nature makes it easy to scale up and adopt for a cloud environment. However, there are some disadvantages, too. Compared with the desktop-based DSS-WM, it is more difficult to develop the web-based application, as it involves more languages and technologies, and other details need to be carefully considered, such as communications between browsers and servers.

As indicated by many studies and practices, stakeholders have significant impacts on the success of developing IT projects or facilities. This is especially so in situations when the funders and users are different individuals or not even in the same organization. As stakeholders may have different interests, it is very important to identify and involve these stakeholders at an early stage of the system implementation. In our case, we have three major groups of stakeholders: The funders, watershed managers and public users, all focusing on different aspects of the DSS-WMRJ. The funders are concerned more about the effectiveness of DSS-WMRJ; the watershed mangers are focussed on the conciseness of the interface and the efficiencies; while the public users worry about the ease of information sharing. To fulfill these requirements, technologies, such as Web GIS, distributed models and cloud technologies, and the agile software development methodology were adopted in the development of DSS-WMRJ, to promote adaptive planning, evolutionary development, early delivery and to encourage rapid and flexible responses to changes. We are currently at an initial stage of the development cycle, and the prototype of DSS-WMRJ that we provided is mainly for the purpose of demonstrating to and communicating with stakeholders, stimulating them to provide more specific and accurate system demands. Therefore, the proposed prototype is not a fully functional one; nevertheless, the DSS-WMRJ will evolve into a fully-fledged tool.

In the future versions, DSS-WMRJ will be improved by a continuous enriching of the watershed management functionalities. The performance of DSS-WMRJ will be focused on, as well. For example, large hydrological models may take hours for a single execution, and this inevitably impedes the goal of real-time decision making support. This problem cannot be solved by simply paralleling the model simulations. It is important to reduce the model's execution time, so as to achieve the goal of real-time decision making. The hydrological processes at HRUs (the most process-intensive parts) and the sub-basin level are independent of each other by design in the modeling concept of SWAT. These processes at HRUs and sub-basins are traditionally computed in a serial manner by a single computer, which requires much computing time. Thus, parallelizing the calculation procedures for HRUs and sub-basins should be effective at reducing the simulation time, as proven by the studies of Yalew et al. [21] and Wu et al. [22], by using grid computing. Another possible solution to reduce the simulation time of SWAT is to divide the single and large SWAT watershed models into smaller ones and route them from the headwater basins to the terminal basin, then parallelize the calculation procedures of the headwater basins. Recently, Sun et al. [23] developed three metamodels (model reduction) to support real-time decision making regarding activities relative to surface water quality in a coastal watershed in Texas, USA. They approximated the SWAT model by a reduced order model in order to speed up the running time in the web environment. We would like to evaluate these two methods in our cloud environment and analyze the trade-offs for them.

Effect of management practices (such as EVFS) are not evaluated in our initial prototype of DSS-WMRJ. Many other studies [24–27] have already proven these management practices to be effective. We will also evaluate the incorporation of management practices in our future version of DSS-WMRJ with a well-calibrated SWAT watershed model. In addition, Hadoop technology is available via Amazon's Elastic MapReduce and Microsoft's HDInsight, thus making it possible to migrate DSS-WMRJ to a public cloud. We will evaluate the model simulation service with one of these public services.

5. Conclusions

A user-friendly and quasi-real-time prototype of DSS-WMRJ was developed by seamlessly integrating an open-source Web GIS tool, Geoserver, a modeling component, SWAT, a cloud computing platform, Hadoop, and other open-source components and libraries. Due to its flexible and innovative features, DSS-WMRJ has some advantages over other decision support systems for watershed management: (1) Quasi-real-time decision making is obtained by utilizing cloud computing technology; (2) An intuitive and user-friendly GUI is provided, which largely enhances the user experience; and (3) It is very economic, as the DSS-WMRJ was almost entirely built on open-source software, and this feature lends to it great prospects of being applied to other watersheds. This is also valuable and informative for building other environmental DSSs.

However, as a prototype of DSS-WMRJ, there are some inadequacies (e.g., the nutrient components of SWAT were not well calibrated and evaluated, due to insufficient monitoring data, and limited management options and functionalities were implemented), and thus, continuous improvement is necessary. In the next version of DSS-WMRJ, more management practices will be incorporated, and the model simulation time will be further reduced by modifying the structure of the SWAT model. We will also evaluate DSS-WMRJ with a well calibrated (including the runoff, sediment and nutrients) model and evaluate the model simulation service with a public cloud, such as Amazon's cloud services.

Acknowledgments: The study was financially supported by the Key Specialized Program of Public Research Institutes of Fujian (Grant No. 2013R04) and the National Natural Science Foundation of China (No. 50979015). We wish to thank the reviewers and editors for their kind and constructive comments.

Author Contributions: Xingwei Chen had the basic idea of the present work and coordinated the research group. Dejian Zhang contributed in designing, developing and testing the prototype web-based decision support system for watershed management and applying it to the case study. Dejian Zhang initiated the manuscript writing, Huaxia Yao made improvements to the English writing.

Conflicts of Interest: The authors declare no conflict of interest.

References

1. Karimi, P.; Bastiaanssen, W.G.M.; Molden, D. Water accounting plus (wa plus)—A water accounting procedure for complex river basins based on satellite measurements. *Hydrol. Earth Syst. Sci.* **2013**, *17*, 2459–2472.
2. Chung, E.-S.; Lee, K.S. Prioritization of water management for sustainability using hydrologic simulation model and multicriteria decision making techniques. *J. Environ. Manag.* **2009**, *90*, 1502–1511.
3. Qi, H.H.; Altinakar, M.S. Vegetation buffer strips design using an optimization approach for non-point source pollutant control of an agricultural watershed. *Water Resour. Manag.* **2011**, *25*, 565–578.
4. Wool, T.A.; Davie, S.R.; Rodriguez, H.N. Development of three-dimensional hydrodynamic and water quality models to support total maximum daily load decision process for the neuse river estuary, north carolina. *J. Water Res. Pl-Asce* **2003**, *129*, 295–306.
5. Park, Y.S.; Park, J.H.; Jang, W.S.; Ryu, J.C.; Kang, H.; Choi, J.; Lim, K.J. Hydrologic response unit routing in swat to simulate effects of vegetated filter strip for south-korean conditions based on vfsmod. *Water* **2011**, *3*, 819–842.

6. Kim, Y.-J.; Kim, H.-D.; Jeon, J.-H. Characteristics of water budget components in paddy rice field under the asian monsoon climate: Application of hspf-paddy model. *Water* **2014**, *6*, 2041–2055.

7. Matthies, M.; Giupponi, C.; Ostendorf, B. Preface—Environmental decision support systems: Current issues, methods and tools. *Environ. Model. Softw.* **2007**, *22*, 123–127.

8. McIntosh, B.S.; Ascough, J.C.; Twery, M.; Chew, J.; Elmahdi, A.; Haase, D.; Harou, J.J.; Hepting, D.; Cuddy, S.; Jakeman, A.J.; *et al.* Environmental decision support systems (EDSS) development—Challenges and best practices. *Environ. Model. Softw.* **2011**, *26*, 1389–1402.

9. Cortes, U.; Sanchez-Marre, M.; Ceccaroni, L.; Roda, I.R.; Poch, M. Artificial intelligence and environmental decision support systems. *Appl. Intell.* **2000**, *13*, 77–91.

10. Mysiak, J.; Giupponi, C.; Rosato, P. Towards the development of a decision support system for water resource management. *Environ. Model. Softw.* **2005**, *20*, 203–214.

11. Cau, P.; Paniconi, C. Assessment of alternative land management practices using hydrological simulation and a decision support tool: Arborea agricultural region, sardinia. *Hydrol. Earth Syst. Sci.* **2007**, *11*, 1811–1823.

12. Hipel, K.W.; Fang, L.; Kilgour, D.M. Decision support systems in water resources and environmental management. *J. Hydrol. Eng.* **2008**, *13*, 761–770.

13. Rao, M.; Fan, G.; Thomas, J.; Cherian, G.; Chudiwale, V.; Awawdeh, M. A web-based GIS decision support system for managing and planning USDA's conservation reserve program (CRP). *Environ. Model. Softw.* **2007**, *22*, 1270–1280.

14. Zeng, Y.; Cai, Y.; Jia, P.; Jee, H. Development of a web-based decision support system for supporting integrated water resources management in Daegu city, South korea. *Expert Syst. Appl.* **2012**, *39*, 10091–10102.

15. Sun, A. Enabling collaborative decision-making in watershed management using cloud-computing services. *Environ. Model. Softw.* **2013**, *41*, 93–97.

16. Lam, C. *Hadoop in Action*, 1st ed.; Manning Publications Co.: Stamford, CT, USA, 2010; p. 336.

17. White, T. *Hadoop: The Definitive Guide*, 2nd ed.; O'Reilly Media: Sebastopol, CA, USA, 2009; p. 625.

18. Arnold, J.G.; Fohrer, N. SWAT2000: Current capabilities and research opportunities in applied watershed modelling. *Hydrol. Process.* **2005**, *19*, 563–572.

19. Arnold, J.G.; Srinivasan, R.; Muttiah, R.S.; Williams, J.R. Large area hydrologic modeling and assessment part I: Model development1. *JAWRA* **1998**, *34*, 73–89.

20. Wang, L.; Chen, X.W. Runoff simulation with calibration and validation of three station in Jinjiang river basin. *Sci. Soil Water Conserv.* **2007**, *5*, 21–26. (In Chinese)

21. Yalew, S.; van Griensven, A.; Ray, N.; Kokoszkiewicz, L.; Betrie, G.D. Distributed computation of large scale SWAT models on the grid. *Environ. Model. Softw.* **2013**, *41*, 223–230.

22. Wu, Y.P.; Li, T.J.; Sun, L.Q.; Chen, J. Parallelization of a hydrological model using the message passing interface. *Environ. Model. Softw.* **2013**, *43*, 124–132.

23. Sun, A.Y.; Miranda, R.M.; Xu, X. Development of multi-metamodels to support surface water quality management and decision making. *Environ. Earth Sci.* **2015**, *73*, 423–434.

24. Chu, T.-W.; Lin, Y.-C.; Shirmohammadi, A.; Huang, Y.-C. BMP evaluation for nutrient control in a subtropical reservoir watershed using SWAT model. In Proceedings of the 2013 the International Conference on Remote Sensing, Environment and Transportation Engineering; (Rsete 2013). Gahegan, M.N., Xiong, N., Eds.; Atlantis Press: Paris, France, 2013; Volume 31, pp. 914–917.

25. Liu, R.; Zhang, P.; Wang, X.; Wang, J.; Yu, W.; Shen, Z. Cost-effectiveness and cost-benefit analysis of BMPs in controlling agricultural nonpoint source pollution in China based on the SWAT model. *Environ. Monit. Assess.* **2014**, *186*, 9011–9022. PubMed]

26. Giri, S.; Nejadhashemi, A.P.; Woznicki, S.; Zhang, Z. Analysis of best management practice effectiveness and spatiotemporal variability based on different targeting strategies. *Hydrol. Process.* **2014**, *28*, 431–445.

27. Dechmi, F.; Skhiri, A. Evaluation of best management practices under intensive irrigation using swat model. *Agric. Water Manag.* **2013**, *123*, 55–64.

MDPI AG

Klybeckstrasse 64

4057 Basel, Switzerland

Tel. +41 61 683 77 34

Fax +41 61 302 89 18

http://www.mdpi.com/

Water Editorial Office

E-mail: water@mdpi.com

http://www.mdpi.com/journal/water

www.ingramcontent.com/pod-product-compliance
Lightning Source LLC
Chambersburg PA
CBHW051924190326

41458CB00026B/6400